Springer Tracts in Civil Engineering

Springer Tracts in Civil Engineering (STCE) publishes the latest developments in Civil Engineering - quickly, informally and in top quality. The series scope includes monographs, professional books, graduate textbooks and edited volumes, as well as outstanding PhD theses. Its goal is to cover all the main branches of civil engineering, both theoretical and applied, including:

- Construction and Structural Mechanics
- Building Materials
- Concrete, Steel and Timber Structures
- Geotechnical Engineering
- Earthquake Engineering
- Coastal Engineering; Ocean and Offshore Engineering
- Hydraulics, Hydrology and Water Resources Engineering
- Environmental Engineering and Sustainability
- Structural Health and Monitoring
- Surveying and Geographical Information Systems
- Heating, Ventilation and Air Conditioning (HVAC)
- Transportation and Traffic
- Risk Analysis
- Safety and Security

Indexed by Scopus

To submit a proposal or request further information, please contact: Pierpaolo Riva at Pierpaolo.Riva@springer.com, or Li Shen at Li.Shen@springer.com

More information about this series at http://www.springer.com/series/15088

Marijana Serdar · Ivan Gabrijel ·
Dirk Schlicke · Stéphanie Staquet ·
Miguel Azenha
Editors

Advanced Techniques for Testing of Cement-Based Materials

Springer

Editors
Marijana Serdar
Faculty of Civil Engineering
University of Zagreb
Zagreb, Croatia

Ivan Gabrijel
Faculty of Civil Engineering
University of Zagreb
Zagreb, Croatia

Dirk Schlicke
Institute of Structural Concrete
Graz University of Technology
Graz, Austria

Stéphanie Staquet
BATir
Université Libre de Bruxelles
Brussels, Belgium

Miguel Azenha
Department of Civil Engineering
University of Minho
Guimarães, Portugal

ISSN 2366-259X ISSN 2366-2603 (electronic)
Springer Tracts in Civil Engineering
ISBN 978-3-030-39737-1 ISBN 978-3-030-39738-8 (eBook)
https://doi.org/10.1007/978-3-030-39738-8

This Springer imprint is published by the registered company Springer Nature Switzerland AG
The registered company address is: Gewerbestrasse 11, 6330 Cham, Switzerland

Preface

The current book was conceived as one of the outputs of the EU-funded network COST Action TU1404 "Towards the next generation of standards for service life of cement-based materials and structures" (www.tu1404.eu), active between 2014 and 2018. This network intended to provide added value through collaborations across the entire field of cement-based materials and structural concrete—ranging from concrete technology, over simulation and structural design, production and building, up to operation, monitoring and maintenance.

Within the Working Group 1 of the Action, which was focussed on experimentation, over 40 laboratories and institutes around the world participated in a joint testing of cementitious materials using standardised and advanced techniques within the extended round-robin programme, called RRT+. The idea behind the book started to develop when it was realised by the editors that there are numerous innovative concrete characterisation methods being developed across these laboratories that deserve to be put in the forefront. The role of leaders of the WG1 was to pinpoint these techniques and try to organise them into this book.

Advanced testing of cement-based materials gives rise to different associations. Some may think of highly specialised techniques in research and development, while others may perceive it as technical assistance towards the mindset shift from applying concrete by its strength class towards establishing a performance-based application of cement-based materials. This book focuses on both the research and development as well as practical application, aiming to serve as idea generator and reference work at the same time. Overall, this book contains seven targeted chapters presenting highly specialised techniques for obtaining individual performance characteristics of cement-based materials. The application fields of the presented techniques range hereby from basic material properties up to structural behaviour as well as from the beginning of mixing over hydration up to the performance during service life.

In particular, Chap. 1 presents the characterisation of fresh cement-based materials with acoustic emission measurements. In Chap. 2, up-to-date developments and achievements in the field of ultrasonic testing of the early-age cement-based materials are presented. In Chap. 3, a novel method for the continuous measurement of elastic modulus evolution during the hardening phase of cement-based materials is presented. Chapters 4–6 are dealing also with the characterisation of cement-based materials at early ages, particularly aiming at viscoelastic properties by means of repeated minute-scale duration loadings (Chap. 4), the development of the autogenous strain and the coefficient of thermal expansion (Chap. 5) and the consideration of the restraining condition in testing concrete since setting time (Chap. 6). Besides these innovations, Chap. 7 presents an experimental setting for the holistic investigation of crack width development in reinforced concrete members under restrained conditions with the superposition of hardening-induced stresses with restraint stresses during service life.

The presented experimental techniques were derived from profound development and verification within the research of the presenting authors. Besides, they were also applied, verified and/or advanced within the scientific work of the COST Action TU1404. The application, verification and/or advancement of the presented techniques refer hereby mainly to the RRT+. It also refers to the numerical benchmark campaign in which the well-grounded database from the RRT+ was further used for sophisticated modelling and simulation with multi-scale approaches at time and spatial levels.

Beside pure technical aspects, the chapters in this book highlight that concrete—a material perceived in the practice and everyday life as rather simple—is a remarkable and complex material bringing constant challenges to scientist and practitioners in grasping all aspects of its properties and behaviour.

The editors are grateful to all authors, who shared their knowledge and described their methods in detail, allowing us to share this in the present book. The editors are also greatly thankful to all participants in the COST Action TU1404, which took their time to test, report, analyse and discuss results obtained during the Action, all on purely volunteering bases. Finally, a remark shall be given regarding the wide international support that was behind all the activities of the COST Action TU1404—and hereby also behind the creation of this book. General support was appreciated from COST, RILEM, fib, JCI and ACI as well as specific support of the RRT+ by the Austrian Railway Association (ÖBB Infra), the Norwegian Road Administration (Statens Vegvesen), Schleibinger Geräte GmbH from Germany, CEVA Logistics and EDF.

Zagreb, Croatia — Marijana Serdar
Zagreb, Croatia — Ivan Gabrijel
Graz, Austria — Dirk Schlicke
Brussels, Belgium — Stéphanie Staquet
Guimarães, Portugal — Miguel Azenha

Contents

Chapter 1
Acoustic Emission Characterization of Fresh Cement-Based Materials

Dimitrios G. Aggelis, Christian U. Grosse and Tomoki Shiotani

Abstract The increase of performance requirements for contemporary cement-based media and structures calls for better control of the material processes. Acoustic emission (AE) is one of the non-invasive techniques that can provide information on the internal condition of the material. This includes the time of crack occurrence, the location as well as the fracture mode. In addition, it can offer valuable insight on the initial, crucial period of curing and hardening that has serious impact on the performance and durability of concrete. The present chapter aims to review the background, offer the reader a basic overview of the application of the technique on fresh cement-based material and at the same time give some new directions for testing of concrete.

Keywords Curing · Monitoring · Sensor · Shrinkage · Settlement · Cracking · Source · Energy · Frequency · Fracture

1.1 Introduction

Acoustic emission is a technique offering specific benefits in the monitoring of cement-based materials and structures. One of the most important AE features is the high sensitivity to very low energy events, which practically means accurate determination of the onset of micro-cracking or other processes occurring in fresh or hardened cement. In addition, it can localize the sources in three dimensions even though no visual contact is possible with the point of the source. Also, it helps to

D. G. Aggelis (✉)
Department of Mechanics of Materials and Constructions, Vrije Universiteit Brussel, Pleinlaan 2, Brussels 1050, Belgium
e-mail: daggelis@vub.be

C. U. Grosse
Non-Destructive Testing Lab, Center for Building Materials, Technische Universität München, Franz-Langinger-Straße 10, 81245 Munich, Germany

T. Shiotani
Graduate School of Business & Graduate School of Engineering, Kyoto University, Kyoto 615-8540, Japan

M. Serdar et al. (eds.), *Advanced Techniques for Testing of Cement-Based Materials*, Springer Tracts in Civil Engineering, https://doi.org/10.1007/978-3-030-39738-8_1

characterize the fracture mode or in general the type of source based on the received waveform shape and different methods of analysis. More specifically:

- The high sensitivity of the method allows monitoring the actual initiation of microcracking or other processes in cement much earlier than other indications, either macroscopic deflection measurements, stress-strain curves or even other monitoring techniques can indicate. This capability also enables AE to monitor processes like settlement of concrete, segregation, shrinkage or bubble formation. It is characteristic that the absolute energy of the AE signals is measured in atto-Joules (10^{-18} J). To realize the order of magnitude of such signal, it can be stated that one single cement grain of diameter 10 μm, falling from a height of 15 cm possesses a dynamic energy in the order of 10^{-14} J being already 4 orders of magnitude larger than the sensitivity of the technique.
- Localization is another basic feature of AE. By applying at least four sensors in a 3D position pattern, the internal cracking system can be quite accurately visualized. This is related to the delay of acquisition of a single acoustic event between the different sensors. Knowing the propagation velocity of the medium (practically the longitudinal wave velocity or pulse velocity) the exact location can be determined. If certain conditions are met, the localization error is contained within maximum 15% of the gauge length (distance between the furthest sensors) even in cases of severe damage where some linear wave paths are blocked. In that case higher number of sensors (>4) would of course be recommended.
- Another advantage is the possibility to characterize the fracture mode or the source type in general. The mode of the crack in a matrix or the type of process that occurs is indicative of the deterioration process, thus providing information on the remaining life of the component or structure. As an example, when a concrete beam is under bending load, the initial cracks are from tension on the matrix. Later the shear diagonal cracks are developed with the debonding between the reinforcement and matrix while friction between the banks of the cracks is also included. On a different material type, fibre composites (can also be textile reinforced cement) first develop matrix crack, while later delaminations between layers occur, debonding of fibres, pull-out and finally even fibre rupture. These distinct mechanisms do not emit the same elastic waveform and therefore, detailed monitoring by AE gives the possibility to characterize the phenomenon and know exactly what is happening in the material. Such techniques are similar to seismology where the waves radiated from an earthquake are analysed in respect to the rupture process in the earth (Grosse and Ohtsu 2008).

In this chapter, after an initial overview of the technique, monitoring by AE is discussed for the early age of cement-based material. In this case AE intends to capture phenomena not only related to cracking as will be extensively mentioned. In this stage, no external stress is applied, as the material is let to cure. However, still thousands of AEs are recorded being related to intrinsic processes that take place in fresh cement, mortar or paste. Possible sources that have been discussed in literature are settlement of cement, segregation, cavitation, bubbles and water migration, formation of hydrates as well as shrinkage cracking. A summary of the basic literature in the field is given, recent advancements are highlighted and insights on the possibility to enhance the understanding are proposed.

1.2 Overview of Acoustic Emission

The AE technique detects and records elastic waves that are radiated by irreversible processes in the material. These processes include deteriorations in any form (mainly cracking, delaminations, fiber pull-out etc.), development of corrosion and related cracking in concrete or other specific irreversible processes, like the ones occurring in fresh concrete. Usually piezoelectric transducers are applied, placed on the surface of the material with suitable coupling (e.g. silicon grease). The piezoelectric elements of these transducers transform transient pressure changes into electric waveforms. After amplification, the signals are driven to the acquisition board where digitization takes place, while the main waveform parameters are analysed (Grosse and Ohtsu 2008; Mix 2005; Grosse et al. 2016). A representation of an AE setup is given in Fig. 1.1, where two possibilities are demonstrated: monitoring of fracture of a concrete beam on the left and monitoring of curing of fresh cement-based material on the right.

An AE waveform is depicted in Fig. 1.2. Following standards, like RILEM 2010, one can extract parameters out of such a waveform that are representative for source types or other influencing conditions like the media in which the wave propagates. Sensor characteristics (frequency response functions) can have additional influences on the waveform.

The waveform starts with a "threshold crossing". The threshold is a predefined voltage set by the user to define the onset of the wave. When background noise allows the setting of a low threshold, then the measured onset is close to the real one. The time between the first threshold crossing (count) of the signal and the time of the maximum peak is called Rise time (RT). Similarly, duration (Dur) is defined as the time between the first and last threshold crossings. Intensity related parameters include Amplitude (A) which is the highest voltage of the waveform. Additionally, energy (MARSE, measured area under the rectified signal envelope) takes into account the content of the waveform between the 1st and last threshold crossings. These parameters can be indicative of the mechanical energy released by

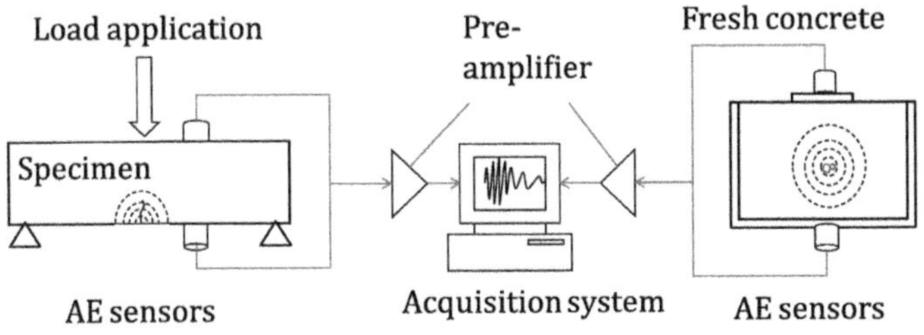

Fig. 1.1 Typical AE setups

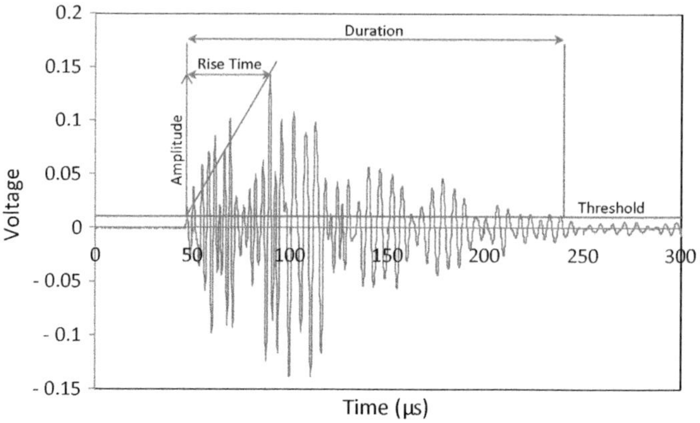

Fig. 1.2 Typical AE and main parameters

the source. Based on the shape of the first part of the signal, the RA value (RT/A) or the inverse of the rising angle has been introduced. The frequency content is also important, and several parameters are used to exploit the frequency information. One is the "average frequency" (AF) which is calculated in time domain as the ratio of the total number of threshold crossings over the duration of the waveform in kHz. Additionally, the peak frequency (PF) is defined as the frequency with the highest magnitude in the FFT domain, while central frequency (CF) is the centroid of the spectrum. Going back to the waveform shape, features like the RT, RA and Duration occupy higher values for shear type of sources, while frequency ones are higher for tension events (Shiotani 2008; Aggelis 2016).

1.3 Related Literature

In the literature related to elastic (or stress) waves, fresh cement-based material has been extensively tested by ultrasound. Wave speed (or ultrasonic pulse velocity) and other parameters show good sensitivity to the development of hydration process, being able to examine the degree of curing. They also provide a reliable evidence of stiffness at any given time of curing (Robeyst et al. 2009a, b; Carette and Staquet 2016; RILEM 2011). A comprehensive review can be found in Trtnik and Gams (2014).

Similarly to other fields, application of AE followed ultrasound testing in fresh cement-based material as well. The difficult point is to separate the AEs according to their source mechanism, something complicated by the number of different processes and their overlap in time. Still, the number of studies and research groups interested in the application of AE for monitoring of the fresh state of the material has recently started to increase as the results show that the processes occurring in the fresh material emit elastic energy that can be monitored by AE sensors.

Different approaches have been used so far to first monitor experimentally and then interpret the results. One of the main aims is the characterization of shrinkage cracking. When fresh paste loses moisture to the environment it starts to shrink. This shrinkage strain results in stresses due to movement restrains of the boundary conditions, and finally to micro- or macro-cracking close to the surface of the material. Since a part of the energy is released in the form of elastic waves, AE use is well motivated as a tool to study shrinkage cracking (McLaskey et al. 2007) and specifically where and when cracking takes place along with other characteristics. In this study exactly this phenomenon of fresh concrete was targeted i.e. shrinkage cracking due to aggregates restrain as well as from differential moisture loss at different depths. This cracking may start from the shrinking paste, but it is likely to propagate through the paste-aggregate interface which is the weakest zone of concrete. Two types of specimen geometries were used, one cubic of 142 mm side and a larger wedge-shaped (380 mm^2 and 190 mm deep). The reason for applying the second type (see Fig. 1.3a) was to avoid possible reflections that did not allow characterization of the different wave modes in the smaller cube. Specimens were equipped with multiple broadband sensors (8 for the cube and 16 for the wedge) and the monitoring lasted at least four days for both specimens. Results were compatible with the drying (and therefore, shrinkage) rate as AE started also with high rate being gradually decreased after the first day of test. Cubic specimen exhibited about double the AE activity of the wedge shape, something attributed to the smaller size and thus attenuation of the material. As far as source localization is concerned, despite the inherent difficulties (variations in propagation velocity, scattering, low signal-to-noise ratio) a standard error of 5–8 mm was achieved as tested by artificial excitations of pencil lead breaks. Most AE events appeared within a few centimetres from the surfaces of the cube and many of them corresponded to visually observed surface cracks. A slight correlation between the frequency content and the age was noted as high frequency emissions were more densely recorded after the first two days (Fig. 1.3b). This was attributed to the decrease of attenuation as the material cured. Examining the waveform shape at the different sensor, the tensile surface crack model was confirmed, as sensor 1 at the surface felt large Rayleigh waves, while for the same events, sensors 2–4 felt much stronger longitudinal waves and no sharply defined Rayleigh waves, Fig. 1.3a (McLaskey et al. 2007).

Shiotani et al. (2003) monitored the shrinkage cracking of cement containing glass spheres as aggregates, using AE sensors of frequency 200–800 kHz on the surface of relatively fresh concrete after demoulding. The specimens were circular with diameter of 235 mm and thickness of 42 mm. The results showed that more activities were recorded from a concrete (paste with glass spheres) than a cement paste, something attributed to shrinkage cracking that was more pronounced due to the additional restrain of the stiff aggregates. Several AE events were localized and validated with microscopy. The localized events for the composite continued for more than 10 h compared to the plain paste. In addition, indices like the average frequency and the *Ib*-value of the emissions indicated the moment when small scale cracking gave its place to larger scale phenomena.

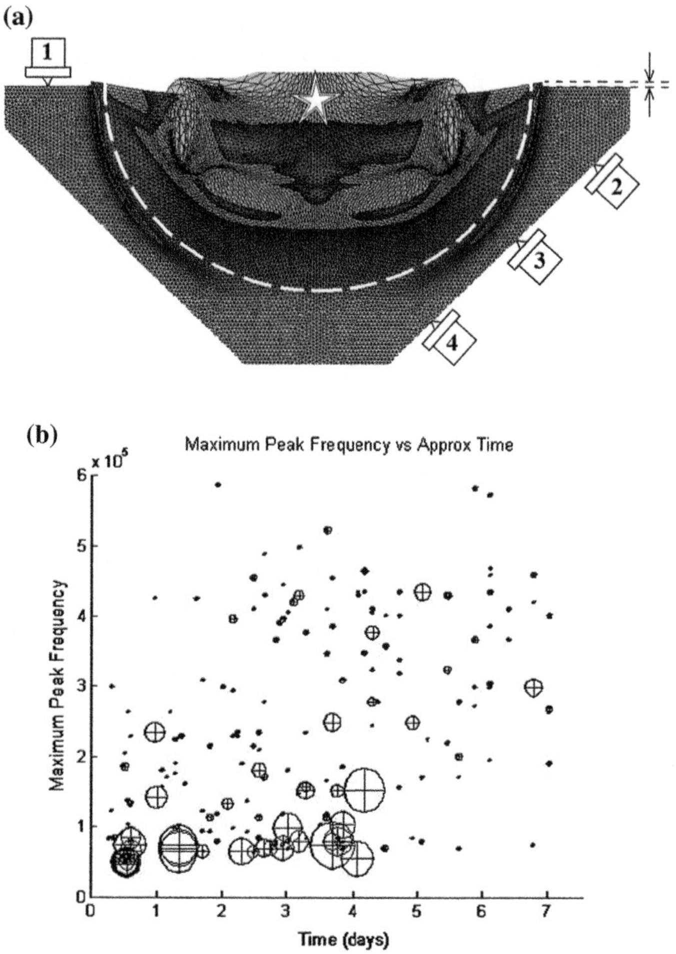

Fig. 1.3 **a** Wedge type specimen with indicative positions of sensors (1–4) and exaggerated displacement field after a tensile cracking event denoted by a star. **b** Peak frequency of AE events versus time (McLaskey et al. 2007)

Still, in the same subject of detecting of early age cracking, another study utilized concrete prisms of different water to cement ratio, strength class and curing conditions (Bacharz et al. 2015). AE monitoring with two sensors at the band of 30–80 kHz started 10 days after mixing (allowing for the selected curing) continued up to approximately 2 months and the authors investigated the possibility to characterize different types of cracking (micro-cracks in paste, micro-cracks in aggregate paste interface and cracks on the concrete surface). Although the background behind the separation of classes is not thoroughly explained, it is reported that the surface micro-cracks produce comparatively high AE energy, while the lowest energy is produced by paste micro-cracking. According to the authors this allows to identify classes and thus characterize these destructive processes in relatively young concrete due to hardening, shrinkage, evaporation in the absence of external mechanical load.

Researchers have tried to acquire AE even earlier, while the material is still in the liquid state in the mould. Most used waveguides inserted into the concrete to transmit signals to the AE sensor, sometimes without details about the material of the waveguide. For example, Van Den Abeele et al. (2009) used two sensors of 375 kHz resonance "attached on top of two protruding bars" to register the signals from the microstructural activity of the cement mortar samples. For all specimens, the main AE activity started after the temperature peaks. Specimens with lower w/c ratio (0.33) exhibited higher amount of activity earlier being in accordance with the speed of hydration compared to higher w/c (0.5). Self-compacting concrete which showed higher temperature, implying more intensive chemical activity, exhibited also more intense AE from the reference concrete of the same w/c at that stage but later during the mechanical setting the activity was weaker. Results were also supported by ultrasonic measurements (longitudinal and shear velocities and nonlinear indices). Despite that AE activity showed good correlation to hydration and mix proportions, it only emerged 15 h after mixing, showing no activity from other very early processes.

Another approach based on waveguide concerned the frequency band 100–500 kHz (Skal's'kyi et al. 2004). The waveguide had variable cross section and was checked that its main resonant frequencies fell in the range of sensitivity of the sensor. In this case AE started to be recorded earlier, between 10 to 20 min after preparation of the 100 mm cubic concrete specimen. The accumulated AE was connected to the development of the crystal structure and when it was mostly completed (i.e. up to 6 h) the AE intensity noticeably decreased. It is stated that the formation of the concrete microstructure and the increase of strength is correlated to the cumulative AE numbers and therefore, AE monitoring can contribute to the optimization of the mechanical characteristics of the final product.

In Chotard et al. (2001) despite that AE was not recorded for the first 3 h, it was shown that the frequency content of emissions changed and specifically increased with curing time by about 50 kHz, something mostly attributed to the change of the medium illustrating the liquid-to-solid state shift. Here, again the main AE activity started during the temperature peak. The authors used a small PMMA plate of 3 mm thickness to couple the sensor to the fresh cement paste. The AE in this study was attributed to the drying process, creation of porosity and formation of hydrates. The sensors were sensitive between 50 and 300 kHz and the specimen had dimensions of $100 \times 100 \times 30$ mm with the last being the height. Amplitude and frequency of AE were proposed to characterize internal mechanisms like hydration, porosity onset, micro-cracking.

Lura et al. (2009) recorded AE from 8 g samples of cement paste with varying w/c ratio. They used a sensor of 375 kHz and received the maximum rate of AE after 8–10 h. Many mechanisms were considered as sources (dissolution of cement grains, formation of hydration products, friction of the shrinking paste with the container walls, shrinkage cracking) but it was concluded that nucleation of bubbles including air and water vapour is the most dominant. As a preparation, the authors monitored AE activity from carbonated liquids (soft drinks) which showed quite high activity for several minutes before losing carbon dioxide. The authors argued that possibly the early age AE in cement could be attributed to formation of hydrates, but the activity did not peak at the end of the dormant period, leading them to the conclusion that possibly the settlement is responsible for the AE at this early stage. Slightly after the setting time as measured by the Vicat needle, low w/c specimens (w/c = 0.3) exhibited a peak of AE activity higher than samples with high w/c (0.4). However, when the open top surface of the small samples was covered by a layer of water, this peak was eliminated, as according to the authors this water penetrated into the paste at setting replacing the chemically used water and reducing thus the possible cavitation phenomena. Also the fact that the AE energy was proportional to the mass of the specimens, indicated that AE comes from the whole volume and not from possible friction with the walls of the mould.

Recently another approach saw publicity (Bardakov and Sagaidak 2016), mainly focusing on predicting the strength of concrete based on the AE acquisition rate recorded at the first hours after mixing. A seemingly metal waveguide connected to a 30–300 kHz sensor was inserted into the fresh concrete. AE was recorded from the first minute and three distinct stages were recognized. Initially AE rate was very high (stage I, first 4–5 h) attributed to "mobility" or mechanical displacement of the mixture. As the water was consumed in hydration, the mobility (and the originating AE) became gradually less until temporarily stopping at 10 h. The authors did not consider other mechanisms as sources (dissolution of cement grains and formation of hydration products) because their emissions were expected with low amplitude and taking into account the high attenuation of fresh concrete, these signals would not be recorded. However, shrinkage cracking is acknowledged as a possible source. The 2nd stage (between approximately 10–15 h) had very limited AE activity something attributed to low energy sources like formation of calcium hydrosilicates which, however, are lost due to attenuation. The final stage showed an increase in AE attributed to growth of crystallites in conditions of limited volume. With the above logic, the AE slope of the 3rd stage was directly indicative of the intensity of structure formation and giving eventually good correlations with the final compressive strength. Correlations were also reasonable with the duration of the 2nd, "silent" stage, which was inversely connected to strength, as according to the authors, longer silence meant delay in hydration formation.

A similar approach was applied in Pazdera et al. (2014) based on relatively broadband sensors placed on steel holders on top of the specimens for 28 days. Here, the monitoring started after a period of 8 h and a similar trend of AE was shown: initially high rate of activities followed by a gradual decrease which came close to the temperature peak. After 24 h the specimens (100 × 100 × 400 mm) were demoulded and the sensors attached again using steel waveguide and wax. AE continued to register in low rates up until the end of the experiment. The main result was that the mix exhibiting the least AE (one containing fly ash), exhibited the highest compressive strength. This according to the authors was attributed to less early age cracking which from one side increases the numbers of emissions and at the same time decreases the strength. In another study of the same group (Hoduláková and Topolár 2016), mortar was tested, showing, according to the authors, that addition of microsilica possibly mitigates early age microcracking as the AE amplitudes were lower than the reference sample.

In Qin et al. (2014) embedded cement-based piezocomposites with sensitivity up to 600 kHz were applied for AE monitoring along with a broadband commercial transducer fixed on a waveguide. The embedded sensor exhibited higher activity for the first 12 h, attributed to better contact with the material, having peak frequencies below 150 kHz. Based on the shape of the cumulative hits curve, three distinct stages were suggested, namely the dormant period of low emissions, due to negligible microstructure development and heavy attenuation, followed by higher AE. At this second stage a strong increase in temperature was also measured showing the intensive hydration phenomena as well as shrinkage induced cracking. After the temperature peak, the AE rate decreased. Then, the authors applied two embedded sensors with distance of 2 m in a large-scale construction (foundation base element of depth 2.5 m) and were able to linearly localize several hundreds of events, attributed to the high temperature gradients that were measured inside the bulk of fresh concrete.

Recently, in Iliopoulos et al. (2016a) AE was monitored from the period just after mixing (10 min) using 150 kHz resonant sensors on the external surface of the metal mould. This approach led to many thousands of hits showing that the monitored volume of the material is much larger than using a single waveguide. In this case, the whole metal mould acted as a waveguide transferring the AE signals to the sensors. The rate of emissions started at a very high value, being gradually decreased until approximately 4–5 h. Concretes with high w/c ratio seemed to exhaust their activity earlier, completing the 85% of their total activity in less than one hour, while for w/c = 0.65, the same proportion was reached at almost 3 h. Concerning the aggregate content, it was shown that high a/c (up to 5) led to reduced emissions compared to cement paste (a/c = 0). The high AE activity for paste could be attributed to the lack of multiple scattering and attenuation of the waves onto the aggregates, but more likely to the higher amount of cement which contributes to the mobility through settlement and is the chemically active compound. Apart from the total number of emissions, in this study some qualitative features used previously for hardened concrete were also examined. Parameters RA and AF show distinct trends, with high w/c mixtures giving higher frequencies in average and shorter signals, something more evident 3–4 h after mixing. AE activity and parameters during the first hours correlated well with the 28 days compressive strength. Specifically, mixes with high a/c ratio that exhibited the highest strength, had also the highest average frequency, lower RA value between 4 and 6 h after mixing and lower AE activity in total when fresh (Iliopoulos et al. 2016a). Excluding one outlier class a general trend is given in Fig. 1.4. The result could be considered reasonable since high amount of early AE activity is connected to increased settlement and therefore, excessive bleeding.

A multi-disciplinary approach was used in Dzaye et al. (2016). Apart from two AE sensors of 150 kHz resonance attached directly on the mould of standard 40 × 40 × 160 mm paste and mortar specimens (Fig. 1.5), capillary pressure measurements and pulse velocity measurements were simultaneously taken on a separate specimen of the same batch. Results showed that initially, as cumulative AE was increasing with high rate, the capillary pressure was also building up. When AE was nearly saturated

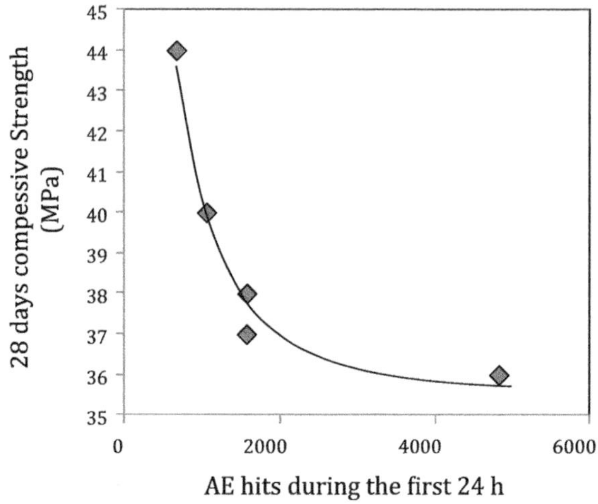

Fig. 1.4 Correlation between compressive strength of hardened concrete and the AE during the fresh state (results from Iliopoulos et al. 2016a)

Fig. 1.5 AE monitoring of fresh paste in standard mould, taken from Dzaye et al. (2016)

for this initial stage and the increase was marginal (approximately at 3 h after mixing of paste) the capillary pressure dropped suddenly indicating that air has entered into the matrix. At the same time, ultrasonic pulse velocity and amplitude started to rapidly develop indicating the setting time (or the point that an interconnected network of hydration products has been built, see Fig. 1.6.

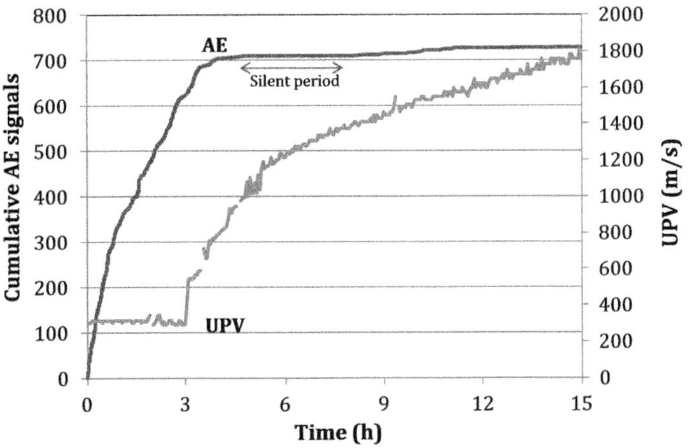

Fig. 1.6 AE cumulative curve and UPV development for a cement paste measured with the test setup of Dzaye et al. (2016)

In addition, measurements in vertical mould of 200 mm height and cross section of 40 × 40 mm with multiple sensors at different heights (Fig. 1.7) revealed that the activity was higher for the bottom sensors showing that the initial hydrostatic pressure is important. Furthermore, and in an effort to isolate AE producing mechanisms, individual aggregates were let to drop from the top of the mold while an air hose was separately used to produce air bubbles which rose from the bottom to the top

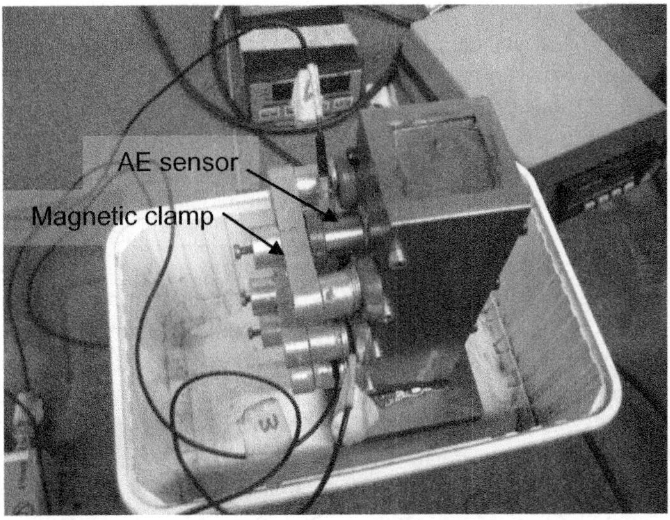

Fig. 1.7 AE monitoring of fresh paste in vertical mould with sensors at different heights (Dzaye et al. 2018)

(Dzaye et al. 2018). It was found that both types of events could be well monitored by the three AE sensors placed along the height of the mold of Fig. 1.7. There was a positive correlation between the aggregate size and the AE energy release from the impact, while the average amplitude due to air bubbles was approximately 6 dB lower than the one from aggregate impacts but was still detectable. An interesting detail is that aggregates impact in cement paste resulted in higher frequency than impact of the same size aggregates in water-filled or empty mold. This was attributed to the lower speed of the particles due to viscosity of the paste and therefore, shorter impact contact time with the inner of the steel mould. The basic conclusion was that both mechanical sources (aggregates impact and air bubble creation/movement) produce waves that are well received by AE systems (Dzaye et al. 2018).

The early-age cracking was targeted in Topolár et al. (2017), as is the reason for severe problems with the durability of concrete. Specifically mortars of different w/c ratio were prepared and two AE sensors of frequency range 100–800 kHz were attached to steel waveguides embedded in the material. The specimens were prismatic with 1 m length and cross section of 100 × 60 mm. There, the AE activity started at 1 h after mixing. The initial hits were attributed to "plastic setting" and early microstructure formation, while the peak of AE rate came at 4 h. Duration and amplitude of AE signals are also presented, without however, showing consistent trends between the different w/c specimens as one of the specimens contained also super-plasticizer. The activity continued with some silent stages up to 72 h when the experiments were terminated.

In another recent study, paste, mortar and concrete prism specimens of dimensions 100 × 100 × 500 mm were monitored by AE sensors 15–20 min after mixing and for 72 h in plastic molds (Thirumalaiselvi and Sasmal 2019). Five broadband sensors with range 100–900 kHz were placed along the top surface of the specimen, and one at the base as reference. Activity started after 2–3 h for all material types. The highest AE rate for paste came between 10 and 20 h (at 10 events/h), while for concrete the activity peaked between 23 and 35 h at approximately 2.5 events/h. These peaks coincided with the heat evolution period in each material. Overall the paste accumulated 120 events, concrete approximately half, while mortar was in between. Attenuation was acknowledged as a crucial factor since simulations showed that it starts at high values at time zero while strongly decreases further (at 22 h and later). Looking at the frequency content, the band 0–100 kHz was always by far the most powerful for the early stage, while later, the band of 100–200 kHz took over in content. According to the presented analysis, four basic types of signals were classified, being related to the CSH formation and emptying of capillary pores, chemical shrinkage while at the end of monitoring, a cluster attributed to micro-cracking was also added, which was stronger for concrete than for paste. As a general comment for this interesting and detailed study, it presents many AE characteristics combined also with "active" wave parameters (pulse velocity and attenuation) however, it does not present any activity at the first 2 h, missing therefore, the process of settlement, which in other studies results in thousands of AE signals within 2 h after mixing. This may be due to the placement of the sensors directly on the surface of concrete, which initially, as acknowledged, is very attenuative.

Assi et al. (2018), after discussing the lack of consensus in literature on the correlation between AE signals and hydration mechanisms, reported the AE application on fresh cement paste with the use of pattern recognition algorithms, combined with analysis of wavelet energy distribution. Specimens were prisms of 38 × 38 × 110 mm cast in Plexiglas molds with two different w/c ratios, i.e. 0.3 and 0.5. One relatively wideband sensor (range from 200–900 kHz) was placed on each specimen, after placing an acetate sheet of the free surface to increase the transmission of waves. Results showed the highest amplitude signals came close to the "middle" stage where the temperature peak was exhibited. While there was strong variation in cumulative energies even of specimens of the same composition, it was seen that a sudden initial increase was noted at the start of the "acceleration" stage, when temperature was rising for all specimens. Main mechanisms considered as AE sources were crystallization, formation of hydrates during the acceleration stage, chemical shrinkage and related micro-cracking and delayed hydration at the deceleration stage. The authors did not consider settlement as possible source maybe because the setup received limited AE activity in the first minutes after mixing.

In the field of concrete admixtures Lefever et al. (2017) monitored the AE activity during the standard shrinkage cracking ring test. Two 150 kHz resonant sensors were placed on the inner side of the metal ring. In the reference mortar specimen, AE started to be recorded after about 4 h in hand with the increase in shrinkage as measured by strain gauges on the steel ring, while after approximately 4 days, the mortar rings were cracked emitting very strong AE, simultaneously with the strain drop. A mix containing super-absorbent polymers (SAPs) followed a different behavior, as it did not exhibit strong shrinkage. This was attributed to the action of SAPs that at some point release water during hydration. At the same time, AE started to be recorded at very high rates (higher than the reference mix) at 7–8 h after mixing indicating the release of water by the SAPs, showing how its sensitivity can be utilized to monitor delicate processes in the microstructure.

Lastly, two interesting cases are reported by Ohtsu (2005). One concerns the identification of consistency of roller-compacted concrete mixes. An AE sensor was placed on the outer surface of the large cylindrical container of 48 cm diameter and height 40 cm ("VS test apparatus"). The contained concrete was submitted to vibration until bleeding water appeared on the surface. AE was continuously acquired for all mixes in two stages, exhibiting a bilinear curve, first with high slope and then with lower. This turning point was well related to the number of vibrations as concrete with higher "VC value" (according to the authors the time until bleed water is observed at the surface) exhibited this transition point earlier.

In a similar case also reported by Ohtsu (2005) AE was monitored by a sensor applied on the external surface to estimate the compaction of the concrete mixes. The concrete volume in the cubic container was of 30 cm side. During compaction by a vibrator, AE was recorded in rates of more than 1000 hits/s until approximately 80 s. Then AE rate decreased and stayed constant to values around 500 hits/s implying according to the authors that compaction was completed, or that "the steady state" was reached. Changing the w/c ratio from 0.45 to 0.60 did not have strong effect on the time to reach the fully compacted stage, but a stiffer mixture tested (called "non-slump concrete") exhibited this transition from high to low AE energy much later.

1.4 Discussion

The above studies show that there is potential in AE monitoring of fresh cement-based material, as it contributes to the understanding of the complicated processes during hydration and also provides possible predictions for the final mechanical properties. However, the explanations of the sources of AE are not consistent and depend mainly on the assumptions allowed by the experience of the researchers and the accompanying measurement conditions.

It is characteristic that some studies find positive correlation between the sum of AE and final strength (Bardakov and Sagaidak 2016), while others find negative, with specimens of low AE activity at the fresh state, possessing higher strength at the hardened state (Pazdera et al. 2014; Iliopoulos et al. 2016a). These results are not necessarily contradictive as the different setups may be sensitive to other processes that take place at different times of curing (e.g. settlement at very fresh age, shrinkage cracking or hydration formation later). However, they indicate that a common understanding would be necessary to advance the field.

The basic mechanisms that are considered as sources in the literature are settlement of cement, migration of water through the capillary pores and air bubbles, cavitation, formation of hydrates, early micro-cracking, friction with the mould due to shrinkage, while cracking on the aggregate-paste interphase has been considered (McLaskey et al. 2007; Pease et al. 2003). These mechanisms may occur at different time zones but they certainly exhibit overlapping windows.

A characteristic point of the monitoring is the temperature peak. In some of the aforementioned studies, activity started or peaked at approximately that point, e.g. (Van Den Abeele et al. 2009; Chotard et al. 2001). However, other studies (Bardakov and Sagaidak 2016; Pazdera et al. 2014; Iliopoulos et al. 2016a; Dzaye et al. 2016; Dzaye et al. 2018; Dzaye et al. 2020; Topolár et al. 2017) showed the large majority of recorded hits earlier than that, leading to the conclusions that significant processes (possibly of lower intensity and thus more difficult to register) occur from as early as the mixing time, much before the chemical reaction of hydration and any heat is developed. This type of mechanical sources has also been studied in relation to AE for characterizing the deformation of granular media due to jamming of grains and frictional slip of particles which release strain energy (Michlmayr and Or 2014). However, differences in the acquisition equipment (including sensor sensitivity), the coupling (with or without waveguide), the specimen size do not allow for robust conclusions relatively to some trends. It could be mentioned though that in most studies different stages are identified, the first of which is characterized by large AE numbers while later AE gradually stops. In the most recent ones that are able to get activity from the first moment of the test, meaning just after mixing, the AE curve looks like the examples of Fig. 1.6, where a period of silence is found between the initial high activity period and the moderate activity at the end (e.g. indicatively 5–8 h in Fig. 1.6).

However, the work is still hardly enough to unambiguously explain the origin of the AE activity and to contribute to specific open questions:

- Can the "setting time" be characterized by AE?
- Can the mixture proportion be reliably investigated?
- Can the influence of different type of admixtures or additives be evaluated?
- Can the suitability of a mixture be assessed in the first hours just by passive AE monitoring (or combination of techniques)?
- Can the final properties be predicted by the AE activity of fresh cement-based material?
- Can other AE parameters (apart from the hit number and possibly energy) seriously assist the characterization?
- Can AE provide feedback to the complex modelling of processes taking place in the microstructure of hydrating cement, basically can the different AE signals be identified according to their source?

Finally, another serious issue is the contribution of wave propagation through this changing medium. A certain excitation will be recorded with very different characteristics after propagating through fresh heavily damping low velocity paste or after propagation through stiff hardened cement (or concrete) what is well known from ultrasound experiments at fresh and hardened concrete (Robeyst et al. 2009a, b; RILEM 2011; Iliopoulos et al. 2016b; Popovics and Popovics 1998). In fact, fresh concrete can be characterized as a "suspension" of particles (cement, sand, aggregates, air bubbles) in water. The viscosity of the liquid matrix and its strong mismatch with the stiff aggregates as well as the cavities induces tremendous amount of damping and scattering attenuation. This has been demonstrated (Popovics and Popovics 1998) and treated through multiple scattering models like the well-known model of Waterman and Truell (1961), in studies aiming at ultrasonically examining the material since the fresh state (Aggelis et al. 2005). In particular the presence and content of air bubbles is crucial (Aggelis et al. 2005; Sayers and Grenfell 1993; Boutin and Arnaud 1995; Robeyst et al. 2008). Even in very small volume percentages (i.e. 0.1%) they are responsible for severe signal distortion compared to de-aired paste and a considerable delay in the rise of longitudinal and shear wave velocities (Zhu et al. 2011). They also cause very high scattering attenuation at moderate frequencies (i.e. below 200 kHz) attributed to the "bubble resonance effect" which is more pronounced in paste rather than mortar (Aggelis et al. 2005; Sayers and Grenfell 1993; Zhu et al. 2011). On the other hand, sand and larger aggregates dominate the attenuation behaviour through scattering above 300 kHz (Aggelis et al. 2005). It should also be mentioned that, especially for the early ages, compaction is very important, as it drastically reduces the air content and accelerates the settlement and better packing of cement particles in the mix (Zhu et al. 2011). Later, with the progress of hydration the material is gradually transformed to solid and thus in terms of wave propagation it has been well treated as a water saturated porous solid medium through Biot theory

(Sayers and Grenfell 1993; Zhu et al. 2011; Biot 1956). The transition between different forms has also a crucial effect on the transmitted frequency content, which for the same initial excitation, extends to much higher range after setting (De Belie et al. 2005; Robeyst et al. 2009a, b).

An also quite important issue is that with the development of material stiffness, other types of waves obtain significance. Initially after mixing the material supports only longitudinal waves due to its liquid form. Later and with the increase of shear modulus, shear and surface (Rayleigh) waves are also supported and actually may take up more of the energy from the excitation (Graff 1991). Therefore, although initially the acquired AE waveforms include only longitudinal wave components, at later ages they may well contain contributions from all different wave modes, with longitudinal dominating the early part of the signal and shear and Rayleigh arriving later within the waveform. This effect may be more straightforward in case the transducers are placed directly on the surface of the material, where the different waves can even be separated based on the waveform shape and content (McLaskey et al. 2007).

All above mentioned propagation effects are coupled to the AE results and therefore, so far, any discussion or conclusion on specific AE values (e.g. amplitude or frequency of emissions) corresponding to a certain mechanism is inconclusive since it is also a function of curing time that defines the damping and the stiffness of the matrix. To support this, the following indicative simulations are presented. The geometry of the simulations may be representative but not directly corresponding to any of the above-mentioned literature studies. The external mould and the waveguide have properties of steel (longitudinal velocity 6000 m/s, shear 3200 m/s, no damping), while the cement paste obtains two different set of properties: one corresponding to fresh (longitudinal velocity 800 m/s, shear negligible, high damping) and the other corresponding to hardened (longitudinal 4000 m/s, shear 2100 m/s, low damping). The excitation is in both cases one cycle of period 1 μs (basic frequency 1 MHz), occurring at the same point, as shown in Fig. 1.8. In the same figure the displacement fields after several μs are seen when the wave has already reached the waveguide. In Fig. 1.9a one can see the difference between the received waveforms through the sensor at the top of the waveguide despite the same excitation. The waveform through hardened material exhibits much higher intensity and has seemingly longer duration due to the limited damping of the materials that allows multiple reflections in this simulation. On the other hand, the waveform through fresh material is much weaker due to the lower elastic properties and the higher damping factor used. Figure 1.9b shows the corresponding power spectra showing not only the tremendous difference in magnitude but the loss of the higher frequencies in the case of fresh matrix. It is indicative that when the material is fresh, the acoustic impedance mismatch between paste and metal is huge, not allowing wave energy to propagate through the waveguide, as it is almost entirely reflected back to the paste (see the arrow in Fig. 1.8a). In Fig. 1.8b, a reflection is still obvious, but a considerable amount of energy enters the waveguide.

Therefore, propagation through the medium should be seriously taken into account when trying to interpret the AE results in this time-dependent material.

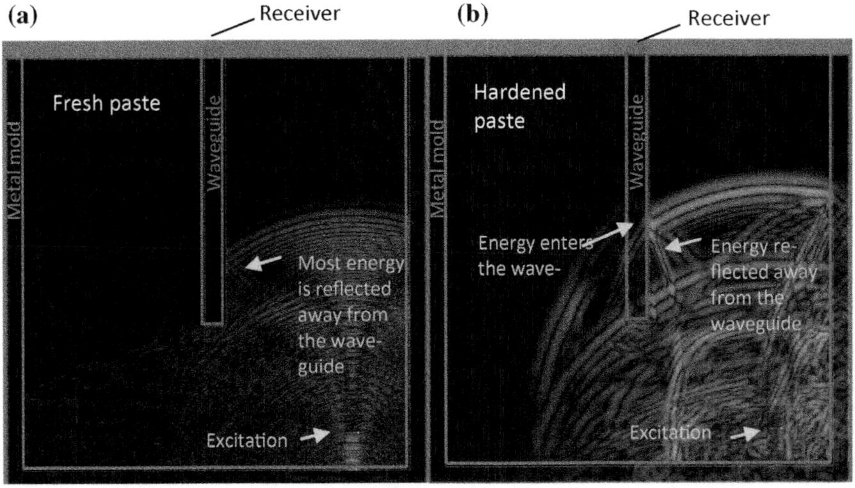

Fig. 1.8 Snapshot of strain field through numerical simulations in hydrating paste. The external dimensions of the mould are 80 × 80 mm and the thickness is 5 mm. The length of the waveguide is 50 mm, **a** fresh (liquid) matrix, **b** hardened matrix

It is certain that combined experimental techniques are necessary. Recently such a combination is realized in a collaborative project between VUB and UGent in Belgium. Since AE may be of chemical (e.g. hydrate formation) as well as mechanical (e.g. settlement, cracking) origin, measurements of temperature can enlighten the ongoing degree of hydration, while the settlement can also provide information on the mobility of the cement particles. Digital Image Correlation is also used to confirm the onset of shrinkage cracking at the surface, while capillary pressure measurements indicate when the paste has dried and air penetrates through micro-cracking. In any case the fresh material is extremely attenuative (much more than hardened concrete) and therefore, the experimental setup should be sensitive enough. Attaching multiple sensors on the metal mould seems to be a good experimental solution to receive the emitted activity. Even if, due to shrinkage, one surface of the specimen is detached from the mould wall, others will remain in contact providing continuous information.

On another notice, localization of AE events in the volume of the material would be very beneficial. This would help to clarify some cases, e.g. if an event is due to the friction with or detachment from the mould it would be localized on the concrete-mould interface. However, the tremendous attenuation of fresh cement paste or mortar would not allow capturing the same event by as many as four sensors to achieve 3D localization. As argued in Iliopoulos et al. (2016a) the attenuation of fresh paste and mortar is of the order of 2 dB/mm depending also on the frequency and composition. This means that for further propagation of 1 cm, the signal would lose 20 dB which is a tremendous loss. Practically it would be extremely difficult to record the same source by multiple receivers, taking into account that the sensors occupy some physical dimension of mm or cm and cannot be placed very close to

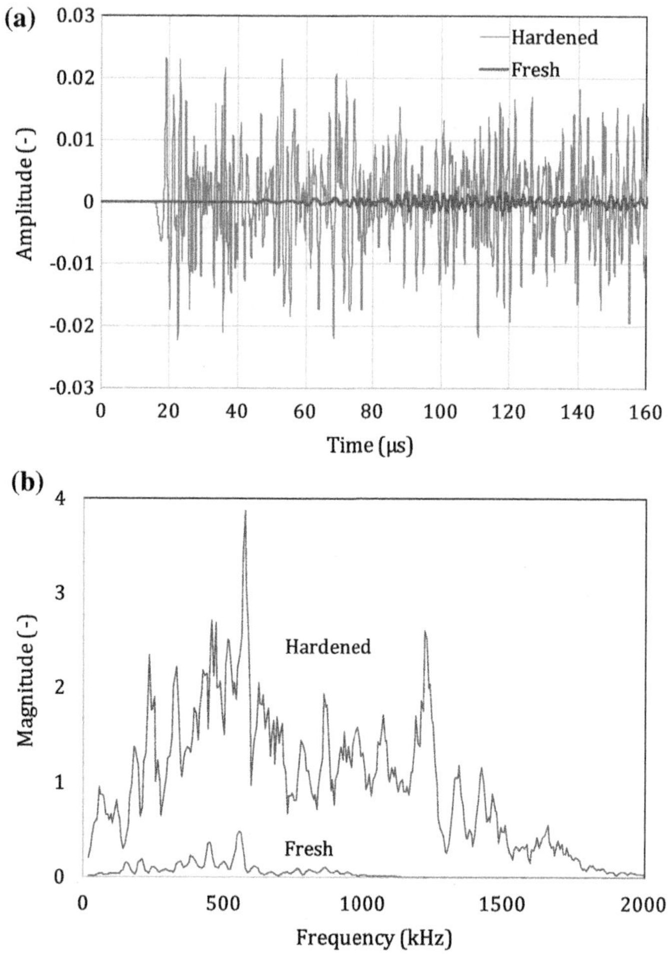

Fig. 1.9 **a** Simulated waveforms on the receiver of the geometry of Fig. 1.8 for hardened and fresh matrix. **b** FFT of the same waveforms

each other. Even if that was possible, then the accuracy would not be sufficient as the "gauge length" would be too small and comparable to errors in placement and signal reception. Furthermore, the pulse velocity in such a medium may present variations for different directions something that would contribute to the increase of errors.

1.5 Conclusions

AE monitoring is a field that can offer much to the characterization of cement-based materials especially at the sensitive stage of curing. If during this early stage the material is controlled, the possibility that later the performance is less than satisfactory is crucially diminished. A lot of meaningful research efforts have emerged recently showing the capability of the AE technique to correlate with different processes of properties. These efforts, however, are all case-specific, meaning that geometries, sensors, coupling types, monitoring periods are different. Some suggestions for future research include:

- Careful selection of the experimental setup including records of environmental conditions (ambient temperature, humidity, etc.) and concrete mixing process (compaction).
- AE experiments using multiple sensors with varying frequency response functions and varying positions, ensuring a good coupling throughout the experiments and taking into account attenuation and acoustic impedance mismatches.
- Designing experiments that would physically isolate as much as possible the different source mechanisms.
- Combination of multidisciplinary measurements to reduce the assumptions on the behaviour of the material and create different correlations with AE to help interpretation.
- Numerical simulation to understand the wave propagation in cementitious materials and to examine in what way does the received wave after propagation through cement and waveguide/mould correspond to the sources.
- A coordination in the field in order to steer the efforts in a common direction.

References

Aggelis, D. G. (2016). Acoustic emission analysis for nde in concrete (Chap. 5). In M. Ohtsu (Ed.), *Innovative AE and NDT Techniques for On-Site Measurement of Concrete and Masonry Structures*. RILEM State-of-the-Art Reports 20, pp. 69–88. https://doi.org/10.1007/978-94-017-7606-6_6.

Aggelis, D. G., Polyzos, D., & Philippidis, T. P. (2005) Wave dispersion and attenuation in fresh mortar: Theoretical predictions vs. experimental results. *Journal of the Mechanics and Physics of Solids, 53*(4), 857–883 (2005).

Assi, L., Soltangharaei, V., Anay, R., Ziehl, P., & Matta, F. (2018). Unsupervised and supervised pattern recognition of acoustic emission signals during early hydration of Portland cement paste. *Cement and Concrete Research, 103,* 216–225.

Bacharz, M., Goszczyńska, B., & Trąmpczyński, W. (2015). Analysis of destructive processes in unloaded early-age concrete with the acoustic emission method. *Procedia Engineering, 108,* 245–253.

Bardakov, V. V., & Sagaidak, A. I. (2016) Forecasting of concrete strength during the hardening process by means of acoustic emission method. PROGRESS in ACOUSTIC EMISSION XVIII.

In T. Shiotani, S. Wakayama, M. Enoki, & S. Yuyama (Eds.), *Proceedings of the 23rd International Acoustic Emission Symposium, the Inauguration Conference of International Institute of Innovative Acoustic Emission & the 8th International Conference on Acoustic Emission JSNDI & IIIAE.*

Biot, M. A. (1956). Theory of propagation of elastic waves in a fluid-saturated porous solid. I. Low-frequency range. *The Journal of the Acoustical Society of America, 28*(2), 168–78 (1956).

Boutin, C., Arnaud, L. (1995). Mechanical characterization of heterogeneous materials during setting. *European Journal of Mechanics. A. Solids, 14*(4), 633–656 (1995).

Carette, J., & Staquet, S. (2016). Monitoring and modelling the early age and hardening behaviour of eco-concrete through continuous non-destructive measurements: Part II. *Mechanical behaviour, Cement and Concrete Composites, 73,* 1–9.

Chotard, T. J., Barthelemy, J., Smith, A., Gimet-Breart, N., Huger, M., Fargeot, D., et al. (2001). Acoustic emission monitoring of calcium aluminate cement setting at the early age. *Journal of Materials Science Letters, 20,* 667–669.

De Belie, N., Grosse, C. U., Kurz, J., & Reinhardt, H.-W. (2005). Ultrasound monitoring of the influence of different accelerating admixtures and cement types for shotcrete on setting and hardening behavior. *Cement and Concrete Research, 35,* 2087–2094.

Dzaye, E., De Schutter, G., & Aggelis, D. G. (2016). Acoustic emission monitoring of fresh cementitious material. In *International RILEM Conference on Materials* (pp. 105–113). Lyngby, Denmark: Technical University of Denmark, 22–24 Aug 2016.

Dzaye, E., De Schutter, G., & Aggelis, D. G. (2018). Study on mechanical acoustic emission sources in fresh concrete. *Archives of Civil and Mechanical Engineering, 18,* 742–754.

Dzaye, E. D., De Schutter, G., & Aggelis, D. G. (2020). Monitoring early-age acoustic emission of cement paste and fly ash paste. *Cement and Concrete Research, 129,* 105964. https://doi.org/10.1016/j.cemconres.2019.105964

Graff, K. F. (1991). *Wave motion in elastic solids*. New York: Dover publications.

Grosse, C. U., & Ohtsu, M. (2008). *Acoustic emission testing*. Berlin: Springer.

Grosse, C., Aggelis, D., & Shiotani, T. (2016). Concrete structures. In M. Ohtsu (Ed.) *Innovative AE and NDT techniques for on-site measurement of concrete and masonry structures*. RILEM State-of-the-Art Report 20 (pp. 5–25). Berlin Heidelberg: Springer. https://doi.org/10.1007/978-94-017-7606-6.

Hodulàková, M., & Topolář, L. (2016). The assessment of the structural changes of fine-grained cement-based composites in the early-age using acoustic testing. *International Review of Applied Sciences and Engineering*. https://doi.org/10.1556/1848.2019.0004.

Iliopoulos, S. N., Khattabi, Y. E., & Aggelis, D. G. (2016a). Towards the establishment of a continuous nondestructive monitoring technique for fresh concrete. *Journal of Nondestructive Evaluation, 35*(3), article no. 37. https://doi.org/10.1007/s10921-016-0355-7.

Iliopoulos, S. N., Aggelis, D. G., & Polyzos, D. (2016b). Wave dispersion in fresh and hardened concrete through the prism of gradient elasticity. *International Journal of Solids and Structures, 78–79,* 149-159 (I.F2015 = 2.081). http://dx.doi.org/10.1016/j.ijsolstr.2015.09.005.

Lefever, G., De Boe, E., Aggelis, D.G., De Belie, N., Snoeck, D., & Van Hemelrijck, D. (2017). Monitoring the reduction in shrinkage cracking of mortars containing superabsorbent polymers. In *Proceedings of the 2nd International RILEM/COST Conference on Early Age Cracking and Serviceability in Cement-based Materials and Structures—EAC2* (pp. 221–226), September 12–14, 2017, Brussels.

Lura, P., Couch, J., Jensen, O. M., & Weiss, J. (2009). Early-age acoustic emission measurements in hydrating cement paste: Evidence for cavitation during solidification due to self-desiccation. *Cement and Concrete Research, 39,* 861–867.

McLaskey, G. C., Glaser, S. D., & Grosse, C. U. (2007). (desiccation). Integrating broad-band high-fidelity acoustic emission sensors and array processing to study drying shrinkage cracking in concrete. In *Proceedings of SPIE—The International Society for Optical Engineering*, 6529 PART 1, art. no. 65290C. https://doi.org/10.1117/12.715055.

Michlmayr, G., & Or, D. (2014). Mechanisms for acoustic emissions generation during granular shearing. *Granular Matter, 16,* 627–640.

Mix, P. E. (2005). *Introduction to nondestructive testing: A training guide*. Hoboken, New Jersey: Wiley.

Ohtsu, M. (2005) Acoustic emission in concrete of early-age (Chap. 5). In H.-W. Reinhardt, C. Grosse (Eds.), *Advanced testing of cement-based materials during setting and hardening*. RILEM Report 31, ISBN: 2912143705, RILEM Publ. S.A.R.L.: Cachan ENS, ca. 341 p.

Pazdera, L., Topolar, L., Korenska, M., Smutny, J., & Bilek, V. (2014). Advanced analysis of acoustic emission parameters during the concrete hardening for long time. In *11th European Conference on Non-Destructive Testing (ECNDT 2014)*, October 6–10, 2014, Prague, Czech Republic. (http://www.ndt.net/events/ECNDT2014/app/content/Paper/384_Pazdera.pdf).

Pease, B., Neuwald, A., & Weiss, J. (2003). The influence of aggregates on early age cracking in cementitious systems. In *Role of Concrete in Sustainable Development* (pp. 329–338), January 2003.

Popovics, S., & Popovics, J. S. (1998). Ultrasonic testing to determine water-cement ratio for freshly mixed concrete. *Cement, Concrete and Aggregates, 20*(2), 262–268.

Qin, L., Ren, H.-W., Dong, B.-Q., & Xing, F. (2014). Acoustic emission behavior of early age concrete monitored by embedded sensors. *Materials, 7,* 6908–6918. https://doi.org/10.3390/ma7106908.

RILEM Technical Committee (Masayasu Ohtsu). (2010). Recommendations of RILEM Technical Committee 212-ACD: acoustic emission and related NDE techniques for crack detection and damage evaluation in concrete: Test method for damage qualification of reinforced concrete beams by acoustic emission. *Materials and Structures, 43*(9), 1183–1186. https://doi.org/10.1617/s11527-010-9639-z.

RILEM Technical Committee (H-W. Reinhardt). (2011). Recommendation of RILEM TC 218-SFC: Sonic methods for quality control of fresh cementitious materials. Testing of fresh concrete by ultrasound transmission. *Materials and Structures, 44,* 1047–1062.

Robeyst, N., Gruyaert, E., Grosse, C. U., & De Belie, N. (2008). Monitoring the setting of concrete containing blast-furnace slag by measuring the ultrasonic p-wave velocity. *Cement and Concrete Research, 38,* 1169–1176.

Robeyst, N., Grosse, C. U., & De Belie, N. (2009a). Measuring the change in ultrasonic p-wave energy transmitted in fresh mortar with additives to monitor the setting. *Cement and Concrete Research, 39,* 868–875.

Robeyst, N., Grosse, C. U., & De Belie, N. (2009b). Monitoring fresh concrete by ultrasonic transmission measurements: Exploratory multi-way analysis of the spectral information. *Chemometrics and Intelligent Laboratory Systems, 95,* 64–73.

Sayers, C. M., & Grenfell, R. L. (1993). Ultrasonic propagation through hydrating cements. *Ultrasonics, 31*(3), 147–153.

Shiotani, T. (2008). Parameter analysis (Chap. 4). In C.U. Grosse, M. Ohtsu (Ed.), *Acoustic Emission Testing* (pp. 41–51). Berlin: Springer (2008).

Shiotani, T., Bisschop, J., & Van Mier, J. G. M. (2003). Temporal and spatial development of drying shrinkage cracking in cement-based materials. *Engineering Fracture Mechanics, 70,* 1509–1525.

Skal's'kyi, V. R., Koval', P. M., Serhienko, O. M., & Lotots'kyi, Y. L. (2004). Investigation of the solidification of concrete according to the signals of acoustic emission. *Materials Science, 40*(5).

Thirumalaiselvi, A., & Sasmal, S. (2019). Acoustic emission monitoring and classification of signals in cement composites during early-age hydration. *Construction and Building Materials, 196,* 411–427.

Topolár, L., Pazdera, L., Kucharczyková, B., Smutný, J., & Mikulášek, K. (2017). Using acoustic emission methods to monitor cement composites during setting and hardening. *Applied Sciences, 7*(5), 451. https://doi.org/10.3390/app7050451.

Trtnik, G., & Gams, M. (2014). Recent advances of ultrasonic testing of cement based materials at early ages. *Ultrasonics, 54,* 66–75.

Van Den Abeele, K., Desadeleer, W., De Schutter, G., & Wevers, M. (2009). Active and passive monitoring of the early hydration process in concrete using linear and nonlinear acoustics. *Cement and Concrete Research, 39,* 426–432.

Waterman, P. C., & Truell, R. (1961). Multiple scattering of waves. *Journal of Mathematical Physics, 2,* 512–537.

Zhu, J., Kee, S.-H., Han, D., & Tsai, Y.-T. (2011). Effects of air voids on ultrasonic wave propagation in early age cement pastes. *Cement and Concrete Research, 41,* 872–881.

Chapter 2
Ultrasonic Techniques for Determination and Monitoring Various Properties of Cementitious Materials at Early Ages

Ivan Gabrijel, Stéphanie Staquet, Markus Krüger, Jérome Carette, Christian U. Grosse and Gregor Trtnik

Abstract Ultrasonic techniques have been developed worldwide aimed to determine and monitor various important properties of cement-based materials. In recent years, advances in computer and experimental equipment have allowed the development of advanced ultrasonic systems. This chapter describes principles of determination of various properties of cement-based materials using ultrasonic techniques. Advantages and limitations of these techniques are briefly presented and discussed. State-of-the-art form of the chapter gives other researchers a comprehensive overview of the developed ultrasonic procedures. Due to their clear physical basis, accuracy, and ease of use, ultrasonic techniques have a great potential to soon become standard testing techniques for determination of setting and other properties of fresh and early-age cement-based materials. The first steps in this direction have already been performed.

Keywords Ultrasound · Cementitious materials · Hydration · Microstructure · Setting

I. Gabrijel (✉)
Department of Materials, Faculty of Civil Engineering, University of Zagreb, Zagreb, Croatia
e-mail: gabrijel@grad.hr

S. Staquet · J. Carette
BATir Department, Université Libre de Bruxelles ULB, Brussels, Belgium

M. Krüger
Institute of Technology and Testing of Construction Materials, Graz University of Technology, Graz, Austria

C. U. Grosse
Civil, Geo and Environmental Engineering, Technische Universität München, Munich, Germany

G. Trtnik
Construction Department, Igmat Building Materials Institute, Polje, Slovenia

Faculty of Civil and Geodetic Engineering, Department of Civil Engineering, Ljubljana, Slovenia

M. Serdar et al. (eds.), *Advanced Techniques for Testing of Cement-Based Materials*, Springer Tracts in Civil Engineering, https://doi.org/10.1007/978-3-030-39738-8_2

2.1 Introduction

Immediately after mixing cement and water, exothermic chemical reactions begin. These reactions are known as hydration and result into solidification process of cement-based materials (CBM). Consequently, various hydration products are formed, representing a solid phase within the material's microstructure.

Hydration and associated process of structure formation can be divided into different time periods (Ramachandran et al. 2002), as it is illustrated in Fig. 2.1. In the first stage, as soon as the cement and water get in contact, a short-lasting *pre-induction stage* starts during which individual cement particles are completely separated by water. This initial stage is followed by a low reactivity *dormant stage*, starting at time t_2. During this stage, several phenomena can be observed, e.g. intensive formation of ettringite needles, early C–S–H gel formation, loss of workability, internal settling process caused by gravity, migration of entrapped air bubbles to the surface of the material, etc. However, the material is still in its liquid form and highly workable. At a certain moment t_{PT} called *percolation threshold*, solid phase becomes interconnected. Due to perpetual development of the hydration products, a transformation of the material from its liquid to solid state occurs, which is generally known as *setting* of CBM. Two characteristic setting times can be defined, namely initial t_I and final t_F setting. This important and high-reactivity stage is known as *acceleration stage*. Generally, the following relations can be given:

$$t_{PT} \cong t_I \cong t_3 \tag{2.1}$$

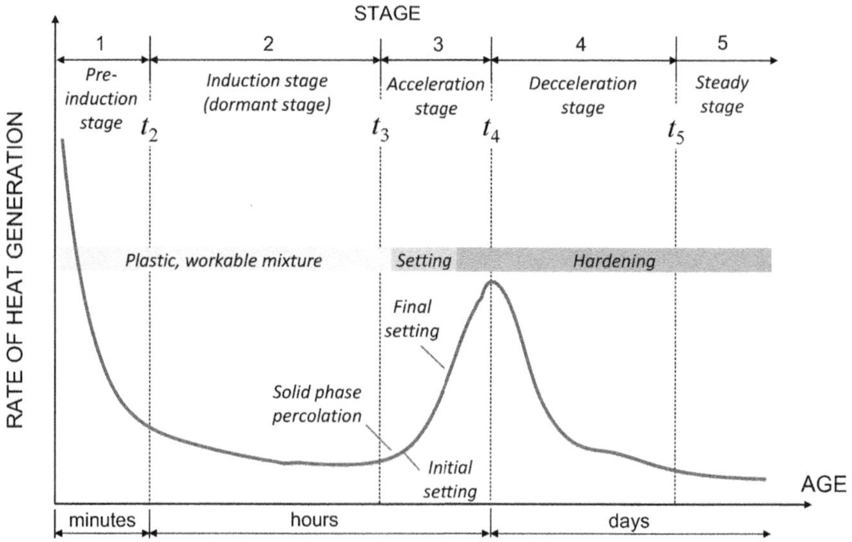

Fig. 2.1 Schematic representation of the overall hydration process evaluated by monitoring rate of heat generation

$$t_F < t_4 \tag{2.2}$$

where t_3 and t_4 stand for the beginning of the *acceleration* and *deceleration stage* of the solidification process, respectively. During the acceleration stage, a rapid increase in both total and connected solid phase within the material's microstructure can be observed which is later reduced during the deceleration stage. At a time t_5, the final (long-lasting) *steady stage* begins. Different stages of the structure formation process are schematically presented in Fig. 2.1.

Various testing techniques exist for determination of different phenomena that occur during the structure formation of CBM. Among others, calorimetric tests, Vicat method, and other penetration resistance techniques are frequently used. Although being widely accepted, these techniques have some limitations and disadvantages, e.g.: (1) the accuracy of these techniques usually depends on the technologist's skills, (2) the techniques can give different results for the same material, (3) the techniques are affected by the presence of (large) aggregates, and (4) they cannot usually be used directly in situ.

To overcome these inconveniences, advanced testing techniques to determine and monitor different phenomena and properties of early age CBM are constantly being developed. Such methods are usually fully automated, easy to perform, applicable directly in situ, and have a clear physical basis. Ultrasonic techniques are among the most frequently used methods and are of great interest. In the following, different ultrasonic techniques together with their advantages and limitations are briefly presented and described.

2.2 Ultrasonic Techniques for Measuring Early Age Properties of CBM

To establish a measurement of CBM properties during setting and hardening using ultrasound, a basic understanding of the wave propagation theory is highly recommended. In the following sub-chapters, some information about relevant issues related to the setup using different emitter-receiver configurations are given starting with an introduction to the mathematic-physical relations.

2.2.1 Theoretical Basics

A description of the physical phenomena of wave propagation would not be complete without an introduction to the different wave types, the derivation of these wave types from a generalized Hook's law and the relation of wave properties to elastic parameters. Following the literature (Landau and Lifschitz 1986; Reinhardt and Grosse 2005) this will establish a foundation to many ultrasonic applications where it is the intention to derive information about material properties of CBM like the Young's modulus, workability, setting, air content and others. The derivations are based on some pre-assumptions and these should be considered to decide, whether a selected measuring method is suitable for the task and if the elastic quantities can be determined with expected accuracy. Some physical and particularly geometrical boundary conditions can prevent relating the easily measurable wave velocities to the elastic modules like bulk modulus, shear modulus, Poisson ratio, and Lame's parameters.

2.2.1.1 Body Waves

The general form of an elastic wave in a solid is the body wave. It propagates as a spherical wave (generated at a point source) through a solid and has two modes which are distinguished by the polarization and which are usually coupled. The wave oscillating in the propagation direction is called *longitudinal* or *compression wave*. Because this wave arrives always as first at the measuring point it is also called P-wave (for primary). The wave which causes a particle movement transverse to the propagation direction is called *transversal* or *shear wave*. It is also called S-wave because it arrives as a secondary wave. In a fluid or in gases, no shear wave exists. In a suspension, the amplitude of these waves is smaller in relation to the compression wave amplitude in a solid. If there are boundaries present in a solid a third wave type emerges, namely surface waves. However, these waves are usually more difficult to be applied in the testing of hardening materials. In Fig. 2.2, propagation of compression and shear waves through a solid body is schematically presented.

2.2.1.2 Derivation of the Wave Equations for Isotropic Elastic Media

In the general case of a solid body, compression and shear waves are coupled. Under certain circumstances, the wave equations can be decoupled what leads to an important simplification. Usually the mechanic behaviour of a solid is described by continuum mechanics (Irgens 2008). A condition for the application of this theory to non-destructive testing is that the wavelength of the acoustic signal is largely greater than the microscopic dimensions (lattice constant) of the objects. Regarding small displacements, the behaviour of an ideal-elastic solid can be described by the generalized Hooke's law:

$$T_{ij} = C_{ijkl} \cdot D_{kl} \tag{2.3}$$

The equation relates the stresses with the deformation of a solid. In each point the elements of the deformation tensor $\boldsymbol{D}$, which is also called Green's deformation tensor, are linked to the stress tensor $\boldsymbol{T}$ by a linear function. $\boldsymbol{T}$ is a tensor of second order which must be symmetric due to the conservation of moments. The link between $\boldsymbol{T}$ and $\boldsymbol{D}$ is given by a tensor of fourth order, the elasticity tensor $\boldsymbol{C}$, the coefficients

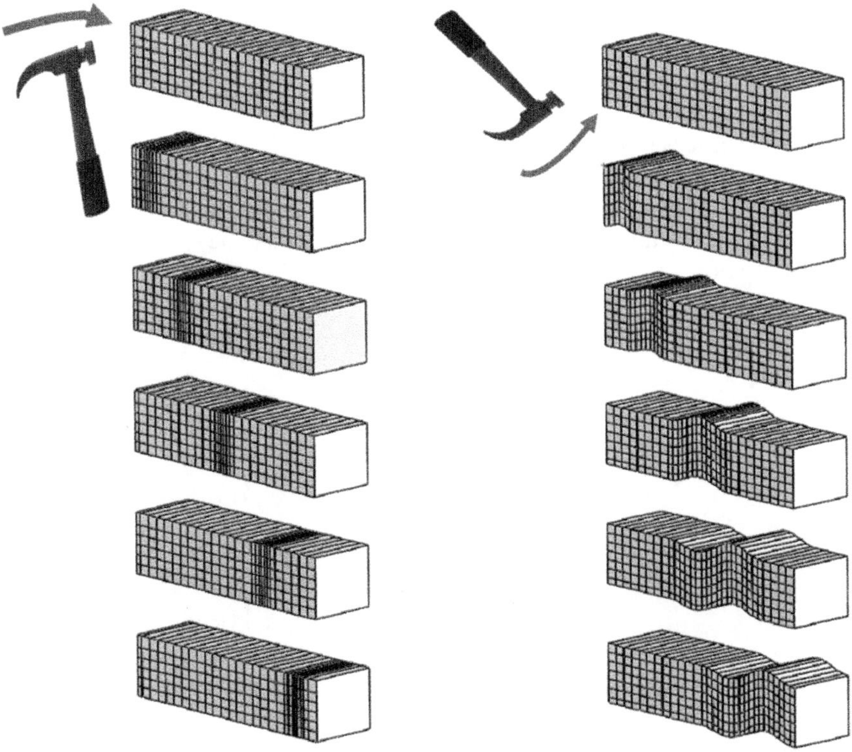

Fig. 2.2 Schematic of propagation of body waves: compressional waves (left) and shear waves (right) (Berckhemer 1990)

of which have the same dimension as the stress. By symmetry considerations (for instance $C_{ijkl} = C_{jikl} = C_{ijlk} = C_{jilk}$) follows (Landau and Lifschitz 1986) that the original 81 elements are reduced to 21 independent elements for a general anisotropic body. If there are symmetries in the stress-strain relation the number of independent components of the tensor and the number of coefficients is reduced again. A further simplification arises for isotropic materials. The number of independent elements of the elasticity tensor can be reduced to two using the following identities: $C_{1313} = ½(C_{1111} - C_{1122})$, $(C_{3333} = C_{1111})$ and $(C_{1133} = C_{1122})$. One can define the following two constants:

$$C_{1122} = \lambda \quad \text{and} \quad \frac{1}{2}(C_{1111} - C_{1122}) = \mu \tag{2.4}$$

The elasticity tensor simplifies to:

$$C_{ijkl} = \lambda\delta_{ij} \cdot \delta_{kl} + \mu\left(\delta_{ik}\delta_{jl} + \delta_{il}\delta_{jk}\right) \tag{2.5}$$

where δ is Kronecker symbol. For a homogeneous isotropic body Eq. 2.3 becomes:

$$T_{ij} = \lambda \Theta \delta_{ij} + 2\mu D_{ij} \tag{2.6}$$

This equation has only two parameters, λ and μ, that describe material behaviour and are called Lamé's constants. Θ is the trace of the deformation tensor ($D_{11} + D_{22} + D_{33}$) and is called *dilatation*. All elastic parameters follow Lamé's constants. E is the effective elasticity modulus, K the bulk modulus, G the shear modulus and ν the Poisson's ratio:

$$E = \mu \frac{3\lambda + 2\mu}{\lambda + \mu} \tag{2.7}$$

$$K = \lambda + \frac{2}{3}\mu \tag{2.8}$$

$$G = \mu = \frac{E}{2(1 - \nu)} \tag{2.9}$$

$$\nu = \frac{\lambda}{2(\lambda + \mu)} \tag{2.10}$$

The further description of the elastic behaviour of a solid can be described using *static equilibrium conditions*:

$$F_i + \frac{\partial T_{ij}}{\partial x_j} = 0 \tag{2.11}$$

A volume element is in equilibrium, if the sum of all body forces F_i vanishes. The transition from the static equilibrium condition (*d'Alembert's principle*) to the dynamic case follows from the addition of Newton's inertia forces. One gets the equation of motion in a generalized linear form:

$$\rho \frac{\partial^2 u_i}{\partial t^2} = F_i + \frac{\partial T_{ij}}{\partial x_j} \tag{2.12}$$

where ρ is the density and u_i is a component of displacement vector $\vec{u}$ in i direction. After introduction of the displacements u with:

$$D_{ij} = \frac{1}{2}\left(\frac{\partial u_j}{\partial x_i} + \frac{\partial u_i}{\partial x_j}\right) \tag{2.13}$$

one gets together with Eq. 2.6:

$$\rho \frac{\partial^2 u_i}{\partial t^2} = F_i + \frac{\partial}{\partial x_j}\left[\lambda \frac{\partial u_k}{\partial x_k}\delta_{ij} + \mu\left(\frac{\partial u_j}{\partial x_i} + \frac{\partial u_i}{\partial x_j}\right)\right] \tag{2.14}$$

In a homogenous medium λ and μ are spatially invariant. This allows the following transformations:

$$\rho \frac{\partial^2 u_i}{\partial t^2} = F_i + \left[\lambda \frac{\partial^2 u_k}{\partial x_j \partial x_k} + \mu\left(\frac{\partial^2 u_j}{\partial x_j \partial x_i}\right) + \mu\left(\frac{\partial^2 u_i}{\partial x_j \partial x_j}\right)\right] \tag{2.15}$$

and

$$\rho \frac{\partial^2 u_i}{\partial t^2} = F_i + (\lambda + \mu)\frac{\partial \Theta}{\partial x_i} + \mu \nabla^2 u_i \quad \text{with} \quad \Theta = \frac{\partial u_j}{\partial x_j} \tag{2.16}$$

Using the transformation $\nabla^2 \vec{u} = (grad\, div\, \vec{u} - rot\, rot\, \vec{u})$ Eq. (2.16) can be generalized for random Cartesian coordinate systems:

$$\rho \frac{\partial^2 \vec{u}}{\partial t^2} = \vec{F} + (\lambda + 2\mu) grad\ div\ \vec{u} - \mu\ rot\ rot\ \vec{u} \tag{2.17}$$

This equation is called *basic equation of elastodynamics.*

2.2.1.3 Decoupling of the Wave Equation

For homogeneous isotropic media the right term of Eq. 2.17 can be split into two independent equations for *compression* and *shear* waves. Another advantage of this splitting is that the two body waves can be handled separately. This is a considerable simplification for practical applications of the wave theory and of value for investigations of hardening materials where shear waves are assumed inexistent right after mixing and emerging during setting and hardening of CBM.

The mathematical separation is achieved through applying the operations of divergence (*div*) and rotation (*rot*), respectively, what results in a vortex-free and a source-free term. The following transformations are limited to small elastic deformations, which occur during wave propagation. Therefore, the force term $\vec{F}$ in Eq. 2.17 vanishes.

First, Eq. 2.17 is used for the generation of the divergence of Θ with respect to x_i. In order to eliminate the third addend with the help of identity $div\ rot\ \vec{u} = 0$:

$$\left[\rho \frac{\partial^2 \Theta}{\partial t^2} = (\lambda + 2\mu)\nabla^2 \Theta\right]_i \tag{2.18}$$

it follows:

$$\frac{\partial^2 \Theta}{\partial t^2} - \frac{(\lambda + 2\mu)}{\rho}\nabla^2 \Theta_P = 0 \tag{2.19}$$

The equation represents an inhomogeneous wave equation in which only the compression part of the displacement appears. Using the "divergence" operation, the rotation terms vanished. Investigating this wave type in more detail (compare to Fig. 2.2, left) one can see that particle motion is parallel to the direction of the wave propagation. This is the reason why this wave type is also called *longitudinal*.

Compression waves are vortex free. The expression $(\lambda + 2\mu)/\rho$ has a dimension of a squared velocity. It describes the propagation velocity of a compressional wave (V_P).

$$V_P = \sqrt{\frac{\lambda + 2\mu}{\rho}} = \sqrt{\frac{E(1-\nu)}{\rho(1+\nu)(1-2\nu)}} \tag{2.20}$$

Lamé's parameters are often expressed by the better-known parameter *modulus of elasticity* E (Young´s modulus) and *Poisson's ratio* ν (Landau and Lifschitz 1986). The Poisson's ratio describes the absolute relation of transversal to longitudinal deformations, i.e. the relative compression with relation to the relative dilatation. It yields

$$\nu = \left|\frac{D_{22}}{D_{11}}\right| = \left|\frac{D_{33}}{D_{11}}\right| = \frac{\lambda}{2(\lambda + \mu)} \tag{2.21}$$

To eliminate the divergence terms using the identity *rot grad* $\Theta = 0$ and applying the mathematical operation "rotation" the following equation is generated:

$$\left[\rho\frac{\partial^2}{\partial t^2}(rot\,\vec{u}) = \mu\nabla^2(rot\,\vec{u})\right]_i \tag{2.22}$$

If one puts $rot\,\vec{u} = 2\vec{\phi}$, Eq. (2.22) yields the form:

$$\begin{aligned}
&\frac{\partial^2\vec{\phi}}{\partial t^2} - \frac{\mu}{\rho}\nabla^2\vec{\phi} = 0 \quad \text{with} \\
&\vec{\phi} = \frac{1}{2}\left(\frac{\partial u_z}{\partial y} - \frac{\partial u_y}{\partial z}\right)\vec{i} + \frac{1}{2}\left(\frac{\partial u_x}{\partial z} - \frac{\partial u_z}{\partial x}\right)\vec{j} + \frac{1}{2}\left(\frac{\partial u_y}{\partial x} - \frac{\partial u_x}{\partial y}\right)\vec{k}
\end{aligned} \tag{2.23}$$

This expression is divergence free. It represents an inhomogeneous wave equation from the pure shear part. The waves cause a particle motion, which is perpendicular to the propagation direction as depicted in Fig. 2.2 (right). This wave type is hence called *transversal* wave. The term μ/ρ stands for the square of the shear wave velocity (V_S).

$$V_S = \sqrt{\frac{\mu}{\rho}} = \sqrt{\frac{E}{2\rho(1+\nu)}} \tag{2.24}$$

Both wave types propagate independently from another in homogeneous media. P- and S-waves are decoupled. If the elastic properties or the density of a medium changes P- and S-waves are immediately coupled. An example concerns boundaries between two homogeneous media. If the elastic properties change only steadily (over a distance, which is large compared to the wavelength) P- and S-waves coupling can be neglected. The shear modulus μ (and hence V_S) vanishes in fluids; i.e. no shear forces are transmitted. Contrary to that, for many solids (like stone) in a first approximation $\lambda \approx \mu$ is valid resulting in

$$V_S = \frac{1}{\sqrt{3}} V_P \tag{2.25}$$

The dependence of the wave velocities from the elastic parameters (Eqs. 2.20 and 2.24) can be used for the determination of E and ν under the condition that the density and the velocities V_P and V_S can be measured. However, in practical applications in non-destructive testing of CBM this can cause difficulties. Particularly the determination of the arrival time (onset) of a S-wave at the sensor can cause problems since they can be camouflaged by the oscillations of the faster P-wave, which is always simultaneously transmitted even if special shear wave transducers are used. The damping of concrete is high compared to other materials (like steel) so that sensitive sensors must be used. A significant high sensitivity can be achieved using sensors with a more resonant behaviour. However, then the onset of S-wave can be distorted by the Coda of the P-wave (strong oscillations) so that the determination of the shear wave velocity is becoming difficult, especially at later stages of hydration. One possibility for determining the onset time of the shear wave is by transformation into a time-frequency domain by means of a continuous wavelet transform (Krüger et al. 2011). The analysis using a continuous wavelet transform is based on the fact, that the S-wave is of lower frequency compared to the P-wave. Further on, it is usually more difficult to generate shear waves in concrete than compression waves. Emitters with special ability to excite shear waves are beneficial in measuring P- and S-wave velocities and elastic parameters can be more easily evaluated. Two examples for ultrasound wave transmission tests with different transducer settings are shown in Fig. 2.3. S-waves are almost invisible in the left graph where P-wave transmitters were used. S-waves are clearly visible in the right graph as soon as the fluid suspension turns into a water-saturated solid what will be explained below.

2.2.1.4 Relations Between Elastic Parameters

In ultrasound testing of fresh CBM it is often useful to derive the elastic parameters related to compression, shear and torsion (*E, G, K*), Poisson's ratio (ν) and Lamé's parameters (μ and λ) from the velocities. For Poisson's ratio the equation

$$\nu = \frac{\frac{1}{2}V_P^2 - V_S^2}{V_P^2 - V_S^2} \tag{2.26}$$

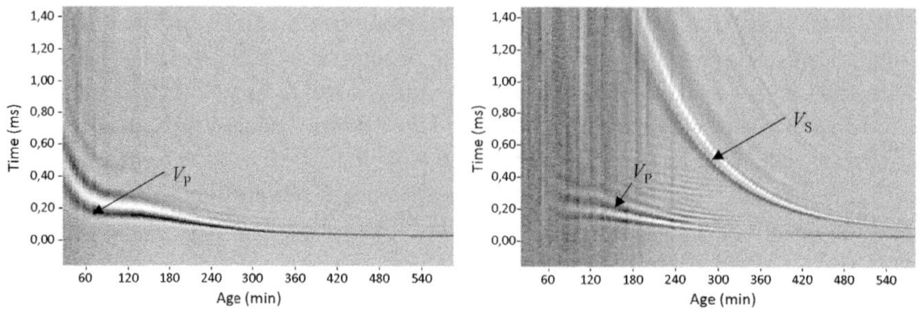

Fig. 2.3 Intensity graphs of normalized ultrasonic signals versus hydration time measured with P-wave transducers (left) and S-wave transducers (right) (cement paste, CEM I 42.5R, w/c = 0.42)

can be found according to the literature (Landau and Lifschitz 1986), while the effective elastic modulus of a specimen can be determined as:

$$E = V_P^2 \rho \frac{(1+\nu)(1-2\nu)}{(1-\nu)} \tag{2.27}$$

The determination of Young´s modulus with Eq. 2.27 can only be used, if the specimen is homogeneous and Poisson's ratio is assumed to be constant during hydration, which is not the case as the investigated mix changes from a suspension to a solid body. An overview of relations between the various elastic parameters is given in the Table 2.1.

Table 2.1 Relation between elastic parameters

	E	G	K	ν	μ	λ
E	–	$E = 2G(1+\nu)$	$E = 3K\cdot(1-2\nu)$	$E = 2\mu(1+\nu)$	$E = \frac{3\lambda+2\mu}{\lambda+\mu}\mu$	$E = \frac{\lambda(1+\nu)(1-2\nu)}{\nu}$
K	$K = \frac{E\mu}{3(3\mu-E)}$	$K = \frac{2G(1+\nu)}{3(1-2\nu)}$	–	$K = \frac{E}{3(1-2\nu)}$	$K = \frac{\lambda(1+\nu)}{3\nu}$	$K = \lambda + \frac{2}{3}\mu$
G	$G = \frac{3KE}{9K-E}$	–	$G = \frac{3K(1-2\nu)}{2(1+\nu)}$	$G = \frac{E}{2+2\nu}$	–	$G = \frac{\lambda}{2\nu} - \lambda$
ν	$\nu = \frac{1}{2} - \frac{E}{6\cdot K}$	$\nu = \frac{\lambda}{2(\lambda+G)}$	$\nu = \frac{3K-2\mu}{2(3K+\mu)}$	–	$\nu = \frac{E}{2\mu} - 1$	$\nu = \frac{\lambda}{3K-\lambda}$
μ	$\mu = \frac{E}{2\cdot(1+\nu)}$	–	$\mu = \frac{3KE}{9K-E}$	$\mu = \frac{\lambda}{2\nu} - \lambda$	–	$\mu = \frac{3}{2}(K-\lambda)$
λ	$\lambda = \frac{3K(3K-E)}{9K-E}$	$\lambda = K - \frac{2}{3}G$	$\lambda = \frac{3K\nu}{1+\nu}$	$\lambda = \frac{\nu\cdot E}{(1+\nu)\cdot(1-2\nu)}$	$\lambda = \frac{2\mu\nu}{1-2\nu}$	–

2.2.1.5 Extension of Wave Equation to Porous Solids

A theory of elastic wave propagation in solids was extended by Biot (1956a, b) for a solid body which consist of two phases: one phase representing purely elastic skeleton and another phase representing interconnected pores throughout the elastic skeleton. If it is assumed that pores are filled with water, stresses in the body can be transferred through solid and liquid phase. For the anisotropic case relation between stresses and strain can be written:

$$\begin{pmatrix} T_{xx} \\ T_{yy} \\ T_{zz} \\ T_{yz} \\ T_{zx} \\ T_{xy} \\ s \end{pmatrix} = \begin{pmatrix} C_{11} & C_{12} & C_{13} & C_{14} & C_{15} & C_{16} & Q_{xx} \\ \cdot & C_{22} & C_{23} & C_{24} & C_{25} & C_{26} & Q_{yy} \\ \cdot & \cdot & C_{33} & C_{34} & C_{35} & C_{36} & Q_{zz} \\ \cdot & \cdot & \cdot & C_{44} & C_{45} & C_{46} & Q_{yz} \\ \cdot & \cdot & \cdot & \cdot & C_{55} & C_{56} & Q_{zx} \\ \cdot & \cdot & \cdot & \cdot & \cdot & C_{66} & Q_{xy} \\ \cdot & \cdot & \cdot & \cdot & \cdot & \cdot & R \end{pmatrix} \begin{pmatrix} D_{xx} \\ D_{yy} \\ D_{zz} \\ D_{yz} \\ D_{zx} \\ D_{xy} \\ e \end{pmatrix} \quad (2.28)$$

where s represents the total normal tension force applied to the fluid part of the faces of the cube (representative volume element of porous solid) and e is the strain component for the fluid. The seven by seven matrix of coefficients constitutes a symmetric matrix with 28 distinct coefficients.

For isotropic case relation between stresses and strains can be written as follows:

$$\begin{pmatrix} T_{xx} \\ T_{yy} \\ T_{zz} \\ T_{yz} \\ T_{zx} \\ T_{xy} \\ s \end{pmatrix} = \begin{pmatrix} (2\mu+\lambda) & \lambda & \lambda & 0 & 0 & 0 & Q \\ \cdot & (2\mu+\lambda) & \lambda & 0 & 0 & 0 & Q \\ \cdot & \cdot & (2\mu+\lambda) & 0 & 0 & 0 & Q \\ \cdot & \cdot & \cdot & 2\mu & 0 & 0 & 0 \\ \cdot & \cdot & \cdot & \cdot & 2\mu & 0 & 0 \\ \cdot & \cdot & \cdot & \cdot & \cdot & 2\mu & 0 \\ \cdot & \cdot & \cdot & \cdot & \cdot & \cdot & R \end{pmatrix} \begin{pmatrix} D_{xx} \\ D_{yy} \\ D_{zz} \\ D_{yz} \\ D_{zx} \\ D_{xy} \\ e \end{pmatrix} \quad (2.29)$$

where μ and λ are Lamé's constants. Coefficient R is a measure of the pressure required on the fluid to force certain volume of the fluid into the elastic skeleton while the total volume remains constant. The coefficient Q is of the nature of coupling between the volume change of the solid and that of the fluid.

The displacement vector of the solid is defined as:

$$\vec{u} = u_x\vec{i} + u_y\vec{j} + u_z\vec{k} \quad (2.30)$$

If the average fluid displacement vector is defined as:

$$\vec{U} = U_x\vec{i} + U_y\vec{j} + U_z\vec{k} \quad (2.31)$$

then the strain in the fluid is defined by the dilatation:

$$e = \frac{\partial U_x}{\partial x} + \frac{\partial U_y}{\partial y} + \frac{\partial U_z}{\partial z} \tag{2.32}$$

The total mass of the fluid-solid per unit volume can be defined as:

$$\rho = \rho_{11} + 2\rho_{12} + \rho_{22} \tag{2.33}$$

The mass of the solid per unit volume (ρ_1) can be defined as:

$$\rho_1 = (1 - \beta)\rho_s \tag{2.34}$$

where β is the porosity and ρ_s is mass density of the solid.

The mass of the fluid per unit volume (ρ_2) can be defined as:

$$\rho_2 = \beta\rho_f \tag{2.35}$$

where ρ_f is mass density of the fluid. Since,

$$\rho = \rho_1 + \rho_2 = (1 - \beta)\rho_s + \beta\rho_f \tag{2.36}$$

it can be shown that:

$$\rho_1 = \rho_{11} + \rho_{12} \quad \text{and} \quad \rho_2 = \rho_{12} + \rho_{22} \tag{2.37}$$

where ρ_{12} is a mass coupling parameter between fluid and the solid.

Now the equations for propagation of elastic waves can be written:

$$\mu\nabla^2\vec{u} + grad[(\lambda + \mu)\Theta + Qe] = \frac{\partial^2}{\partial t^2}\left(\rho_{11}\vec{u} + \rho_{12}\vec{U}\right) \tag{2.38}$$

$$grad[Q\Theta + Re] = \frac{\partial^2}{\partial t^2}\left(\rho_{12}\vec{u} + \rho_{22}\vec{U}\right) \tag{2.39}$$

These 6 equations for the 6 unknown components of the displacements $\vec{u}$ and $\vec{U}$ determine the propagation of elastic waves. In the above equations it was assumed that the material is statistically isotropic so that for all cross sections we always observe the same ratio of the fluid area to solid area. Because of this assumption dilatational waves can be uncoupled from the rotational waves. Applying the operations of divergence *(div)*

$$div\,\vec{u} = \Theta \quad \text{and} \quad div\,\vec{U} = e \tag{2.40}$$

we obtain

$$\nabla^2[(\lambda + 2\mu)\Theta + Qe] = \frac{\partial^2}{\partial t^2}(\rho_{11}\Theta + \rho_{12}e) \tag{2.41}$$

$$\nabla^2[Q\Theta + Re] = \frac{\partial^2}{\partial t^2}(\rho_{12}\Theta + \rho_{22}e) \tag{2.42}$$

These two equations govern the propagation of dilatational waves.

By applying the rotation (*rot*) operation to Eqs. 2.38 and 2.39

$$rot\,\vec{u} = 2\vec{\phi} \quad \text{and} \quad rot\,\vec{U} = 2\vec{\Phi} \tag{2.43}$$

we obtain

$$\frac{\partial^2}{\partial t^2}\left(\rho_{11}\vec{\phi} + \rho_{12}\vec{\Phi}\right) = \mu\nabla^2\vec{\phi} \tag{2.44}$$

$$\frac{\partial^2}{\partial t^2}\left(\rho_{12}\vec{\phi} + \rho_{22}\vec{\Phi}\right) = 0 \tag{2.45}$$

These two equations govern the propagation of pure rotational waves.

From Eqs. 2.44 and 2.45 wave velocity of rotational waves is:

$$V_S = \sqrt{\frac{\mu}{\rho_{11}\left(1 - \frac{\rho_{12}^2}{\rho_{11}\rho_{22}}\right)}} \tag{2.46}$$

The velocity of dilatational waves can be derived from Eqs. 2.41 and 2.42. Since the expression for velocity of dilatational waves is too long to be represented using 4 elastic constants (μ, λ, R, Q) it is convenient to introduce additional parameters. Therefore, we can introduce reference velocity V_C defined by:

$$V_C^2 = \frac{H}{\rho} \quad \text{with} \quad H = \lambda + 2\mu + R + 2Q \tag{2.47}$$

$$a_{11} = \frac{\lambda + 2\mu}{H} \quad a_{22} = \frac{R}{H} \quad a_{12} = \frac{Q}{H} \tag{2.48}$$

$$\gamma_{11} = \frac{\rho_{11}}{\rho} \quad \gamma_{22} = \frac{\rho_{22}}{\rho} \quad \gamma_{12} = \frac{\rho_{12}}{\rho} \tag{2.49}$$

The a_{ij} parameters define the elastic properties of the material while the γ_{ij} parameters define its dynamic properties. With these parameters Eqs. 2.41 and 2.42 can be written:

$$\nabla^2[a_{11}\Theta + a_{12}e] = \frac{1}{V_c^2}\frac{\partial^2}{\partial t^2}(\gamma_{11}\Theta + \gamma_{12}e) \tag{2.50}$$

$$\nabla^2[a_{12}\Theta + a_{22}e] = \frac{1}{V_c^2}\frac{\partial^2}{\partial t^2}(\gamma_{12}\Theta + \gamma_{22}e) \tag{2.51}$$

Solutions to these equations can be written in the form:

$$\Theta = C_1 \exp[i(kx + \omega t)] \tag{2.52}$$

$$e = C_2 \exp[i(kx + \omega t)] \tag{2.53}$$

where k is wavenumber and ω is angular frequency. The phase velocity of these waves is:

$$V = \frac{\omega}{k} \tag{2.54}$$

Velocity can be calculated by substituting parameters Θ and e in Eqs. 2.50 and 2.51 with solutions for these parameters from Eqs. 2.52 and 2.53. Additionally, if we define:

$$z = \frac{V_C^2}{V^2} \tag{2.55}$$

we obtain

$$z(a_{11}C_1 + a_{12}C_2) = \gamma_{11}C_1 + \gamma_{12}C_2 \tag{2.56}$$

$$z(a_{12}C_1 + a_{22}C_2) = \gamma_{12}C_1 + \gamma_{22}C_2 \tag{2.57}$$

Eliminating C_1 and C_2 yields an equation for z,

$$\left(a_{11}a_{22} - a_{12}^2\right)z^2 - (a_{11}\gamma_{22} + a_{22}\gamma_{11} - 2a_{12}\gamma_{12})z + \left(\gamma_{11}\gamma_{22} - \gamma_{12}^2\right) = 0 \tag{2.58}$$

This equation has two roots corresponding to two velocities of propagation and shows that there are two dilatational waves: high velocity dilatational wave (V_1) and low velocity dilatational wave (V_2).

$$V_1^2 = \frac{V_C^2}{z_1} \quad \text{and} \quad V_2^2 = \frac{V_C^2}{z_2} \tag{2.59}$$

Equations for elastic wave propagation can be further extended to a case when dissipation of wave energy occurs, if it is assumed that there is a movement of the fluid through the pores relative to the solid skeleton. Dissipation of energy is dependent on the type of flow through the pores which can be laminar (Poiseuille type), transitional or turbulent. Assumption of laminar flow is valid until the Reynolds number reaches a certain critical value. This assumption also breaks down when we exceed a certain characteristic frequency. According to Biot (1956a), for a porous material we may assume that laminar flow breaks down for frequencies higher than

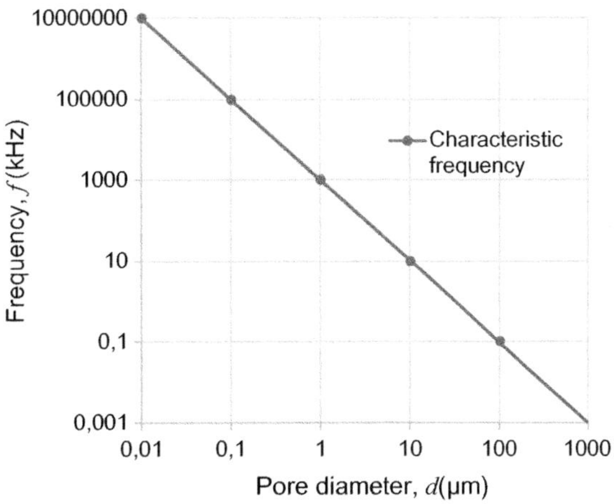

Fig. 2.4 Characteristic frequency as a function of pore diameter calculated according to Eq. 2.60 for water ($\upsilon = 0.00127$ Pa s)

$$f_t = \frac{\pi \upsilon}{4\rho_f d^2} \tag{2.60}$$

where υ is the fluid viscosity and d is the pore diameter. When the frequency of oscillatory motion is higher than the characteristic frequency, wavelength becomes small enough to interact with a fluid in the pores. In Fig. 2.4 f_t is calculated for pore diameter varying from 0.01 µm to 1 mm according to Eq. 2.60 for the case when water is the fluid in the pores.

If it is assumed that the flow of the fluid through the pores is laminar (low frequency range) than dissipation can be introduced by expanding Eqs. 2.38 and 2.39:

$$\mu \nabla^2 \vec{u} + grad[(\lambda + \mu)\Theta + Qe] = \frac{\partial^2}{\partial t^2}\left(\rho_{11}\vec{u} + \rho_{12}\vec{U}\right) + b\frac{\partial}{\partial t}\left(\vec{u} - \vec{U}\right) \tag{2.61}$$

$$grad[Q\Theta + Re] = \frac{\partial^2}{\partial t^2}\left(\rho_{12}\vec{u} + \rho_{22}\vec{U}\right) - b\frac{\partial}{\partial t}\left(\vec{u} - \vec{U}\right) \tag{2.62}$$

where b is related to Darcy's coefficient of permeability k by:

$$b = \frac{\upsilon \beta^2}{k} \tag{2.63}$$

In a case where the assumption of a laminar flow in not valid (high frequency range) than the coefficient b is replaced by $bF(\kappa)$ in Eqs. 2.61 and 2.62 where $F(\kappa)$ is a function of frequency and viscosity of the fluid.

2.2.1.6 From Suspension to Porous Solid

Suspensions are mixtures of solid particles and a fluid. Simple suspensions are binary mixtures which contain one type of solid material and one type of fluid. Monodisperse suspensions are simple suspensions where solid particles vary only slightly in diameter while polydisperse suspensions have a broader range of particle sizes. In suspensions, solid particles are not mutually interconnected. Therefore, behaviour of the suspension will be similar to a fluid regarding the wave propagation. This behaviour is also greatly influenced by the concentration of solid particles, i.e. the ratio of solid to fluid volume. Dense suspensions with a high concentration of solid particles can be treated as unconsolidated porous solid material.

Neat cement paste which is a mixture of cement and water is therefore a polydisperse suspension in which fluid phase contains a certain volume of air introduced during the mixing process. In plastic mixtures air exists in the form of spherical bubbles (Powers 1968) covering the range of bubble diameters from 10 to 4000 μm. A 1% volume concentration of air bubbles in water produces a compressibility that is about 200 times larger than that of the water alone (Temkin 2005). Because of its high compressibility, another interesting effect arising from the presence of air bubbles is the bubble resonance (see Fig. 2.5). This effect may be present during ultrasonic wave propagation because natural frequency of air bubbles entrapped within cement paste corresponds to the frequencies of ultrasonic waves usually applied for testing. As an example, natural frequency for air bubbles in water at 15 °C is 6.02×10^5 Hz for air bubbles with diameter equal to 10 μm and 5.59×10^4 Hz for air bubbles with diameter equal to 100 μm (Temkin 2005). Both effects influence the propagation velocity and attenuation of acoustic waves.

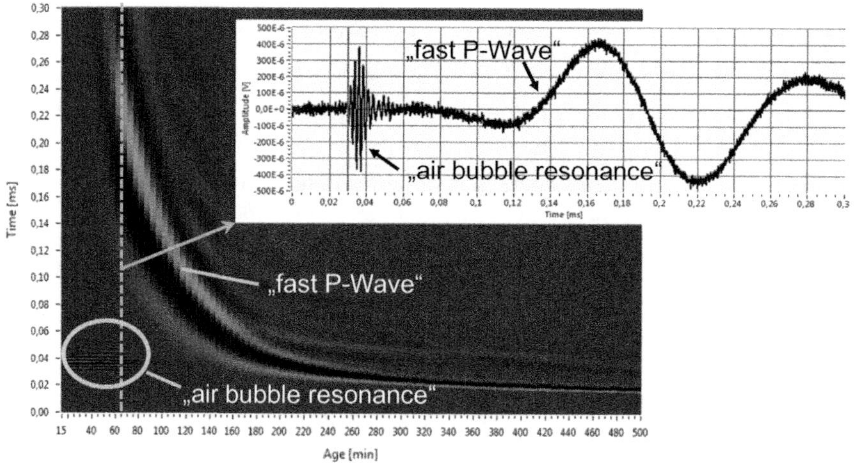

Fig. 2.5 Intensity graph of normalized ultrasonic signals versus hydration time with resonance caused by air bubbles and fast P-wave (cement paste, CEM I 42.5R, W/C = 0.36)

Neat cement pastes with w/c ratio in the range of 0.3–0.7 will have the volume content of solid particles in the range 30–50% so from that perspective it can be treated as suspension while it is in the fresh (liquid) state. The effective wave velocity and attenuation of an ultrasonic wave of angular frequency ω in suspension may be obtained from the real and imaginary parts of the complex wave number k, describing wave propagation in suspension (Sayers and Grenfell 1993).

$$k^2 = \omega^2 \overline{\beta} \rho_{\text{eff}}(\omega) \Rightarrow V^2 = \frac{1}{\overline{\beta} \rho_{\text{eff}}(\omega)} \tag{2.64}$$

In Eq. 2.64 $\bar{\beta}$ is the mean compressibility of the suspension and $\rho_{\text{eff}}(\omega)$ is an effective density given by equation:

$$\overline{\beta} = (1 - \varphi_s)\beta_f + \varphi_s \beta_s \tag{2.65}$$

$$\rho_{eff}(\omega) = \overline{\rho} - \frac{\varphi_s (1 - \varphi_s)\left(\rho_s - \rho_f\right)^2}{M(\omega) + \left[\varphi_s \rho_f + (1 - \varphi_s)\rho_s\right]} \tag{2.66}$$

$$\overline{\rho} = (1 - \varphi_s)\rho_f + \varphi_s \rho_s \tag{2.67}$$

where β_f, ρ_f, β_s, ρ_s are the compressibility and density of the fluid and the solid respectively, $\bar{\rho}$ is the mean density and φ_s is the volume fraction of solid. Harker et al. (1991) gave the following expression for $M(\omega)$ for suspensions:

$$M(\omega) = \frac{\rho_f S}{1 - \varphi_s} \tag{2.68}$$

The term S contains the hydrodynamic interactions between suspended particles and the host fluid and is given by:

$$S = \frac{1}{2} \frac{(1 + 2\varphi_s)}{(1 - \varphi_s)} + \frac{9}{4} \frac{\delta}{r} + i \frac{9}{4} \left(\frac{\delta}{r} + \frac{\delta^2}{r^2} \right) \tag{2.69}$$

$$\delta = \sqrt{\frac{2\upsilon}{\omega \rho_f}} \tag{2.70}$$

where r is the radius of suspended particles and υ is viscosity of the fluid.

As the hydration continues, cement particles become connected so a porous elastic frame will be created which can be characterized by a bulk modulus K and a shear modulus μ. When the cement paste is in the fresh state K and μ can be assumed to be $K = \mu = 0$. When K and μ start to increase the structure of the cement paste can be treated as a porous solid, therefore wave propagation can be approximated using Biot's theory. For high frequency limit ($\omega \to \infty$) Johnson and Plona (1982) derived following solutions for three wave velocities:

$$V_S^2 = \frac{\mu}{(1-\beta)\rho_s + (1-\alpha^{-1})\beta\rho_f} \tag{2.71}$$

$$V_{1,2}^2 = \frac{D \pm \left[D^2 - 4(\rho_{11}\rho_{22} - \rho_{12}^2)(PR - Q^2)\right]^{1/2}}{2(\rho_{11}\rho_{22} - \rho_{12}^2)} \tag{2.72}$$

In these equations V_S, V_1 and V_2 are phase velocities of shear wave, fast compressional wave and slow compressional wave respectively, α is tortuosity and β is porosity. For the case in which solid frame is a single component homogenous medium elastic coefficients P, Q, R and D from Eq. 2.72 can be expressed through porosity, bulk modulus of fluid (K_f), bulk modulus of solid (K_s), bulk modulus of elastic frame (K_b) and shear modulus of both the elastic frame and of the composite (μ).

$$P = \frac{(1-\beta)[1-\beta - K_b/K_s]K_s + \beta K_s K_b/K_f}{1-\beta - K_b/K_s + \beta K_s/K_f} + \frac{4}{3}\mu \tag{2.73}$$

$$Q = \frac{[1-\beta - K_b/K_s]\beta K_s}{1-\beta - K_b/K_s + \beta K_s/K_f} \tag{2.74}$$

$$R = \frac{\beta^2 K_s}{1-\beta - K_b/K_s + \beta K_s/K_f} \tag{2.75}$$

$$D = P\rho_{22} + R\rho_{11} - 2Q\rho_{12} \tag{2.76}$$

In the limiting case where $K_b = \mu = 0$ shear wave and slow longitudinal wave velocities vanish as may be seen from Eqs. 2.71 and 2.72.

For low frequency range the slow compressional wave becomes diffusive and the shear wave velocity and fast longitudinal wave velocity are given by (Sayers and Grenfell 1993):

$$V_S^2 = \frac{\mu}{(1-\beta)\rho_s + \beta\rho_f} \tag{2.77}$$

$$V_1^2 = \frac{P + R + 2Q}{(1-\beta)\rho_s + \beta\rho_f} \tag{2.78}$$

In cement paste solid volume is composed of at least two distinct phases: hydration products and unhydrated cement and therefore bulk modulus K_s and K_b are dependent on the ratio of these two phases. In mortar or concrete aggregate particles are the third solid component that will also have influence on the value of bulk modulus. For such multiphase systems it is necessary to carry out some sort of homogenization to obtain the effective values of bulk modulus $K_{s,eff}$ and $K_{b,eff}$.

2.2.1.7 Relation Between Heterogeneity, Frequency and Attenuation

Despite of perfect crystals all materials exhibit inhomogeneities. For technical materials like CBM it depends on the length scale if it is going to be regarded as homogeneous or not. For applications of ultrasound, the size of inhomogeneities must be put into relation to the wavelength. If the wavelength is larger than the obstacle (aggregate, pore) the wave is almost undistorted in its propagation through the material—aside from attenuation effects. The relation between wavelength λ, compressional wave velocity V_P and frequency f is given by the following equation:

$$f = \frac{V_P}{\lambda} \tag{2.79}$$

Concrete contains aggregates with a maximum size of 16 mm or even 32 mm. In the hardened state this corresponds (according to the above equation) to a frequency between 250 and 125 kHz at $V_P = 4000$ m/s. Using Eq. 2.79 one can estimate the suitable measuring frequencies before testing. Otherwise, the smallest geometrical dimensions of the discontinuities in a material can be determined. The attenuation of a material limits the propagation of waves through this material resulting in a sort of spatial boundary. A wave cannot be detected in an unlimited (or large) distance from the source. Voids and air pores are a governing factor for attenuation concerning ultrasound wave propagation—this is certainly a limiting factor for CBM in the fresh state and for suspensions. Knowing the attenuation factor (damping) of a material one can predict the maximum propagation distance to a transmitter/receiver from an emitter source. Another source of amplitude attenuation of ultrasound acoustic wave exists. The spherical divergence (Militzer and Weber 1987) (*geometrical spreading*) describes the attenuation of the energy or intensity I due to the fact that the area of the wave front increases with distance r. Due to the conservation of energy the decay of the intensity of spherical waves must be inverse proportional to the surface of the sphere. Due to the relation between the intensity I and the amplitude A with $I = A \cdot A^*$ the following expression is obtained:

$$I_K \sim \frac{1}{r^2} \quad \Rightarrow \quad A_K \sim \frac{1}{r} \quad (for\ spherical\ waves) \tag{2.80}$$

To summarize, the dissipative damping is the attenuation of the intensity due to dispersion at inhomogeneities and absorption by inelastic processes and internal friction (transformation of kinetic and potential energy of the wave into heat). The amplitude attenuation depends on the travel distance r in an exponential way and can be described by $A \sim e^{-\vartheta r}$. The damping constant ϑ depends on the material and the frequency.

If one considers both damping types, the following expression for the amplitude is obtained:

$$\widehat{A}_K(r) = \frac{r_0}{r} A_{K,0} e^{-\vartheta(r-r_0)} \tag{2.81}$$

Mounting of sensors to a specimen is addressing issues of *coupling*. In cases where the damping of a material over time (due to hardening) is evaluated, the coupling of a sensor has a decisive influence. Since most ultrasound measurements at CBM use mounted sensors to a surface, it should be mentioned that it is difficult to establish identical coupling conditions for two different measurements, what is the basis for the comparison of damping values. A systematic error must be determined using Eq. 2.81 at practical applications of ultrasound in CBM.

2.2.1.8 Interaction of Waves with Boundaries and Interfaces

Scattering is the consequence of the combined processes of reflection, refraction and diffraction at surfaces caused by differences in density and geometry of inhomogeneities in the medium. If the inhomogeneities are small compared to wavelength, diffraction effects dominate the scattering (Rayleigh scattering). If the inhomogeneities are large compared to wavelength, reflection effects dominate the scattering (geometrical scattering).

Geometrical boundaries of a specimen and transitions to other mediums, layers or internal interfaces have a special importance for the wave propagation. If an elastic wave hits an interface at which the wave velocity changes, the phenomena of refraction and/or reflection occurs. In contrast to pure acoustic or electro-magnetic waves, which propagate as pure compressional or shear waves, both wave types occur as coupled body waves in solids. Thus, refraction and reflection becomes more complicated because the transition of one wave type into the other is possible and *mode splitting* (*mode conversion*) occurs (Shutilov 1988; Rikitake et al. 1987). Only two important details shall be discussed concerning the angle of refraction and the amplitude. Figure 2.6 depicts schematically a compression wave with the amplitude A_P, which impinge an interface at the transition zone between two materials. In the case that the material properties are not identical (*medium I* $\neq$ *medium II*) their *acoustic impedance* is different. In addition to the transmitted (T_{PP}) and reflected (R_{PP})

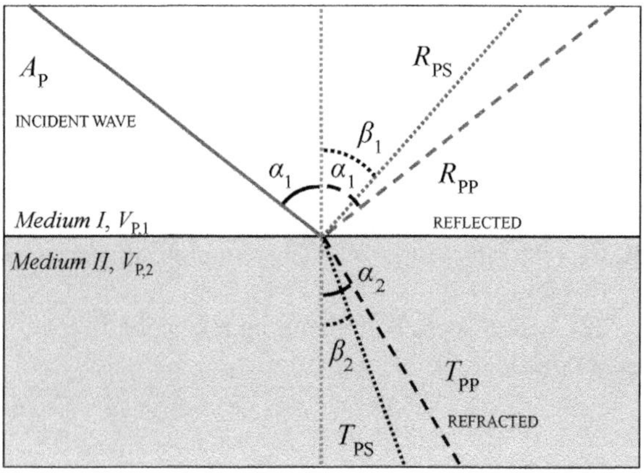

Fig. 2.6 Reflection and transmission of an incident P-wave with $V_{P,2} > V_{P,1}$

compression wave, a transmitted (T_{PS}) and reflected (R_{PS}) shear wave is generated. Under the condition, that *medium II* has a larger propagation velocity for body waves than *medium I* angles α_1, α_2, β_1 and β_2 occur as shown in Fig. 2.6.

According to Snell's law the following relations are valid

$$\frac{\sin\alpha_1}{V_{P,1}} = \frac{\sin\beta_1}{V_{S,1}} = \frac{\sin\alpha_2}{V_{P,2}} = \frac{\sin\beta_2}{V_{S,2}} \tag{2.82}$$

In this context, one can establish R_{PP} and R_{PS} to be *reflection coefficients* and T_{PP} and T_{PS} to be *transmission coefficients*, respectively. Analogue to this are the conditions for an impinging S-wave. One can show that the laws of geometrical optics can be translated to acoustic waves. The following derivations follow principally the Fresnel equations of classical optics. However, contrary to the Fresnel equations the acoustic waves have different modes and different phase velocities (refraction indices).

For the amplitude and phase of incoming waves there are analytical solutions. Their universal derivation for general interfaces is lengthy. The main effects can be described using an incoming P-wave with the amplitude A_1 which hits a free surface ($z = 0$). Then the problem is reduced to two dimensions. For the incoming P-wave the displacements in y-direction are irrespective. At the surface the stress components T_{zz} and T_{zx} (see Eq. 2.6) vanishes in the direction of the z-axis. As shown above, a P-wave generates a reflected P-wave and a reflected S-wave. Using Snell's law, one receives for the reflection coefficients R_{PP} and R_{PS}

$$R_{PP} = -\frac{\left(\cos^2 2\beta - \gamma^2 \sin 2\alpha \cdot \sin 2\beta\right)}{\cos^2 2\beta + \gamma^2 \sin 2\alpha \cdot \sin 2\beta} \tag{2.83}$$

$$R_{PS} = -\frac{2\gamma \sin 2\alpha \cdot \cos 2\beta}{\cos^2 2\beta + \gamma^2 \sin 2\alpha \cdot \sin 2\beta} \tag{2.84}$$

with

$$\gamma = \frac{V_S}{V_P} \quad \text{and} \quad \sin\beta = \gamma \cdot \sin\alpha \tag{2.85}$$

For the vertical ($A_\perp$) and horizontal ($A_\parallel$) components of the displacements one can write at $z = 0$.

$$A_{P,\perp} = (1 - R_{PP})\cos\alpha - R_{PS}\sin\beta = \frac{2\cos\alpha \cdot \cos 2\beta}{\cos^2 2\beta + \gamma^2 \sin 2\alpha \cdot \sin 2\beta} \tag{2.86}$$

$$A_{P,\parallel} = (1 + R_{PP})\sin\alpha - R_{PS}\cos\beta = \frac{2\cos\alpha \cdot \sin 2\beta}{\cos^2 2\beta + \gamma^2 \sin 2\alpha \cdot \sin 2\beta} \tag{2.87}$$

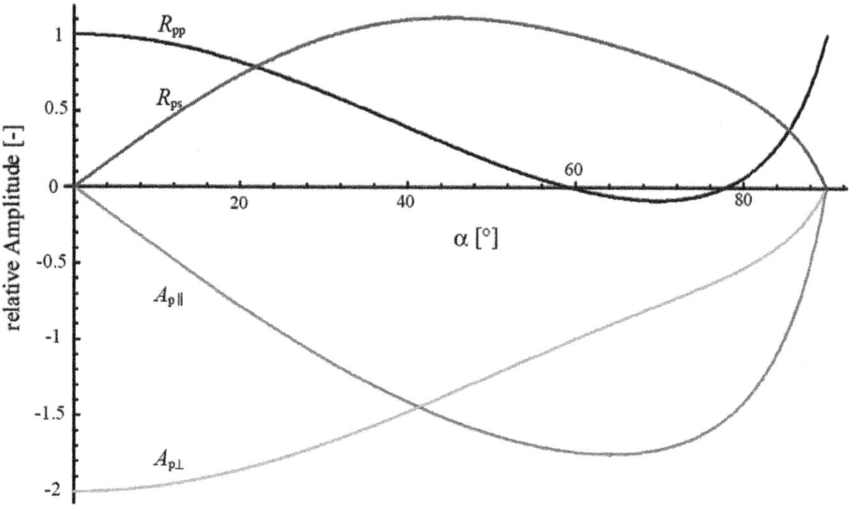

Fig. 2.7 Relative amplitudes of the reflected waves R_{PP} and R_{PS} as well as for $A_\perp$ and $A_\parallel$ depending on the incidence angle α

One can illustrate this fact by plotting these amplitudes and the reflection coefficients as a function of the incidence angle α as it is demonstrated in Fig. 2.7 (Grosse 1996).

If one considers a solid medium instead of a free surface the energy of the waves is transmitted partly at the surface into this medium. Again, mode conversion occurs. An important rule is that the amplitudes of transmitted and reflected waves depend strongly on the incidence angle. The theory for the amplitude distribution of an incoming P- and S-wave and the resulting reflected and refracted body waves is available in analytical form for the case $V_{P,1} > V_{P,2}$ and also for $V_{P,1} < V_{P,2}$. In the case of hardening materials, one would calculate a quantitative energy balance. It is important to state for further considerations that a significant part of wave energy is converted at the interface between two materials with different acoustic impedance due to reflection losses and mode splitting. It is principally not important if this is caused by an air-filled discontinuity or at a sudden change of the body wave velocity. Only the magnitude of the energy losses and the reflected energy are determined by the reflection and transmission coefficients, respectively. However, these findings assume that the wavelength of the incoming wave is one or several order of magnitudes smaller than the dimension of the discontinuity (interface). In case of air pores or other smaller discontinuities the described effects of reflection and transmission is negligible; diffraction and attenuation occur instead.

2.2.2 *Different Types of Ultrasonic Techniques*

2.2.2.1 Ultrasonic Wave Transmission Techniques

The basic principle of the ultrasonic wave transmission method is based on generating ultrasonic pulse and measuring the time t needed for a pulse to travel through the test specimen from transmitter (Tx) to receiver (Rx). Based on a known distance L between the transducers, the velocity of ultrasonic longitudinal waves V_P can be simply calculated according to Eq. 2.88.

$$V_P = \frac{L}{t} \tag{2.88}$$

During setting and hardening process of CBM, the velocity V_P is constantly increasing with different intensity and thus presents a promising parameter for monitoring solidification process of the material.

When performing transmission measurements, maintaining a good acoustic contact between the transducers and the CBM during the whole measurement is of paramount importance. The use of a special coupling agent is usually enough to avoid decoupling. Transducers can also be partly or even fully embedded into the test material to additionally improve the contact and to avoid a potential disconnection between the transducers and the specimen due to the shrinkage of the material. Instead of P-wave transducers, it is also possible to use S-wave transducers, if the aim is to conduct further analysis based on shear waves.

Different experimental setups exist to perform ultrasonic wave transmission technique. However, a vast majority of such instruments are based on a direct transmission technique, i.e. when the transducers are placed on two parallel opposite sides of the test specimen and ultrasonic pulse travels directly through the specimen. Although semi-direct or indirect techniques can also be used, the main differences between transmission techniques originate mostly from the shape and material of the mould to be used and the transducers' frequencies.

Transducers of frequencies between 20 and 500 kHz are usually used, depending on the distance between the transducers and type of the material to be tested. For monitoring early age properties of CBM, high frequency transducers are less recommended due to their high attenuation and small wavelength (Keskin et al. 2011). Short distance (L) between the transducers is often preferable in order to obtain stronger signal but may result into a so-called near-field effect at low V_P during very initial stages of the solidification period. Consequently, the selection of an optimal distance between the transducers (i.e. determining an appropriate size of the specimen) is quite challenging. This is influenced by the type and frequency of the transducers and the material's composition (e.g. the size and amount of aggregate). Based on the experience from several investigations, it can be assumed that the distance between the transducers should be in minimum three times the aggregate size. However, for frequencies of 54 and 82 kHz, $L = 15$ cm and $L = 10$ cm is usually recommended, respectively (Keskin et al. 2011).

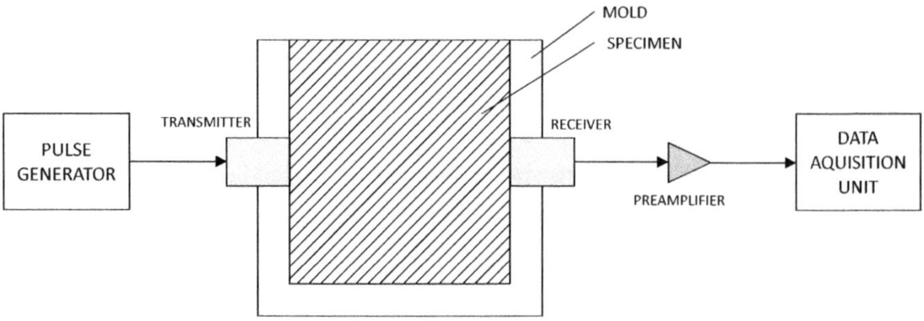

Fig. 2.8 Schematic representations of a typical experimental setup (transmitter/receiver circuit) used to perform ultrasonic wave transmission measurements

In Fig. 2.8, a block scheme of a typical experimental setup (transmitter/receiver circuit) used to perform ultrasonic wave transmission measurements is schematically presented.

An example of a reliable ultrasonic transmission test setup is the FreshCon system that was developed at the Institute of Construction Materials (IWB), University of Stuttgart and was further on commercialized by a spin-off company of the University of Stuttgart (see Fig. 2.9). The system can be used with differently sized containers for concrete and mortar either to get best test results or to reduce waste.

Figure 2.9 shows the experimental test setup that is used for P- and S-wave measurement on cementitious pastes, mortars and concrete (a. PC with DAQ; b. high voltage pulser; c. ultrasonic transducer; d. test container; e. preamplifier). Additionally, in Fig. 2.10 photo of a test container is shown. This test setup is an improved version compared to the one proposed by a RILEM recommendation worked out by RILEM Technical Committee TC 218-SFC (RILEM Technical Committee 2011). Besides the usage of a container equipped with P-wave transducers with a center frequency of 500 kHz an additional container with two broadband S-wave transducers with a center frequency of 250 kHz are used. A sensor distance of approximately 50 mm was found to be sufficient for both containers. The sensors are flexible mounted into the mould and are coupled by a replaceable thin polyimide film (d = 25 μm) directly

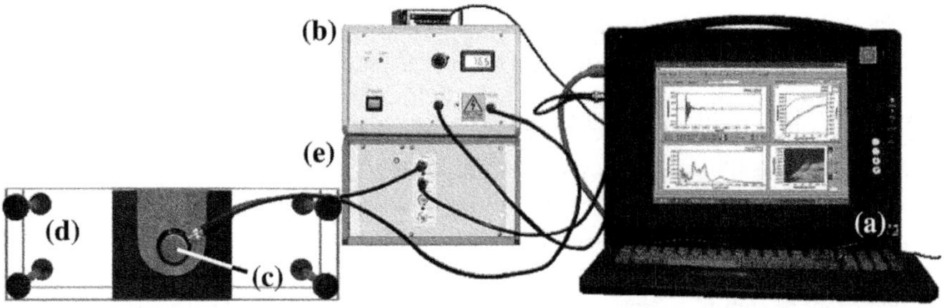

Fig. 2.9 FreshCon instrument, used to perform ultrasonic wave transmission measurements

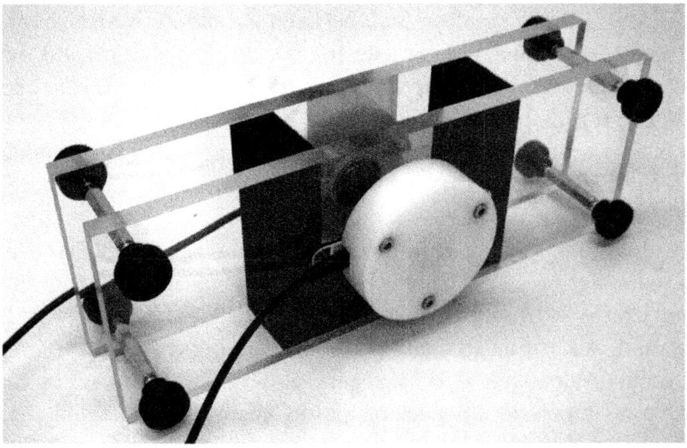

Fig. 2.10 FreshCon test container used to perform transmission measurements with shear wave and compressional wave transducers

through the mix to allow for best signal transmission and minimized risk of sensor decoupling even if shrinkage is observed. A detailed system description is given in Krüger and Lehmann (2010).

Aside the FreshCon instrument, moulds made of different materials are also used. A successful application of steel e.g. (Ye 2003; Ye et al. 2001, 2003, 2004; Krauss and Hariri 2006; Kamada et al. 2005), extruded polystyrene foam (Trtnik 2009; Trtnik et al. 2008, 2009; Trtnik and Gams 2013), aluminium foam (Aggelis et al. 2005; Liu et al. 2011; Zhang et al. 2012), and wood (Keskin et al. 2011) has been reported. However, in order to avoid ultrasonic pulse to travel through the mould during the very early hydration process and to avoid sensor decoupling by shrinkage, the test container must be carefully designed (see Fig. 2.10).

2.2.2.2 Ultrasonic Wave Reflection Techniques

At the interface between two materials with different acoustic impedances Z_1 and Z_2, a part of ultrasonic signal is transmitted into a test material 1 and another part is reflected into a buffer material 2. The principle of ultrasonic wave reflection technique for monitoring solidification process of CBM is based on this well-known phenomenon. During setting and hardening of CBM, the value of the acoustic impedance Z_1 of CBM is constantly changing. If Z_2 is constant, the following dimensionless ratio can be defined:

$$r(t) = \frac{Z_1(t) - Z_2}{Z_1(t) + Z_2} \tag{2.89}$$

Consequently, a dimensionless ratio between Z_1 and Z_2 is also time dependent and thus presents a promising parameter for measuring hydration and formation of structure of CBM. Generally, shear wave reflection coefficient $r(t)$ or reflection loss $RL(t)$ [or $WRL(t)$] is defined for this case.

Even though various materials can be effectively used as a buffer material when reflection technique is applied, quartz (Valič 2000; Jupe et al. 2012) or steel (Öztürk et al. 1999, 2004, 2006; Voigt 2004; Voigt et al. 2005a, b, 2006; Subramaniam et al. 2000, 2002; Sun 2005) are often utilized. Additionally, poly-methyl-methacrylate (PMMA), (Gimet et al. 1999; Chotard et al. 2001a, b; Subramaniam et al. 2005), polystyrene (Chung 2010; Suraneni 2011) and acrylic glass (Öztürk et al. 2004) were effectively used. According to a comprehensive study performed by Voigt (2004), the buffer medium does not affect the general development of $r - t$ (or $RL - t$) curves but has an important influence on the absolute values of r or RL, i.e. the range in which r (or RL) changes. The $r - t$ curve assigned to the PMMA buffer material shows the largest bandwidth followed by quartz and steel. Consequently, various phenomena that occur during the solidification process of CBM can be identified more clearly when buffer material made of PMMA is used. PMMA (or quartz) are therefore usually recommended as a buffer material when performing ultrasonic wave reflection technique.

Various experimental setups are being successfully used to perform different types of ultrasonic wave reflection method. Such setups differ mainly in the buffer material as has been briefly indicated in the previous paragraph. Usually, in order to perform the measurement a small amount of test material (CBM) has to be put on a special element made of the buffer material (e.g. Valič 2000; Jupe et al. 2012). Unfortunately, due to the small testing specimen, such procedure is generally not appropriate for the materials with large aggregate size. Alternatively, incorporation of small buffer elements into a larger test sample is frequently utilized. Steel plates are generally used for such purpose (Öztürk et al. 1999, 2004, 2006; Voigt 2004; Voigt et al. 2005a, b, 2006; Subramaniam et al. 2000, 2002; Sun 2005). For the case of simplicity, the bottom part of steel (Öztürk et al. 1999, 2006; Akkaya et al. 2003), PMMA (Gimet et al. 1999; Subramaniam et al. 2005; Chotard et al. 2001a, b) or polystyrene (Chung 2010; Suraneni 2011) moulds are seldom used as the buffer medium. In Fig. 2.11, three typical transmitter/receiver circuits are schematically presented.

2.2.2.3 Other Ultrasonic Techniques

There are methods closely related to ultrasound that can be used to evaluate the hardening and setting behaviour of CBM. The low-frequency vibroscope technique was developed in the late 1990s at the Ecole Nationale des Travaux Publics de l'Etat in Vaulx-en-Velin near Lyon, France, by Laurent Arnaud to be used for similar applications (Arnaud and Grosse 2005). There have also applications of the Impact Echo technique been reported by Stephen Pessiki at Lehigh University, USA (Grosse et al. 2005). According to the best knowledge of the authors of this book chapter, the

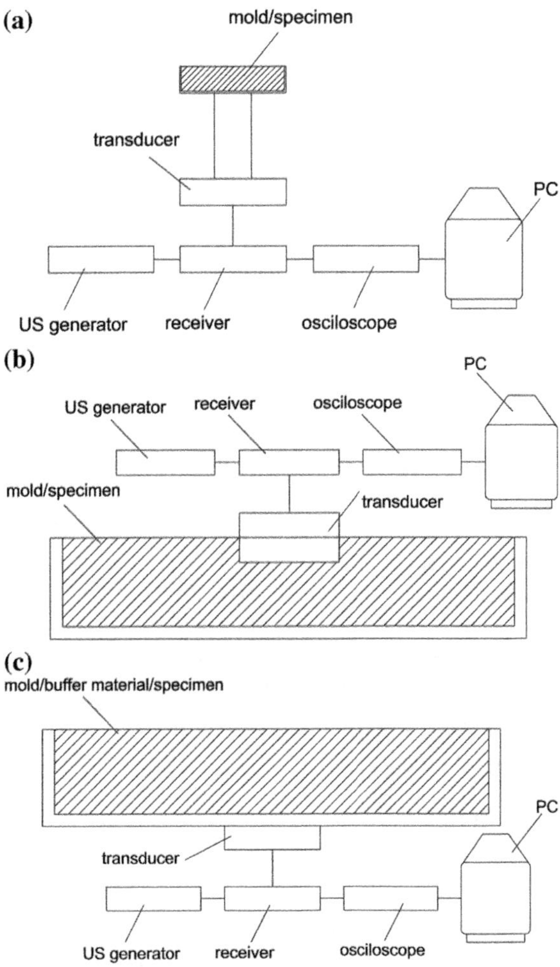

Fig. 2.11 Schematic representations of three typical experimental setups (transmitter/receiver setups) used to perform ultrasonic wave reflection measurements: **a** setup with a special element made of buffer material; **b** setup with a small buffer element incorporated in a larger test specimen; **c** setup with a mould as a buffer material

two mentioned techniques have not been further developed. A description of these techniques and a few others can be found in RILEM Report 31 (2005).

Acoustic emission (AE) techniques are similar to ultrasound with the difference that no active ultrasound generation is used but changes of the material produce elastic waves radiated from the source. Typically, AE techniques are used to record the formation of crack or crack propagation, but other type of changes can be recorded as well. The high sensitivity of the method enables AE to monitor processes like setting of concrete, segregation, shrinkage or bubble formation as well. More details can be found in this book in Chap. 1.

2.3 Monitoring of CBMs at Early Ages

2.3.1 Determination of Different Time Periods Within Structure Formation Process

As has been previously briefly explained in Sect. 2.1, the intensity of the structure formation of CBM at early ages is not constant but highly time-dependent process. Consequently, several stages can be defined during the structure formation process based on the intensiveness of formation of different hydration products. The evolution of V_P in time ($V_P - t$ curve) is strongly and directly influenced by the amount of the total and connected solid phase in the microstructure of CBM and is therefore an excellent parameter to follow this process. When the amount of the total and connected solid phase is increasing due to the increase of the amount of different hydration products, V_P is also increasing. This phenomenon is exploited to monitor solidification process of CBM at early ages.

Three stages (Chotard et al. 2001a; Ye 2003; Ye et al. 2003; Lee et al. 2004; Kamada et al. 2005; Robeyst et al. 2007; Darquenes et al. 2009; Khokhar et al. 2009; Mikulic et al. 2011; Liu et al. 2011; Zhang et al. 2012), or five stages (Feylessoufi et al. 2001; Trtnik 2009, 2012; Trtnik et al. 2009, 2013), during the structure formation process of CBM are usually defined by different researchers based on the $V_P - t$ evolution. These stages can be divided as follows:

- Stage I—Dormant stage
- Stage II—Period of rapid V_P increase
  - II-a—initial stage
  - II-b—acceleration stage
  - II-c—deceleration stage
- Stage III—Later stage or steady stage

Figure 2.12 shows two examples of a measurement on concrete mixture and cement paste mixture where different stages of V_P evolution are identified. Both mixtures in Fig. 2.12 are made with the same cement type and measurements are made with the same measuring system using 54 kHz transducers.

Start of the Stage II is usually very easy to identify because a steep increase in V_P begins. By plotting the rate at which V_P develops (dV_P/dt) a subdivision of the second stage is made on stages II-a, II-b and II-c in Fig. 2.12a. Inflection point PT1 defines the beginning of the stage II-b (acceleration stage) and inflection point PT2 defines the beginning of the stage II-c (deceleration stage). In case presented in Fig. 2.12b. inflection points PT1 and PT2 can't be identified so it is not possible to subdivide Stage II. From these two examples it is evident that type of subdivision can be influenced by the material tested. From mixtures tested without and with aggregate it can be seen that increase of the amount of aggregates always leads to a smoother increase of V_P, so that the different acceleration stages II-a, II-b and II-c become less visible.

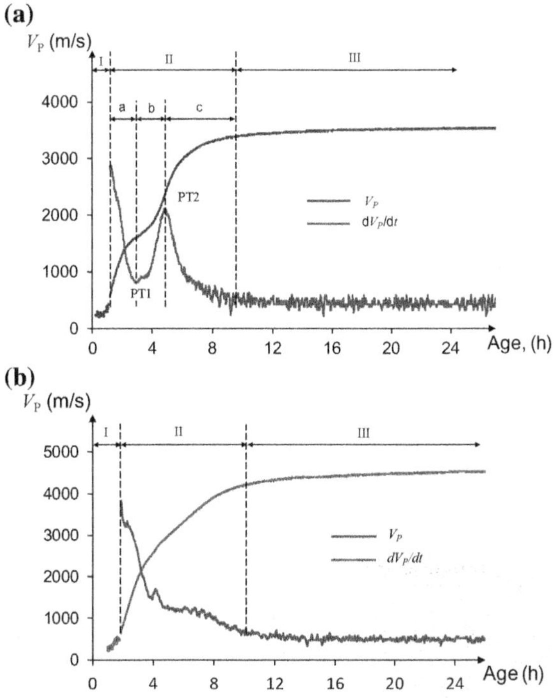

Fig. 2.12 Representation of different stages during formation of structure of CBMs based on the V_P – t curve determined experimentally for: **a** cement paste with w/c ratio 0.3 and made with cement type CEM I: division into three and five stages; **b** concrete with w/c ratio 0.4 and made with CEM I and maximum aggregate size 16 mm

First stage: I—Dormant stage

During the dormant stage transmission is governed by liquid phase and strong attenuation of ultrasonic signals occurs due to the air entrapped in the material (Keating et al. 1989; Sayers and Dahlin 1993; Arnaud and Thinet 2003; Aggelis and Philippidis 2004; Aggelis et al. 2005; Robeyst et al. 2008). Consequently, very low initial values of V_P (Reinhardt and Grosse 1996; Ye 2003; Ye et al. 2003; Kamada et al. 2005; Mikulić et al. 2005; De Belie et al. 2008; Liu et al. 2011; Zhang et al. 2012), usually even lower than V_P through air (i.e. 340 m/s) or even no V_P (Trtnik et al. 2008, 2009; Mikulić et al. 2011; Trtnik and Gams 2013; Cheng et al. 2013) can be observed during this initial period.

Second stage: II—Period of rapid increase of VP

Different hypotheses describe initial increase in V_P during the *Stage II-a*. Many researchers do not attribute it to setting only. The formation of ettringite needles (Reinhardt and Grosse 2004), early C–S–H formation (Robeyst et al. 2011), workability loss and thixotropy (Robeyst et al. 2011), internal settling process caused by gravity (Reinhardt and Grosse 2004; Mikulic et al. 2011; Aggelis and Grammenou 2011), increasing number of contacts between particles (Feylessoufi et al. 2001), and/or entrapped air bubbles migrating to the surface of the material (Ye et al. 2004; Aggelis and Grammenou 2011) are usually used to explain this initial steep increase in V_P. During the *Stage II-b*, the most intensive development of hydration products associated with the most intensive development of V_P in time can be noticed. This important stage finishes at the inflection point PT2, which defines the beginning of the deceleration stage. The rate of V_P evolvement decreases during this stage.

Third stage: III—Later stage or steady stage

Deceleration stage is followed by a long-lasting steady stage, during which most of the hydration products are interconnected. As a result, a very small increase in V_P can be observed during this final stage.

Similarly to the V_P, shear reflection coefficient r or reflection loss RL can also be effectively used as a parameter to monitor formation of structure of CBM at early ages. However, since S-waves do not propagate through the fluid-like viscous suspension, they are not sensitive to the formation of different "non-setting" or "pre-setting" phenomena described in the previous section. Consequently, before the initial setting time t_I, an almost constant value of r (or RL) is usually reported (Sun et al. 2004a, b; Sun 2005; Subramaniam et al. 2005; Trtnik 2009; Trtnik et al. 2009, 2013; Subramaniam and Wang 2010), during the dormant period. The initial increase in r (or RL) is therefore often defined as a beginning of the acceleration stage (Ye et al. 2003; Subramaniam et al. 2005; Trtnik et al. 2008; Trtnik and Gams 2013). Usually, acceleration stage is quite long, and many processes associated with the formation of structure of CBM occur. At the end of this stage, the intensive development of r (or RL) in time stops due to the decrease in the intensity of solidification process and the long-lasting steady stage begins.

2.3.2 Determination of Setting Process

Setting is usually defined as a period during which the material is changing its aggregate state from the initial, viscous suspension to the solid material. This short period defined by the initial t_I and final t_F setting times is considered as one of the most important periods within the process of structure formation. Consequently, an accurate determination and monitoring of setting period is of paramount importance.

Basic and the most direct approach to define setting period using ultrasonic transmission technique is to simply define threshold values of V_P which correspond to t_I and t_F characteristic times. The values of V_P in the range between 800 (Lee et al. 2004)—2700 (Van der Winden 1990) m/s and 1200 (Lee et al. 2004)—3180 m/s (Van der Winden 1990) have been frequently suggested to define initial and final setting of different CBM, respectively. As can be seen, a relatively wide range of threshold values is reported to define setting times because V_P is strongly influenced by the testing equipment and materials composition. In the case of larger amount of aggregate, larger aggregate size, and lower air content in the material, lower V_P values for the initial setting are generally reported. While a larger amount of aggregates will increase the overall wave velocity because they increase the total volume

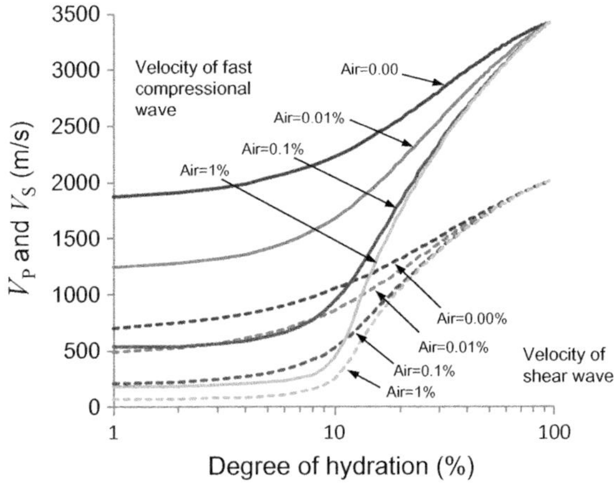

Fig. 2.13 Theoretical analysis of wave velocity for cement paste with w/c ratio = 0.4 according to Biot's theory: variation with degree of hydration and air void contents (percentage of the volume of cement paste)

of solid structure, small changes in the air content strongly influences the absolute value of V_P (see Fig. 2.13) especially during the very early hydration period. Whether a three-stage or five-stage evolution of V_P is recognized by ultrasonic transmission is also predominantly influenced by the mixture proportion. A high amount of solid aggregates will dominate V_P evolution often leading to a three-stage type, whereas tests on pure cementitious pastes that contain certain amounts of air show five stage type of V_P evolution.

Due to the wide interval representing V_P values at initial and final set, different alternative criteria have been proposed to determine these characteristic values more accurately. Such procedures are usually based on some characteristic changes and characteristic points on $V_P - t$ curves. The time of the inflection point PT1 on the $V_P - t$ diagrams (Voigt et al. 2005a; Robeyst et al. 2007, 2008; Trtnik et al. 2008; Trtnik 2009; Darquenes et al. 2009; Khokhar et al. 2009), the time when a significant increase in V_P occur (Garnier et al. 1995; Reinhardt et al. 2000; Ye et al. 2003; Zhang et al. 2010, 2012), or the time of the intersection of three straight lines tangent to the $V_P - t$ curve (Chotard et al. 2001a, b) is usually used to define the initial setting time t_I. Determination of final setting time is often less evident. It can be defined as the time when the decelerating stage on the $V_P - t$ curves starts (Robeyst et al. 2008; Khokhar et al. 2009), the time of maximum rate of change of V_P (Zhang et al. 2012), etc.

Figure 2.14a shows the evolution of V_P and V_S of a cementitious paste composed of Portland cement and limestone (50% CEM I 52, 5R, 50% limestone powder, w/c = 0.43). While the evolution of V_P in principle follows the five-stage division described in the previous chapter the evolution of V_S is of the three-stage type. From determined wave velocities the dynamic elastic properties can be calculated as it is described in Eqs. 2.26 and 2.27, Table 2.1 and plotted in Fig. 2.14b. In Krüger et al. (2016) it is assumed that the dynamic shear modulus of about 0.1 GPa is a good and reproducible indicator for initial setting independently from the mixture composition. However, in Krüger et al. (2018) it is as well discussed that the dynamic

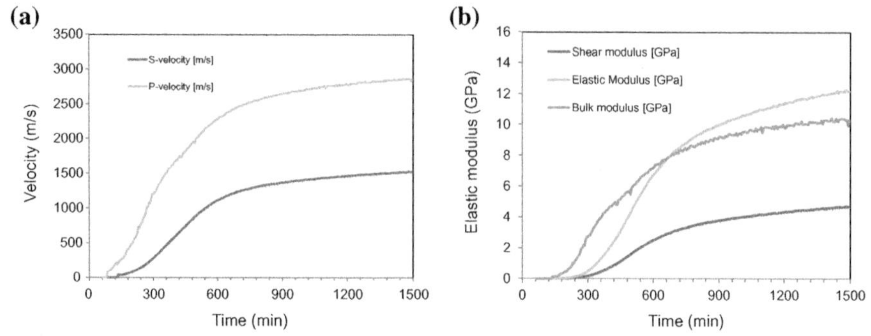

Fig. 2.14 Evolution of compressional and shear wave velocity (left) and dynamic compressional elastic, shear and bulk modulus (right) of a cementitious paste (Krüger et al. 2016)

shear modulus to determine initial setting is a function of packing density and thus not constant.

Based on the combined measurement of compression and shear wave transmission velocity, Carette and Staquet (2015) decomposed the setting process of cement-based materials as represented in Fig. 2.15. Several stages were successfully identified for many distinct mortar mixes presenting various binder natures (presence of limestone filler, blast-furnace slag, fly-ash), different w/c ratios and air contents (Carette and Staquet 2015, 2016c). During stage 1, a progressive increase in V_P is first observed, until it reaches a maximum rate of increase (δV_P^{max}). This point corresponds to the

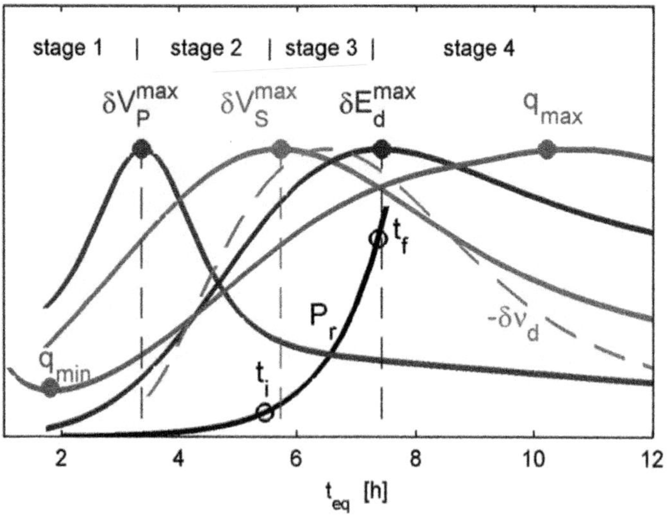

Fig. 2.15 Four stage decomposition of the setting process of mortar, based on shear (green curve) and compression (red curve) wave velocity and dynamic Young's modulus (blue curve). Penetration resistance (black curve) and heat flow (grey curve) measurements are also shown

end of the dormant period, marking the end of the first stage (stage I in Fig. 2.12b). In the second stage, the mechanical properties start to develop as indicated by the initiation of penetration resistance (P_r) and elastic modulus (E_d). This corresponds to a fast increase in the volume of hydrates, leading to a progressive solid percolation. This increase follows until several particles are connected, forming continuous solid paths through the saturated medium. This moment, defined as initial setting, corresponds to the time where the maximum rate of increase of V_S is observed (δV_S^{max}). The third stage spreads from initial to final setting times and is characterized by a fast increase in P_r, corresponding to a strong increase in E_d. This stage ends when the $E_d - t$ curve has reached its inflection point (δE_d^{max}). This time, defined as the final setting time, marks the initiation of a slower formation of hydration products, resulting in significant increase of the mechanical properties with a moderately decreasing rate. Stage 4 corresponds to the hardening process and is not part of the setting process itself. It is interesting to note that the setting process (stage 3) is also characterized by the quickest evolution of the dynamic Poisson's ratio ($\delta \nu_d$), due to the fast transition between a fluid with particles in suspension and a densifying viscoelastic material. Interestingly, this methodology can also be applied to the scale of concrete materials (Carette and Staquet 2016c).

As has been indicated in the previous sections, shear waves do not propagate through liquid substances. If it is assumed that liquid phase dominates the viscoelastic behaviour in fresh state, shear waves cannot propagate through CBM before the initial setting time t_I or are even very hard to measure (only at mixes that show certain yield stress resp. thixotropy). This phenomenon is usually considered when reflection technique is used to define setting period of CBM. Initial setting time is therefore defined as the time of the first significant increase on the $r - t$ curve. Contrary to the V_P, initial setting is often reported to occur at the same or very similar level of r, regardless of the material's composition (Valič et al. 1999; Sun et al. 2004a, b). However, other researchers have indicated that t_I cannot be defined at the same r level for different materials because it is influenced by the material's composition (Trtnik et al. 2012; Trtnik et al. 2013). Alternatively, an inflection point on the $r - t$ (Subramaniam et al. 2005) diagrams or the time when noticeable changes in rate of r occur (Voigt et al. 2004; Wang et al. 2010), can be used as an accurate indicator of the initial setting time of an arbitrary CBM.

Similarly, to the ultrasonic transmission technique, definition of final setting time based on r (or RL) values is less straightforward. Most researchers suggest defining a specific, material dependent value to define the final setting of CBM which can be determined after an adequate calibration. An example for such a correlation for ultrasonic wave transmission can be found in Grosse (2010).

2.3.2.1 Determination of Setting Based on Waveform Analysis

The previous sections indicate that it is rather difficult and often inaccurate to define both t_I and t_F characteristic times of arbitrary CBM based on V_P and r values. Such principle is usually greatly influenced by the material's composition, namely amount, type, and size of (coarse) aggregate, etc. Advances in computer and equipment technology have allowed the development of more comprehensive and advanced ultrasonic techniques to define setting phenomena, which considerably eliminates such discrepancies. Monitoring of the whole spectra of P-waves is often used for such purpose. During the very early stages of the solidification process, low frequencies with relatively small amplitudes are dominating (Boumiz et al. 1996; Reinhardt and Grosse 1996; Grosse and Reindardt 2000, 2001; De Belie et al. 2006; Robeyst et al. 2007, 2009a, b; Darquenes et al. 2009; Trtnik and Gams 2013; Gams and Trtnik 2013) in the spectra (Fig. 2.16). When enough hydration products are formed and interconnected, resulting in a sufficient amount of connected solid phase in the material's microstructure, high frequencies gradually appear (Darquenes et al. 2009; Trtnik and Gams 2013; Gams and Trtnik 2013). Several studies have shown that the time when high frequencies begin to develop can be recognized very accurately and can be directly correlated to the setting phenomena. In order to continuously monitor changes that occur in a frequency spectrum of P-waves during structure formation process, Trtnik and Gams (2013) defined a dimensionless *TG* factor as a ratio between maximum amplitudes of low and high frequencies in a frequency spectrum of P-waves. Several studies have shown that *TG* can be very accurately contributed to various setting phenomena if properly considered (Trtnik and Gams 2013; Gams and Trtnik 2013). The initial increase in *TG* factor from its minimum value was found to be closely related to the initial setting time. Similarly, the time when *TG* factor reaches its final plateau corresponds very well with the final setting time of the material. These phenomena were found to be independent from the material's composition.

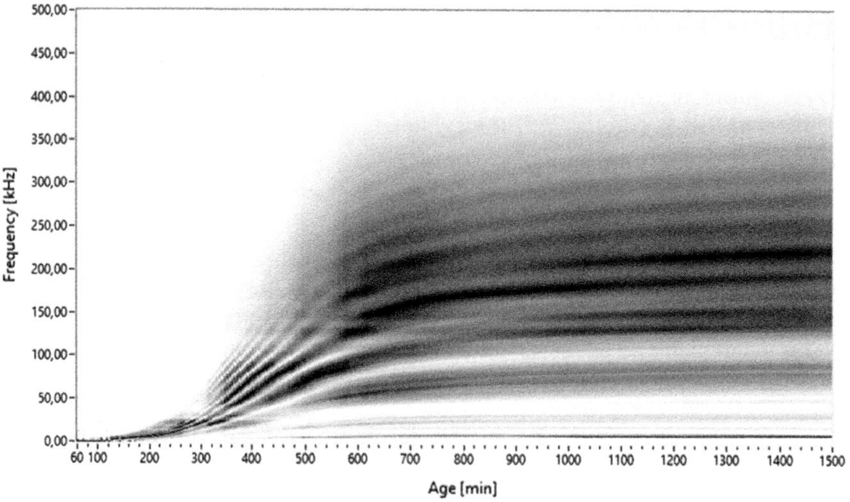

Fig. 2.16 Evolution of frequency spectrum during hydration (broadband 500 kHz P-wave transducer), 50% CEM I 52,5R, 50% limestone powder, w/c = 0.43

In addition to *TG* factor, other advanced parameters can also be effectively and accurately used to define setting process of CBMs. The change of the received ultrasonic wave energy in time was found to be clearly related to the setting phenomena of CBM (De Belie et al. 2005; Robeyst et al. 2009a, b). Thresholds values were proposed to determine initial and final setting of the material (Robeyst et al. 2009a, b). However, the efficiency of this method is undermined by the possible decrease in energy caused by the sample shrinkage or swelling, related to thermal or autogenous origins (Carette 2015). Gabrijel et al. (2010, 2011; Gabrijel and Pintar 2018) applied parametric analysis of transmitted ultrasonic waveforms, typically used in acoustic emission technique to identify setting in cement paste and concrete. They found that some parameters exhibit rapid change during setting and could be useful in estimation of setting period.

2.3.3 Monitoring of Microstructural Properties

In addition to the initial and final setting times, many other important phenomena that appear during the early age formation of structure of CBM can be effectively determined using ultrasonic reflection or transmission technique if properly considered. Such phenomena include e.g. percolation threshold and penetration resistance, degree of connectivity of solid phase, degree of hydration and hydration products, mechanical properties, temperature development, shrinkage, effect of different admixtures and additives, etc.

As has been briefly explained in previous sections, percolation threshold of solid phases is an important microstructural parameter. It is defined as a critical degree of hydration, when enough hydration products are present to form an interconnected solid percolation path. At this critical moment t_{SP}, the transformation of the material from liquid suspension to solid material begins. Consequently, shear waves start to propagate through the material and significant changes can be observed also on $V_P - t$ curve. Based on this explanation, the time of the percolation threshold of solid phase t_{SP} is usually considered as the time when r (or *RL*) starts to increase, (Feylessoufi et al. 2001; Lacouture et al. 2003; Voigt 2004) or when the *Stage II-b* starts on the $V_P - t$ curves. Researchers also define the percolation threshold as the initial setting time t_I of specific CBM.

Some studies have shown that both V_P, V_S, r and *RL* are not only connected to the percolation threshold but also strongly follows the trend of the increase of penetration resistance due to the increasing connectivity of solid phase within the material's microstructure. A strong linear relationship between Vicat penetration resistance and r values was reported during the whole setting period of cement pastes (Trtnik et al. 2012; Trtnik et al. 2013). Using Hymostruc numerical model (Van Breugel 1991; Lokhorst 1999), some other parameters which can be effectively used to define and describe solidification process of CBM have been proposed. Such parameters include e.g. bridge particles and specific contact area of solid phase. A unique correlation between V_P and the bridge volume (Ye 2003) and unique bilinear relationship between RL and specific contact area of solid phase (Voigt et al. 2005a, b; Sun 2005) was reported in the case of cement pastes. A power trend line between RL and gel-space ratio, i.e. parameter which describes a proportion of the available space within the hydrating CBM occupied by the gel has also been established (Voigt 2004).

Being one of the essential parameters to define the intensity and progress of the hydration process, a degree of hydration α is an attractive parameter to be correlated to various ultrasonic parameters. A strong linear relation between α and RL exists for different CBM (Sun et al. 2004a, b; Sun 2005; Voigt et al. 2006). Moreover, relationships between RL and V_P and decrease of capillary porosity, amount of hydrated cement, CH content, degree of C_3S hydration, porous volume fraction, and crystallized hydrates have been proposed by different researchers.

Since temperature development or heat release is often used to describe and monitor hydration process of CBM, various attempts to correlate ultrasonic parameters and temperature development of CBM during early age formation of structure process have been made. Several studies (e.g. Voigt 2004) have shown that V_P increases at a much higher rate than adiabatic heat release T_A at early ages. This can be attributed to various non-setting phenomena described earlier in this section. However, some researchers found out that important changes on $T_A - t$ curves occurred earlier than the changes on $V_P - t$ curves (Gimet et al. 1999; Chotard et al. 2001a, b). On the contrary, an excellent agreement between characteristic points on $V_P - t$ curves and the development of the material's in situ temperature T in time has been reported by the others. A good relationship between adiabatic heat release T_A and RL was reported and was found to be independent from the material's composition (Voigt 2004). The time when the rate of V_P reaches its maximum was found to occur at the time of the maximum rate of T increase for an arbitrary cement paste (Trtnik and Turk 2013).

Finally, it was observed that the sensitivity of cement hydration reactions to temperature changes (through the maturity method and activation energy parameter) can be assessed by the ultrasonic wave transmission method (Carette and Staquet 2016a). It is observed that the pre-setting and setting process is less sensitive to temperature changes than the hardening process. This sensitivity can be determined by performing ultrasonic transmission tests at various cure temperature. It is also shown than blast-furnace slag or limestone filler do not affect this sensitivity to temperature.

2.3.4 Ultrasonic Testing and Early Age Mechanical Properties

A significant amount of studies has been performed in order to correlate V_P and mechanical characteristics of different CBM during the last two decades. However, such studies usually refer to the hardened materials and are therefore beyond the scope of this section. Even though connecting V_P and r (or RL) to early age compressive strength is less common, some studies exist which indicate that V_P and r (or RL) can also be used as an effective parameter to describe the development of the initial compressive strength of CBM if properly considered. However, relatively large discrepancies can be noticed between different studies in this case. The time when RL starts to increase from its initial constant value was found to be a promising indicator of the first significant increase of the material's compressive strength (Voigt and Shah 2006). A bilinear (Byfors 1980; Voigt and Shah 2003), power law (Voigt et al. 2006), and even strong linear (Valič and Vuk 2000; Akkaya et al. 2003) trend was suggested between r and early age compressive strength. Due to the previously described phenomena which are not related to setting, connecting V_P with early age compressive strength is more challenging.

2.3.4.1 Determination of the Static Elastic Modulus from Transmission Techniques

The difference between the dynamic elastic modulus (E_d) determined from ultrasonic transmission and the statically determined elastic modulus (E_s) comes from the non-linear (often considered quadratic) stress-strain behaviour of concrete (Shkolnik 2005). In Eq. 2.90 σ is the stress, ε the strain, E_0 is the elastic modulus for a negligible loading (close to E_d), and a is a non-linear material parameter. The value of this last parameter is significantly material-dependent. For instance, experiments performed on 316LN steel show that the static and dynamic elastic moduli reach the same value, and a is therefore close to 0 (Ledbetter 1993).

$$\sigma = E_0\varepsilon - a\varepsilon^2 \tag{2.90}$$

In early age concrete difference between E_d and E_s is also affected by the creep behaviour. It is thus expected that more creep potential leads to a higher difference between static and dynamic elastic modulus. Cement hydration leads to a significant decrease of the viscoelastic properties at early age. It is therefore expected that the non-linear parameter decreases through hydration, suggesting that the $E_s - E_d$ relationship also follows a non-linear evolution (Fig. 2.17).

Most studies performed on this matter suggest that a linear equation relates both properties (Eq. 2.91) (Klieger 1957; Shkolnik 2005). The reason for this approximation is that these studies are performed on specimens with advanced hardening processes. More recent studies performed at very early age show that this linear trend cannot be applied. From a theoretical perspective, as soon as E_d increases, there is a corresponding increase of E_s. A more accurate exponential relationship between both properties satisfies this condition (Eq. 2.92) (Carette and Staquet 2016b). According to a study by Han and Kim (2004), a power relationship would be even more satisfying in regard of very early age results (Eq. 2.93). It is also shown that the value of parameters p and q is not affected by cement type or w/c ratio.

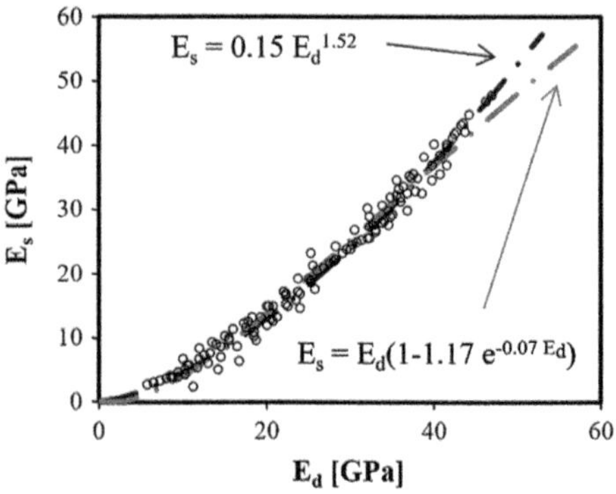

Fig. 2.17 Relationship between static and dynamic elastic modulus (Carette and Staquet 2016b)

$$E_S = pE_d - q \tag{2.91}$$

$$E_S = pE_d^q \tag{2.92}$$

$$E_S = E_d\left(1 - pe^{-qE_d}\right) \tag{2.93}$$

Since there is no unambiguous definition of "dynamic" or "static" elastic modulus, the relationship between both measurements is expected to depend on both, their corresponding loading strain rate and the loading amplitude (Delsaute et al. 2015). Additionally, the viscosity of the CBM is of upmost importance. The amount and nature of aggregates might therefore affect these relationships, especially at early age. These values are highly affected by the content and nature of aggregates (Carette 2015).

2.3.4.2 Determination of Mechanical Strength from Transmission Techniques

The relationship between compressive (f_c) strength and V_P has been known for long time to be exponential. The parameters m and n (m' and n' for S-waves) of such empirical relationship can be calibrated for most CBM. However, their value is highly affected by the cement composition, w/c ratio or the presence of aggregates. This type of relationship has been validated up to the earliest age (from initial setting to several days) (Carette and Staquet 2016a, b, c). It is generally observed that tensile strength (f_t) can be estimated through similar methodologies.

$$f_{c/t}(V_P) = m \cdot e^{nV_P} \tag{2.94}$$

$$f_{c/t}(V_S) = m' \cdot e^{n'V_S} \tag{2.95}$$

It is also shown (Fig. 2.18) that using V_S as an indicator to estimate strength is more reliable than V_P. The possible reason for that is that shear wave is more sensitive to the aggregate interlocking mechanism, which strongly contributes to the concrete strength. As a result, the effect of w/c variations or type of cement composition (i.e. presence of limestone filler or blast furnace slag) will have a lower impact on the V_S—*strength* relationship than on the V_P—*strength* relationship. It is worth to be noted that V_s is also less sensitive to air content compared to V_P.

Ultimately, the aggregate nature and content as well as the air content are the most influential parameters, as they highly affect the wave velocity, while having a limited impact on the tensile or compressive strength (Fig. 2.19). For this reason, it is only recommended to use ultrasonic indicators for strength estimations if the relationship between both has been previously established for the given composition, or if small variations in the air volume and aggregate content and nature are applied.

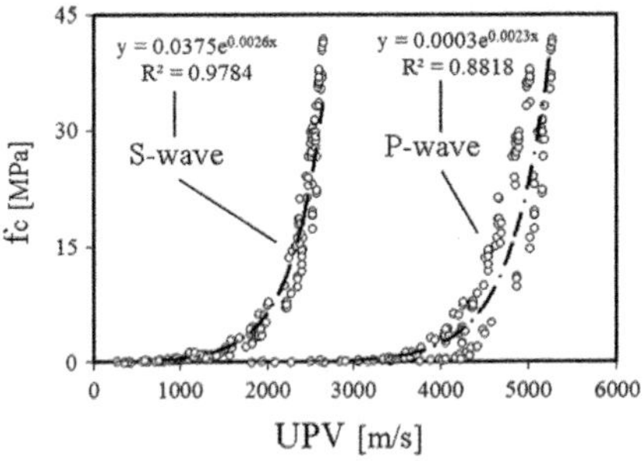

Fig. 2.18 Evolution of compressive strength (left) and tensile strength (right) as a function of V_P and V_S for concrete mixes containing various type of cements and w/c ratio from initial setting to several days (Carette 2015)

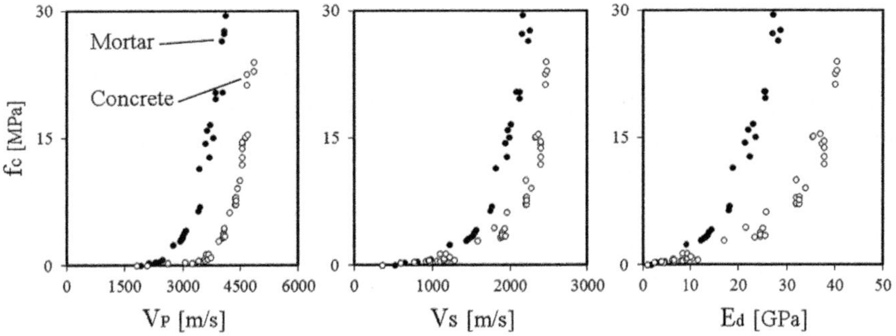

Fig. 2.19 Evolution of compressive strength as a function of V_P and V_S and E_d for concrete and corresponding mortar mixes (Carette and Staquet 2016c)

2.4 Conclusion

In this chapter up to date developments and achievements in the field of ultra-sonic testing of the early age CBM have been presented. At the beginning of the chapter, theoretical aspects and phenomena are briefly presented, followed by some most important experimental achievements. Two ultrasonic techniques, measuring longitudinal and shear waves in transmission or/and reflection modes are discussed in more detail due to their popularity. It is shown that ultrasonic techniques can provide useful insight into structural changes of cementitious materials from the period of placing to a fully hardened state if properly considered.

One of the focus areas of many researchers is to monitor setting behaviour of cementitious materials, characterised as a transition phase during which material changes its aggregate state from liquid to solid. Experimental results show that ultrasonic methods can be used as an effective alternative to standard penetration resistance techniques and can provide a finer representation of the fluid-solid transformation process. However, such measurement results do not unambiguously show distinct points during solidification process which are generally defined as initial and final setting. This led to different criteria for setting times that have been proposed based on measurement of ultrasonic parameters. One may also raise the question if tests like the Vicat needle test give the "right" value for initial and final setting. Vicat needle test is designed to be conducted with a standard consistency that guarantees a comparable plasticity at the beginning of the test. This means that Vicat setting times obtained from mixes with consistency different from the standard consistency might not be interpreted correctly to represent the initial setting time of the material.

In transmission measurements focus is mainly on the measurement of ultrasonic pulse velocity. However, advances in computer and equipment technology during the recent years have allowed the development of more sophisticated and complex ultrasonic techniques, which are based on measuring the whole spectra of ultrasonic signals. Using this technique, one can more accurately define both initial and final set as well as some other important phenomena that appear during solidification process of CBM. Moreover, development of a measuring system for simultaneous measurement of V_P and V_S enabled estimation of dynamic elastic constants of early age CBM under the assumption of homogenous elastic medium. Several research campaigns have found that application of shear waves in transmission and reflection modes gives stronger correlation with early age compressive strength development compared to longitudinal waves.

This chapter shows that due to their clear physical basis, accuracy, and ease of use, ultrasonic techniques have a great potential to become standard testing techniques for determination of setting and other properties of (fresh and hardening) CBM in the (near) future. The first steps in this direction have already been performed.

References

Aggelis, D. G., & Grammenou, D. (2011). Characterization of fresh mortar with chemical admixtures based on stress wave dispersion. In *Proceedings of the International Symposium on Nondestructive Testing of Materials and Structures*, Istanbul (pp. 459–464).

Aggelis, D. G., & Philippidis, T. P. (2004). Ultrasonic wave dispersion and attenuation in fresh mortar. *NDT & E International, 37*(8), 617–631.

Aggelis, D. G., Polyzos, D., & Philippidis, T. P. (2005). Wave dispersion and attenuation in fresh mortar: Theoretical predictions vs. experimental result. *Journal of the Mechanics and Physics of Solids, 53*(4), 857–883.

Akkaya, Y., Voigt, T., Subramaniam, K. V., & Shah, S. P. (2003). Nondestructive measurement of concrete strength gain by an ultrasonic wave reflection method. *Materials and Structures, 36*(8), 507–514.

Arnaud, L., & Grosse, C. U. (2005). Setting phenomena observed in heterogeneous materials by means of the Vibroscope techniques. In H.-W. Reinhardt & C.U. Grosse (Ed.), *Report 31: Advanced testing of cement-based materials during setting and hardening*. Report of RILEM TC 185-ATC, RILEM Publications S.A.R.L.: Cachan ENS (pp. 191–238).

Arnaud, L., & Thinet, S. (2003). Mechanical evolution of concrete during setting. *Materials and Structures, 36*(7), 355–364.

Berckhemer, H. (1990). Grundlagen der Geophysik. Wissenschaftliche Buchgesellschaft.

Biot, M. A. (1956a). Theory of propagation of elastic waves in a fluid-saturated porous solid. I. Low-frequency range. *The Journal of the Acoustical Society of America, 28*(2), 168–178.

Biot, M. A. (1956b). Theory of propagation of elastic waves in a fluid-saturated porous solid. II. Higher-frequency range. *The Journal of the Acoustical Society of America, 28*(2), 179–191.

Boumiz, A., Vernet, C., & Cohen Tenoudji, F. (1996). Mechanical properties of cement pastes and mortars at early ages—Evolution with time and degree of hydration. *Advanced Cement Based Materials, 3*(3–4), 94–106.

Byfors, Y. (1980). *Plain concrete at early ages*. Swedish Cement and Concrete Research Institute.

Carette, J. (2015). *Towards early age characterisation of eco-concrete containing blast-furnace slag and limestone filler*. Ph.D. Thesis. Université Libre de Bruxelles.

Carette, J., & Staquet, S. (2015). Monitoring the setting process of mortars by ultrasonic P-wave and S-wave transmission velocity measurement. *Construction and Building Materials, 94,* 196–208.

Carette, J., & Staquet, S. (2016a). Monitoring and modelling the early age and hardening behaviour of eco-concrete through continuous non-destructive measurements: Part I. Hydration and apparent activation energy. *Cement & Concrete Composites, 73,* 10–18.

Carette, J., & Staquet, S. (2016b). Monitoring and modelling the early age and hardening behaviour of eco-concrete through continuous non-destructive measurements: Part II. Mechanical behaviour. *Cement and Concrete Composites, 73,* 1–9.

Carette, J., & Staquet, S. (2016c). Monitoring the setting process of eco-binders by ultrasonic P-wave and S-wave transmission velocity measurement: Mortar vs concrete. *Construction and Building Materials, 110,* 32–41.

Cheng, X., Qin, L., Zhong, Q. Q., Huang, S. F., & Li, Z. J. (2013). Temperature and boundary influence on cement hydration monitoring using embedded piezoelectric transducers. *Ultrasonics, 53*(2), 412–416.

Chotard, T., Gimet-Brert, N., Smith, A., Fargeot, D., Bonnet, J. P., & Gault, C. (2001a). Application of ultrasonic testing to describe the hydration of calcium aluminates cement at the early age. *Cement and Concrete Research, 31*(3), 405–412.

Chotard, T., Gimet-Brert, N., Smith, A., Fargeot, D., Bonnet, J. P., & Gault, C. (2001b). Characterisation of calcium aluminate cement hydration at young age by ultrasonic testing. In *Proceedings of the International Conference on Calcium Aluminate Cements (CAC)* (pp. 155–167). Edinburgh: The University Press.

Chung, C. (2010). *Ultrasonic wave reflection measurements on stiffening and setting of cement paste*. Ph.D. Thesis. Urbana-Champaign.

Darquenes, A., Staquet, S., Espion, B., Germain, O., & Pierre, C. (2009). Comparison between different techniques for monitoring setting and hardening of concrete. In *7th International Symposium on Nondestructive testing in Civil Engineering—NDTCE'09*. Nantes.

De Belie, N., Grosse, C. U., & Baert, G. (2008). Ultrasonic transmission to monitor setting and hardening of fly ash concrete. *ACI Materials Journal, 105*(3), 221–226.

De Belie, N., Grosse, C. U., Kurz, J., & Reinhardt, H. W. (2005). Ultrasound monitoring of the influence of different accelerating admixtures and cement types for shotcrete on setting and hardening behavior. *Cement and Concrete Research, 35*(11), 2087–2094.

De Belie, N., Grosse, C. U., Robeyst, N., & Gruyaert, E. (2006). Evolution of different ultrasound transmission parameters to monitor microstructure development in fresh concrete or mortar. In *Proceedings of the Second International Symposium on Advances in Concrete Through Science and Engineering*, Quebec City. RILEM Publications SARL.

Delsaute, B., Boulay, C., Granja, J., Carette, J., Azenha, M., Dumoulin, C., et al. (2015). Testing concrete e-modulus at very early ages through several techniques: An inter-laboratory comparison. *Strain, 52*(2), 91–109.

Feylessoufi, A., Tenoudji, F. C., Morin, V., & Richard, P. (2001). Early ages shrinkage mechanisms of ultra-high-performance cement-based materials. *Cement and Concrete Research, 31*(11), 1573–1579.

Gabrijel, I., Mikulić, D., & Bijelić, N. (2010). Ultrasonic characterization of cement composites during hydration. *Technical Gazette, 17,* 493–497.

Gabrijel, I., Mikulić, D., & Milovanović, B. (2011). Application of ultrasonic measurements for determination of setting and hardening in cement paste. *Journal of Civil Engineering and Architecture, 3,* 278–283.

Gabrijel, I., & Pintar, M. (2018). Estimation of setting time in concrete through modelling of ultrasonic parameters. In *Proceedings of the SynerCrete'18: Interdisciplinary Approaches for Cement-based Materials and Structural Concrete: Synergizing Expertise and Bridging Scales of Space and Time*, Funchal (pp. 695–700).

Gams, M., & Trtnik, G. (2013). A new US procedure to determine setting period of cement pastes, mortars, and concretes. *Cement and Concrete Research, 53,* 9–17.

Garnier, V., Corneloup, G., Spraurel, J. N., & Perfumo, J. C. (1995). Setting time study of roller compacted concrete by spectral analysis of transmitted ultrasonic signals. *NDT & E International, 28*(1), 15–22.

Gimet, N., Fargeot, D., Gaillard, J. M., Smith, A., Gault, C., & Bonnet, J. P. (1999). Ultrasonic assessment of Portland cement at the early stages of hydration. *Journal of Material Science Letters, 18*(16), 1335–1337.

Grosse, C. U. (1996). *Quantitative zerstörungsfreie Prüfung von Baustoffen mittels Schallemissionsanalyse und Ultraschall*. Ph.D. thesis. University of Stuttgart.

Grosse, C. U. (2010). Ultrasound testing for early age quality control of concrete. In *Proceedings of the 13th International Conference on Structural Faults & Repair*. Edinburgh: Engineering Technics.

Grosse, C. U., Beutel, R., & Pessiki, S. (2005). Impact-echo. In H.-W. Reinhardt & C.U. Grosse (Ed.), *Report 31: Advanced testing of cement-based materials during setting and hardening—Report of RILEM TC 185-ATC* (pp. 239–253). RILEM Publications S.A.R.L.

Grosse, C. U., & Reindardt, H. W. (2000). Ultrasound technique for quality control of cementitious materials. In *Proceedings of the 15th Conference on Nondestructive Testing*, Rome.

Grosse, C. U., & Reinhardt, H. W. (2001). Fresh concrete monitored by ultrasound methods. *Otto-Graf-Journal, 12,* 157–168.

Han, S. H., & Kim, J. K. (2004). Effect of temperature and age on the relationship between dynamic and static elastic modulus of concrete. *Cement and Concrete Research, 34,* 1219–1227.

Harker, A. H., Schofield, P., Stimpson, B. P., Taylor, R. G., & Temple, J. A. (1991). Ultrasonic propagation in slurries. *Ultrasonics, 29,* 427–438.

Irgens, F. (2008). Continuum Mechanics. Springer.

Johnson, D. L., & Plona, T. H. (1982). Acoustic slow waves and the consolidation transition. *The Journal of the Acoustical Society of America, 72*(2), 556–565.

Jupe, A. C., Wilkinson, A. P., & Funkhouser, G. P. (2012). Simultaneous study of mechanical property development and early hydration chemistry in Portland cement slurries using X-ray diffraction and ultrasound reflection. *Cement and Concrete Research, 42*(8), 1166–1173.

Kamada, T., Uchida, S., & Rokugo, K. (2005). Nondestructive evaluation of setting and hardening of cement paste based on ultrasonic propagation characteristics. *Journal of Advanced Concrete Technology, 3*(3), 343–353.

Keating, J., Hannant, D. J., & Hibbert, A. P. (1989). Correlation between cube strength, ultrasonic pulse velocity and volume change for oil well cement slurries. *Cement and Concrete Research, 19*(5), 715–726.

Keskin, O. K., Yaman, I. O., & Tokyay, M. (2011). Effects of experimental parameters in monitoring the hydration of cement mortars by ultrasonic testing. In *Proceedings of the International Symposium on Nondestructive Testing of Materials and Structures*, Istanbul (pp. 437–444).

Khokhar, M. I. A., Staquet, S., Roziere, E., & Loukili, A. (2009). Ultrasonic monitoring of setting of green concrete containing high cement substitution by mineral additions. In *7th International Symposium on Nondestructive testing in Civil Engineering—NDTCE'09*, Nantes.

Klieger, P. (1957). Long-time study of cement performance in concrete. *Journal of American Concrete Institute, 54,* 481–504.

Krauss, M., & Hariri, K. (2006). Determination of initial degree of hydration for improvement of early-age properties of concrete using ultrasonic wave propagation. *Cement & Concrete Composites, 28*(4), 299–306.

Krüger, M., Bergar, R., David, G. A., & Juhart, J. (2018). Influence of the mixture composition of cementitious pastes with optimized packing density on setting and hardening studied by ultrasound investigation. In *Proceedings of the SynerCrete'18: Interdisciplinary Approaches for Cement-based Materials and Structural Concrete: Synergizing Expertise and Bridging Scales of Space and Time*, Funchal (pp. 243–248).

Krüger, M., Bregar, R., David, G. A., & Juhart, J. (2016). Non-destructive evaluation of eco-friendly cementitious materials by ultrasound. In *Proceedings of the International RILEM Conference: Materials, Systems and Structures in Civil Engineering, Segment on Service Life of Cement-Based Materials and Structures*, Lyngby (pp. 503–512).

Krüger, M., Grosse, C. U., & Lehmann, F. (2011). Automated shear-wave techniques to investigate the setting and hardening of concrete in through-transmission. In *Proceedings of the International Symposium on Nondestructive Testing of Materials and Structures* (V. 6). RILEM bookseries. Istanbul: Springer Netherlands (pp. 431–436).

Krüger, M., & Lehmann, F. (2010). *SmartPick—User manual, Rev. 1.50.* Smartmote.

Lacouture, J. C., Johson, P. A., & Cohen-Tenoudji, F. (2003). Study of critical behaviour in concrete during curing by application of dynamic linear and nonlinear means. *Journal of the Acoustical Society of America, 113*(3), 1325–1332.

Landau, L. D., & Lifschitz, E. M. (1986). Theory of elasticity. In *Course of theoretical physics* (Vol. 7, 3rd ed.). New York: Pergamon Press, Oxford.

Ledbetter, H. (1993). Dynamic vs. static Young's moduli: A case study. *Materials Science and Engineering: A, 165*(1), 9–10.

Lee, H. K., Lee, K. M., Kim, Y. H., Yim, H., & Bae, D. B. (2004). Ultrasonic in-situ monitoring of setting process of high-performance concrete. *Cement and Concrete Research, 34*(4), 631–640.

Liu, Z., Zhang, Y., Jiang, Q., Sun, G., & Zhang, W. (2011). In situ continuously monitoring the early age microstructure evolution of cementitious materials using ultrasonic measurements. *Construction and Building Materials, 25*(10), 3998–4005.

Lokhorst, S. J. (1999). Deformational behaviour of concrete influenced by hydration related changes of the microstructure. *Ph.D.* thesis. University of Delft.

Mikulić, D., Milovanović, B., & Gabrijel, I. (2011). Analysis of thermal properties of cement paste during setting and hardening. In *Proceedings of the International Symposium on Nondestructive Testing of Materials and Structures, Istanbul* (pp. 465–472).

Mikulić, D., Sekulić, D., Štirmer, N., & Bjegović, D. (2005). Application of ultrasonic methods for early age concrete characterization. In *Proceedings of the 8th International Conference of the Slovenian Society for Non-Destructive Testing: Application of Contemporary Non-Destructive Testing in Engineering*, Portorož (pp. 99–108).

Militzer, H., & Weber, F. (1987). *Angewandte Geophysik. Seismik* (Vol. 3). Wien: Springer.

Öztürk, T., Kroggel, O., & Grübl, P. (2004). The influence of temperature on the hydration process of concrete evaluated through ultrasound technique. In *Proceedings of the International RILEM Symposium: Advances in Concrete through Science and Engineering, Evanston*, Illinois (pp. 213–223).

Öztürk, T., Kroggel, O., Grübl, P., & Popovics, J. S. (2006). Improved ultrasonic wave reflection technique to monitor the setting of cement-based materials. *NDT & E International, 39*(4), 258–263.

Öztürk, T., Rapoport, J., Popovics, J. S., & Shah, S. P. (1999). Monitoring the setting and hardening of cement-based materials with ultrasound. *Concrete Science and Engineering, 1*(2), 83–91.

Powers, T. (1968). *The properties of fresh concrete*. New York: Wiley.

Ramachandran, V. S., Paroli, R. M., Beaudoin, J. J., & Delgado, A. H. (2002). *Handbook of thermal analysis of construction materials*. Noyes Publications/William Andrew Publishing.

Reinhardt, H. W., & Grosse, C. U. (1996). Setting and hardening of concrete continuously monitored by elastic waves. *NDTnet, 1*(7).

Reinhardt, H. W., & Grosse, C. U. (2004). Continuous monitoring of setting and hardening of mortar and concrete. *Construction and Building Materials, 18*(3), 145–154.

Reinhardt, H. W., & Grosse, C. U. (2005). Mechanical wave propagation—Basics. In H-W. Reinhardt & C.U. Grosse (Eds.), *Report 31: Advanced testing of cement-based materials during setting and hardening—Report of RILEM TC 185-ATC* (pp. 83–96). RILEM Publications S.A.R.L.

Reinhardt, H. W., Grosse, C. U., & Herb, T. (2000). Ultrasonic monitoring of setting and hardening of cement mortar—A new device. *Materials and Structures, 33*(9), 581–583.

Rikitake, T., Sato, R., & Hagiwara, Y. (1987). *Applied mathematics for earth scientists. Mathematical approaches to geophysics*. Dorcrecht: Kluwer Academic Pub.

RILEM Report 31. (2005). *Advanced testing of cement-based materials during setting and hardening*. RILEM Publ. S.A.R.L.

RILEM Technical Committee 218. (2011). Recommendation of RILEM TC 218-SFC: Sonic methods for quality control of fresh cementitious materials—Testing of fresh concrete by ultrasound transmission. *Materials and Structures, 44*, 1047–1062.

Robeyst, N., Grosse, C. U., & De Belie, N. (2009a). Measuring the change in ultrasonic p-wave energy transmitted in fresh mortar with additives to monitor the setting. *Cement and Concrete Research, 39*(10), 868–875.

Robeyst, N., Grosse, C. U., & De Belie, N. (2009b). Monitoring fresh concrete by ultrasonic transmission measurements: Exploratory multi-way analysis of the spectral information. *Chemometrics and Intelligent Laboratory Systems, 95*(1), 64–73.

Robeyst, N., Grosse, C. U., & De Belie, N. (2011). Factors affecting the monitoring of the early age setting of concrete by ultrasonic P-waves. In *Proceedings of the International Symposium on Nondestructive Testing of Materials and Structures*, Istanbul (pp. 423–430).

Robeyst, N., Gruyaert, E., & De Belie, N. (2007). Ultrasonic monitoring of setting and hardening behavior of concrete and mortar with blast furnace slag cement. In *12eme Congres International de la chimie du ciment*, Montreal.

Robeyst, N., Gruyaert, E., Grosse, C. U., & De Belie, N. (2008). Monitoring the setting of concrete containing blast-furnace slag by measuring the ultrasonic p-wave velocity. *Cement and Concrete Research, 38*(10), 1169–1176.

Sayers, C. M., & Dahlin, A. (1993). Propagation of ultrasound through hydrating cement paste at early times. *Advanced Cement Based Materials, 1*(1), 12–21.

Sayers, C. M., & Grenfell, R. L. (1993). Ultrasonic propagation through hydrating cements. *Ultrasonics, 31*, 147–153.

Shkolnik, I. (2005). Effect of nonlinear response of concrete on its elastic modulus and strength. *Cement and Concrete Composites, 27*(7–8), 747–757, *27*(7–8), 747–757.

Shutilov, V. (1988). *Fundamental physics of ultrasound*. London: CRC Press.

Subramaniam, K. V., Lee, J., & Christensen, B. J. (2005). Monitoring the setting behavior of cementitious materials using one-sided ultrasonic measurements. *Cement and Concrete Research, 35*(5), 850–857.

Subramaniam, K. V., Mohsen, J. P., Shaw, C. K., & Shah, S. P. (2002). Ultrasonic technique for monitoring concrete strength gain at early age. *ACI Materials Journal, 99*(5), 458–462.

Subramaniam, K. V., Rapoport, J. R., & Shah, S. P. (2000). Nondestructive technique for monitoring strength gain in concrete structures. *Cementing the Future, Newsletter of the Center for Advanced Cement-Based Materials, 12*(1), 4–5.

Subramaniam, K. V., & Wang, X. (2010). An investigation of microstructure evolution in cement paste through setting using ultrasonic and rheological measurements. *Cement and Concrete Research, 40*(1), 33–44.

Sun, Z. (2005). *Monitoring the early-age properties of cementitious materials with ultrasonic wave reflection method at macro-and micro-structural levels*. Ph.D. Thesis. Northwestern University.

Sun, Z., Ye, G., Voigt, T., Shah, S. P., & Van Breugel, K. (2004a). Early age properties of Portland cement pastes investigated with ultrasonic shear waves and numerical simulations. In *Proceedings of the first International RILEM Symposium on Advances in Concrete Through Science and Engineering*, Evanston.

Sun, Z., Ye, G., Voigt, T., Shah, S. P., & Van Breugel, K. (2004b). Microstructural aspects of cement hydration—ultrasonic waves and numerical simulation. In *5th international Ph.D. symposium in Civil Engineering*. London.

Suraneni, P. (2011). *Ultrasonic wave reflection measurements on self-compacting pastes and concretes*. M.Sc. Thesis. Urbana-Champaign.

Temkin, S. (2005). *Suspension acoustics*. Cambridge University Press.

Trtnik, G. (2009). *The use of ultrasonic method to monitor the setting and hardening process of concrete*. PhD. Thesis. University of Ljubljana.

Trtnik, G. (2012). Sensitivity of ultrasonic method to detect a presence of plasticizers in cement pastes. *Gradbeni vestnik, 61*, 37–45.

Trtnik, G., & Gams, M. (2013). The use of frequency spectrum of ultrasonic P-waves to monitor the setting process of cement pastes. *Cement and Concrete Research, 43*(1), 1–11.

Trtnik, G., & Turk, G. (2013). Influence of superplasticizers on the evolution of ultrasonic P-wave velocity through cement pastes at early age. *Cement and Concrete Research, 51*, 22–31.

Trtnik, G., Turk, G., Kavčić, F., & Bokan Bosiljkov, V. (2008). Possibilities of using the ultrasonic wave transmission method to estimate initial setting time of cement paste. *Cement and Concrete Research, 38*(11), 1336–1342.

Trtnik, G., Valič, M. I., Kavčić, F., & Turk, G. (2009). Comparison between two ultrasonic methods in their ability to monitor the setting process of cement pastes. *Cement and Concrete Research, 39*(10), 876–882.

Trtnik, G., Valič, M. I., Kavčić, F., & Turk, G. (2012). Ultrasonic wave reflection measurements of cement pastes at early ages. *Advances in concrete technology and sustainability issues: supplementary papers*. Prague.

Trtnik, G., Valič, M. I., & Turk, G. (2013). Measurements of setting process of cement pastes using non-destructive ultrasonic shear wave reflection technique. *NDT & E International, 56*(6), 65–75.

Valič, M. I. (2000). Hydration of cementitious materials by pulse echo USWR. Method, apparatus and application examples. *Cement and Concrete Research, 30*(10), 1633–1640.

Valič, M. I., Stepišnik, J., Gabrijelčić, M., Vuk, T., & Reščić, L. (1999). Setting times measurements of Portland cements with pulsed USWR method. *KZT, 33*(1–2), 83–86.

Valič, M. I., & Vuk, T. (2000). A study of the influence of additives and admixtures on the hydration of cement with the pulsed USWR method. *Materials and Tehnology, 34*(5), 249–255.

Van Breugel, K. (1991). *Simulation of hydration and formation of structure of hardening cement-based materials*. Ph.D. Thesis. University of Delft.

Van der Winden, N. G. (1990). Ultrasonic measurement for setting control of concrete. In *Proceedings of the International Workshop: Testing During Concrete Construction* (pp. 122–137). Mainz: Chapman and Hall.

Voigt, T. (2004). *The application of an ultrasonic shear wave reflection method for nondestructive testing of cement-based materials at early ages*. Ph.D. Thesis. University of Leipzig.

Voigt, T., Grosse, C. U., Sun, Z., Shah, S. P., & Reinhardt, H. W. (2005a). Comparison of ultrasonic wave transmission and reflection measurements with P- and S-waves on early age mortar and concrete. *Materials and Structures, 38*(8), 729–738.

Voigt, T., Malonn, T., & Shah, S. P. (2006). Green and early age compressive strength of extruded cement mortar monitored with compression tests and ultrasonic techniques. *Cement and Concrete Research, 36*(5), 858–867.

Voigt, T., & Shah, S. P. (2003). Nondestructive monitoring of setting and hardening of Portland cement mortar with sonic methods. In *Proceedings of the International Symposium Non-Destructive Testing in Civil Engineering*, Berlin.

Voigt, T., & Shah, S. P. (2006). Chemical shrinkage and calcium hydroxide content of early age Portland cement monitored with ultrasonic shear wave reflections. In *Proceedings of the International Symposium Dedicated to Professor Surendra P. Shah: Measuring, Monitoring and Modeling Concrete Properties*. Alexandroupolis.

Voigt, T., Sun, Z., & Shah, S. P. (2004). Health monitoring of early age concrete. In *Proceedings of the 4th International Conference on Concrete Under Severe Conditions: Environment and Loading*. Seoul.

Voigt, T., Ye, G., Sun, Z., Shah, S. P., & Van Breugel, K. (2005b). Early age microstructure of Portland cement mortar investigated by ultrasonic shear waves and numerical simulation. *Cement and Concrete Research, 35*(5), 858–866.

Wang, X., Subramaniam, K. V., & Lin, F. (2010). Ultrasonic measurement of viscoelastic shear modulus development in hydrating cement paste. *Ultrasonics, 50*(7), 726–738.

Ye, G. (2003). *Experimental study and numerical simulation of the development of microstructure and permeability of cementitious materials*. Ph.D. Thesis. Delft University of Technology.

Ye, G., Lura, P., Van Breugel, K., & Fraaij, A. (2004). Study on the development of the microstructure in cement-based materials by means of numerical simulation and ultrasonic pulse velocity measurement. *Cement & Concrete Composites, 26*(5), 491–497.

Ye, G., Van Breugel, K., & Fraaij, A. (2001). Experimental study on ultrasonic pulse velocity evaluation of the microstructure of cementitious materials at early age. *HERON, 46*(3), 161–167.

Ye, G., Van Breugel, K., & Fraaij, A. (2003). Experimental study and numerical simulation on the formation of microstructure in cementitious material at early age. *Cement and Concrete Research, 33*(2), 233–239.

Zhang, J., Weissinger, E. A., Peethamparan, S., & Scherer, G. W. (2010). Early hydration and setting of oil well cement. *Cement and Concrete Research, 40*(7), 1023–1033.

Zhang, Y., Zhang, W., She, W., Ma, L., & Zhu, W. (2012). Ultrasound monitoring of setting and hardening process of ultra-high performance cementitious materials. *NDT & E International, 47*(4), 177–184.

Chapter 3
Elastic Modulus Measurement Through Ambient Response Method

José Granja and Miguel Azenha

Abstract Concrete endures strong evolution in its mechanical behaviour since casting, passing from a solid suspension to a structural material. It is extremely important to understand and be able to predict the structural behaviour of concrete since the beginning of the hardening process for a good structural design, particularly in regard to the development of self-induced stresses (due to heat of hydration and shrinkage). In order to collect data for the relevant phenomena that have been mentioned, it is important to have methods that allow continuous monitoring of the evolution of mechanical properties of concrete since very early ages, both in laboratory environment and 'in-situ'. In such concern, an experimental method called EMM-ARM (Elasticity Modulus Measurement through Ambient Response Method) was proposed in 2009, which is based on the modal identification of a composite beam during the curing period of concrete, allowing the continuous measurement of the E-modulus of the tested material (e.g. concrete, mortar, cement paste) since casting. This chapter provides an overview of EMM-ARM principles, evolution through the last decade, as well as several examples of application and validation.

Keywords Elastic modulus · Continuous monitoring · Cement-based materials · Early age · Modal identification

3.1 Introduction

The measurement of stiffness in cementitious materials has represented a relevant challenge within both research and practitioner communities for decades, especially with its assessment at early ages (Powers 1938; Jones 1949; RILEM 1975; Boulay and Colson 1979; Chengju 1989; Boumiz et al. 1996; Jin and Li 2001; Bentz 2008, Boulay et al. 2013b). Several distinct methodologies have been proposed throughout the years that allow more or less direct assessment of stiffness properties. One important

J. Granja (✉) · M. Azenha
Department of Civil Engineering, School of Engineering, ISISE—Institute for Sustainability and Innovation in Structural Engineering, University of Minho, Campus de Azurém, 4800-058 Guimarães, Portugal
e-mail: granja@civil.uminho.pt

M. Serdar et al. (eds.), *Advanced Techniques for Testing of Cement-Based Materials*, Springer Tracts in Civil Engineering, https://doi.org/10.1007/978-3-030-39738-8_3

group of such methods pertains to those based on resonance, first developed by Powers (1938).

An important dynamic property of any elastic system is its natural frequency of vibration. As an example, if a vibrating beam with known dimensions is considered, its natural frequency of vibration is mostly related to its geometry, support conditions, density and elastic modulus of the beam's material. Indeed, the elastic modulus of a material can be determined from the measurement of the natural vibration frequency of beams using existing mathematical relationships between the two parameters, provided that all other parameters are known. These relationships were derived for homogeneous solid media, isotropic and perfectly elastic. They may however be also applied to heterogeneous systems, such as concrete, when the sample sizes are large relative to the size of the heterogeneities. Therefore, for flexural vibrations of a simply supported uniform beam, the elastic modulus of the material of which it is composed can be obtained by Eq. (3.1).

$$E = \left(f \cdot \frac{2 \cdot L^2}{\pi}\right)^2 \cdot \frac{\rho \cdot A}{I} \tag{3.1}$$

where E is Elastic modulus of the tested material (Pa), f is the resonance frequency of the fist mode of vibration (Hz), L is the free span of the beam (m), ρ is the material's density (kg/m^3), A is the cross-sectional area (m^2) and I is the second moment of area of the composite cross-section around a centroidal axis perpendicular to the bending plane (m^4).

The classical resonance-based method is already standardized by ASTM (2002) and it has already been extensively used by several authors (Giner et al. 2011; Hassan and Jones 2012; Zhao et al. 2014; Lee et al. 1997; Kolluru et al. 2000; Wang and Subramaniam 2011). Nevertheless these methods have some downsides that prevented their wide spread use as tools for continuous stiffness monitoring (Granja 2016).

Based on the above-mentioned reasoning, and in attempt to overcome the limitations of the classical resonance method, a new method for monitoring the E-modulus evolution was developed called EMM-ARM (Elasticity Modulus Measurement through Ambient Response Method).

3.2 EMM-ARM Testing Apparatus

The EMM-ARM testing technique is a variant of the classical resonance method and has been initially proposed by Azenha (2009). Through a continuous modal identification of a composite beam (comprising the mould filled with the material to be tested) with known geometry and support conditions, it is possible to obtain the evolution of the flexural resonant frequency of the first mode of vibration. From this evolution, it is then possible to directly and quantitatively estimate the evolution of

E-modulus of the tested material, without any kind of user induced ambiguity in the data processing methodology.

The test setup varies depending on the material to be tested. The composite beam can be simply supported, to test concrete mortar, stabilized soil or any hardening material with similar grain sizes, or a cantilever to test materials with smaller grain size, such as cement paste and structural adhesives.

3.2.1 Concrete Testing

The setup of the method for concrete testing includes a 1.1 m long cylindrical PVC tube (acting as mould) with external/internal diameters of 110/98 mm (Granja and Azenha 2017), as shown schematically in Fig. 3.1. The mould is placed horizontally through two horizontal rods (with 12 mm in diameter) that laterally trespass the composite beam through its cross-sectional centre. The beam is positioned over two specially devised steel supports, one fixed and one pivoted, which allows fine-tuning of its transversal inclination as to compensate any potential misalignment between the supporting rods of the beam). This configuration allows the structure to behave as simply supported with a free span of one meter. The structural vibrations are measured with a high precision accelerometer (sensitivity: 10 V/g; range: ± 0.5 g) attached to the mid-span of the simply supported beam.

3.2.2 Mortar Testing

Since the original proposed method for concrete testing (Azenha et al. 2010), the technique has been adapted to test other materials like mortars, cement pastes (Azenha et al. 2012a, Maia et al. 2011a, b, 2012a, b; Garijo et al. 2019), stabilized soils (Azenha et al. 2011; Silva 2010; Silva et al. 2013, 2014) and other non-cementitious materials (Granja et al. 2015; Fernandes et al. 2014, 2015). The variant for mortar testing differs from the concrete version solely in regard to the dimensions of the testing beam. The cross section and the total length of the tube used are significantly reduced, by comparison. In this version, the mould is a 0.55 m long PVC tube with internal and external diameters of 44 and 50 mm respectively and a free span of 0.5 m. Figure 3.2 depicts the experimental setup for this variant of the method.

3.2.3 Cement Paste Testing

The variant for cement paste testing differs from the concrete/mortar versions in two main aspects (Azenha et al. 2012a): (*i*) the cross section of the tube used is greatly reduced and (*ii*) the structural system of the beam is changed to become cantilevered.

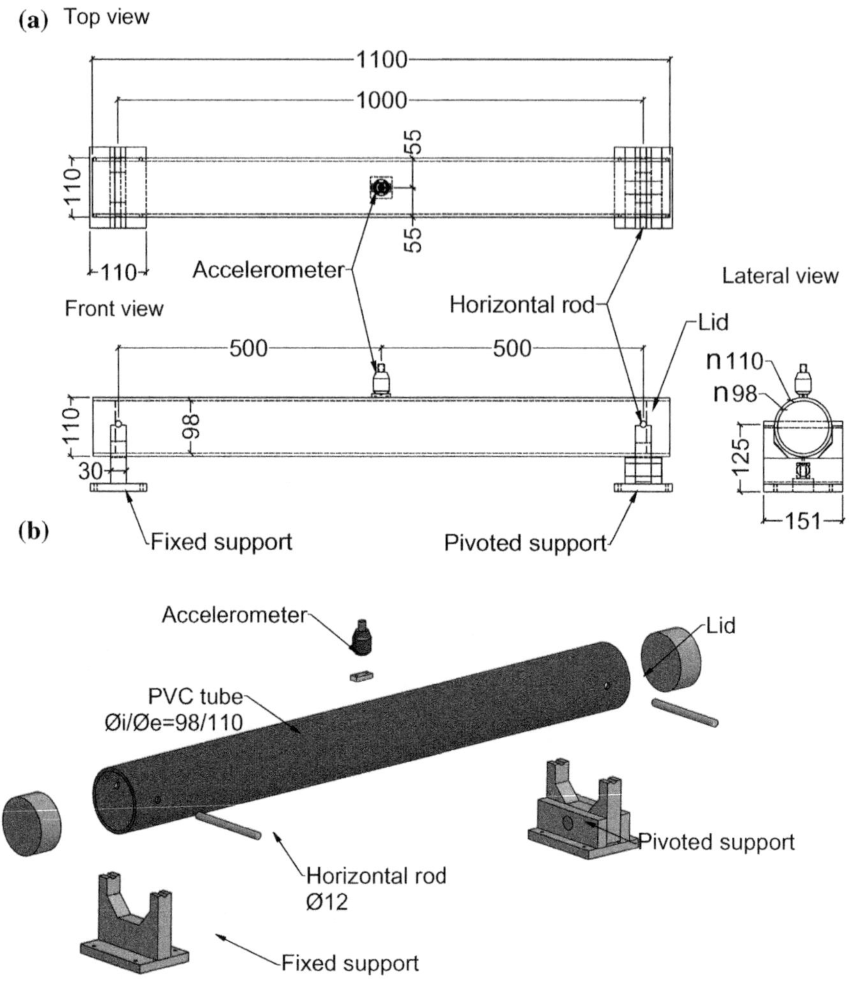

Fig. 3.1 EMM-ARM concrete test apparatus: **a** scheme; **b** 3D exploded view (units: mm)

The experimental setup for this variant of the method is shown in Fig. 3.3. The mould is a hollow tubular beam made of acrylic, with outside/inside diameters of 20/16 mm. The composite beam consists of a 550 mm long acrylic tube to be filled with fresh cement paste, with extremity caps made of polypropylene. The composite beam is then fixed in the horizontal position, operating as a cantilevered structural system with a span of 450 mm. The specimen is fixed using a metal clamping device with inner diameter matching the outer diameter of the beam. The metal clamping device is then rigidly connected to a rigid base to ensure complete fixation.

To monitor the structural vibrations, a lightweight accelerometer (mass and sensitivity equal to 23.25 g and 1 V/g, respectively) is attached to the free end of the cantilevered beam.

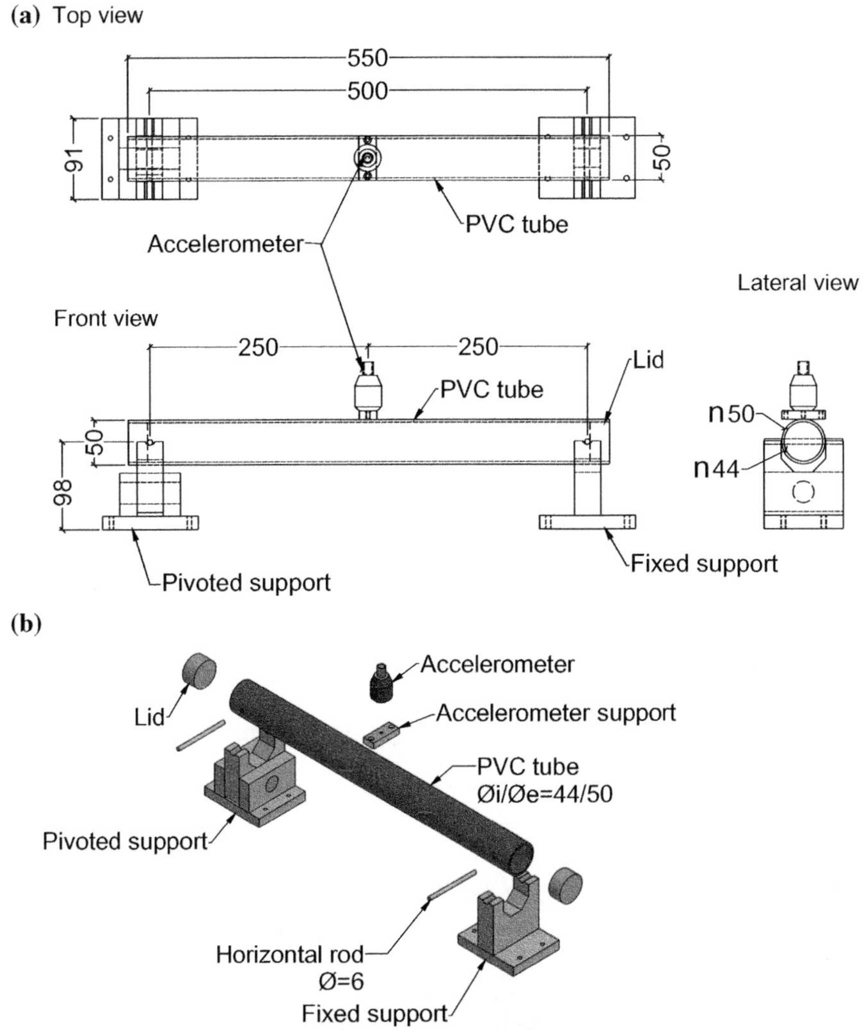

Fig. 3.2 EMM-ARM concrete test apparatus: **a** scheme; **b** 3D exploded view (units: mm)

3.3 Frequency Identification

EMM-ARM is based the modal identification of a composite beam to estimate the stiffness of the material under test. The identification of modal parameters through dynamic testing was originally developed in the fields of mechanics and aerospace engineering (Juang 1994). Modal testing is an experimental technique used to derive the modal model of a linear time-invariant vibratory system (He and Fu 2001). The theoretical basis of the technique is secured upon establishing the relationship between the vibration response at one location and excitation at the same or another

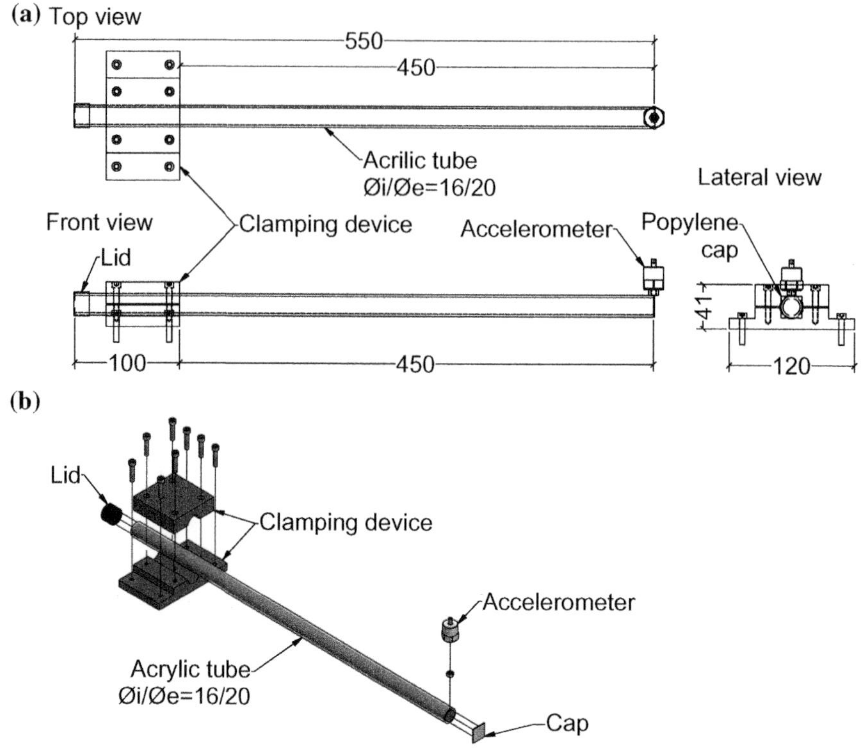

Fig. 3.3 EMM-ARM final assembly for cement pastes tests: **a** scheme; **b** 3D exploded view (units: mm)

location as a function of excitation frequency. Modal analysis involves three constituent phases: data collection, system identification and modal parameter extracting from the identified system.

For the dynamic monitoring structures, and depending on the source of excitation, two different groups of techniques are currently used: experimental modal analysis—EMA (He and Fu 2001; Ewins 2000) and operational modal analysis—OMA (Rainieri and Fabbrocino 2014; Reynders 2012; Peeters and De Roeck 2001; Overschee and Moor 1996). In the current state of development, the EMM-ARM uses the two techniques to estimate the modal parameters, depending on the version used: OMA techniques for the slender beams (e.g. for the cement paste apparatus); and EMA techniques for the other cases.

3.3.1 Operational Modal Analysis (OMA)

OMA is based on the measurement of a structure's response to dynamic excitations caused by environmental factors such as wind and urban traffic, which can be defined as the operational conditions. The dynamic response of the structure is acquired through the measurement of accelerations, velocities or displacements in relevant points of the structure. From the acquired signals the dynamic system is identified, using stochastic modal analysis techniques, and the modal parameters can be estimated, namely the eigenfrequencies, damping coefficients and mode shapes (Reynders 2012).

OMA techniques have the important advantage of not requiring explicit excitation of structures (Ren and Zong 2004). However, the structural responses induced by environmental actions have small amplitudes, which requires the use of very sensitive equipment to monitor the response of the structure (Rodrigues 2004). Fundamentally two sets of OMA methods can be considered (Reynders 2012): the first group comprises the signal analysis methods, where the measured responses of the structural systems are analysed and related to each other based on their transformation to the frequency domain—*frequency domain methods*; the second group corresponds the methods were the acquired repose time series are analysed directly in the time domain—*time domain methods*.

With these techniques, the EMM-ARM beam is considered to be excited by the environmental noise (e.g.: wind, people walking, noises/vibrations from construction site, etc.) which can conceptually be assumed to have an average behaviour of white noise, i.e. a stochastic process with constant spectral intensity in all frequencies. The resulting vibrations are then acquired in the vertical direction through an accelerometer attached to the composite beam. It should be noted that for the application of these techniques, the composite beams need to have a high slenderness in order to be highly excitable, so that the input provided by the environmental noise is enough to induce vibrations that can be detected by the accelerometer, allowing the test to be carried out without the need for explicitly exciting the composite beam. Nonetheless, in order to increase the amplitude of vibration of the beam and facilitate the resonance frequency detection, a fan should be placed in the vicinity of the experiment and blowing air towards the beam.

In regard to data processing, Fig. 3.4 shows a flowchart with a brief overall description of a possible framework to estimate the stiffness of the tested specimen along the curing process. First the vibrations are acquired in packages of 900s (Fig. 3.4a). From the recorded accelerograms (Fig. 3.4b), the collected data is converted from the time domain to the frequency domain through the Welch procedure (Welch 1967), thus resulting in the normalized power spectrum density (NPSD) of each measured package of data (Fig. 3.4c)—see further details in Azenha et al. (2012b). The NPSDs can then be included side-by-side in a coloured frequency *vs* time surface, with colouring that is proportional to the intensity of the power spectra (Fig. 3.4d). Then, the resonance frequencies of the first vibration mode are identified through the highest peak in each amplitude spectrum (Fig. 3.4e). In order to obtain a continuous evolution of

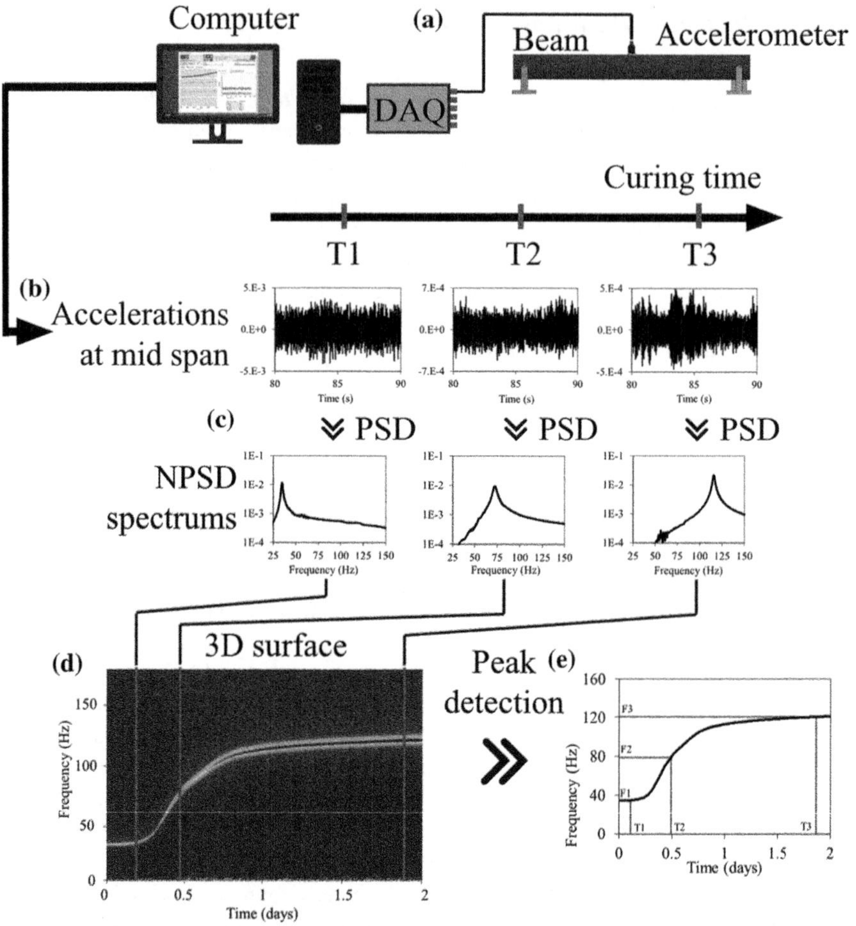

Fig. 3.4 Non-parametric data processing and stiffness estimation flow chart

the resonance frequency of the composite beam, this whole process is repeated every 60 min.

In more recent publications (Granja and Azenha 2017; Granja 2016) the EMM-ARM was used with more advanced OMA techniques like the SSI (Subspace System Identification) in time domain. This technique allowed to increase significantly the accuracy of the frequency identification.

3.3.2 Experimental Modal Analysis (EMA)

The EMA tests essentially consist in applying a controlled excitation to the structure under study with low intensity (so that the produced vibration levels do not affect the integrity of the structure) and measuring its response. Since in these tests, both the excitation forces and the structure response are measured it becomes possible to perform the evaluation of the impulse response functions (IRF) and the frequency response function (FRF). Detailed information about these methodologies may be found in (Caetano 1992; Ewins 2000; He and Fu 2001; Pintelon et al. 1994; McKelvey et al. 1996; Juang and Suzuki 1988; Juang and Pappa 1985; Chen et al. 1993).

3.3.2.1 Excitation

Even though forced vibration techniques can increase the accuracy of the modal identification of the composite beam, it has to be taken into account that the induction of an external vibration during the curing period of the cementitious materials may potentially affect the development of the mechanical properties, particularly in the vicinity of the setting time when bonds are being formed in the cement matrix that can be damaged due to the vibrations. This issue was studied by several authors (Dunham et al. 2007; Hong and Park 2015; Fernandes et al. 2011) and it was found that excessive vibration during the curing process can affect the elastic modulus by almost 10%. Also in the work of Hong and Park (2015) it was found that even small vibrations (with maximum accelerations of 14.4 mg at 5 Hz) can have a relevant impact in the concrete curing process and affect its mechanical properties. Based on these evidences it was established that the use of forced vibrations techniques should be quite limited in terms of vibration intensity applied to the material, as to make the impact of such vibration negligible on the setting and hydration processes (Granja and Azenha 2017).

In addition to the excitation intensity issue, it is also necessary to take into consideration that the application of Eq. (3.1) presented above demands a structural system comprised of a simply supported beam, but does not include any potential interaction issues with an actuator system. Therefore, the actuator must not introduce any added vertical stiffness to the system, otherwise a spring would have to be considered (and dully quantified) in the boundary conditions used to derive the equation.

In this way, a custom electromagnetic actuator was developed, specially designed to be able to apply a very small dynamic force to the testing beam without physical contact between the actuator and the beam. A representation of this custom-made electromagnetic actuator is shown in Fig. 3.5. The force is applied to the beam through a magnet (made of neodymium N50 with remanence: $B_r = 1.4T$; $\varnothing = 8$ mm; height = 20 mm) that is physically attached to the bottom of the beam at mid span. On the opposite side, the magnet is surrounded by a coil (made with copper wire with 0.22 mm in diameter and 3140 turns; $\varnothing_{int} = 10$ mm $\varnothing_{ext} = 25$ mm). When subjected to a given current intensity, this coil creates a magnetic field dependent on

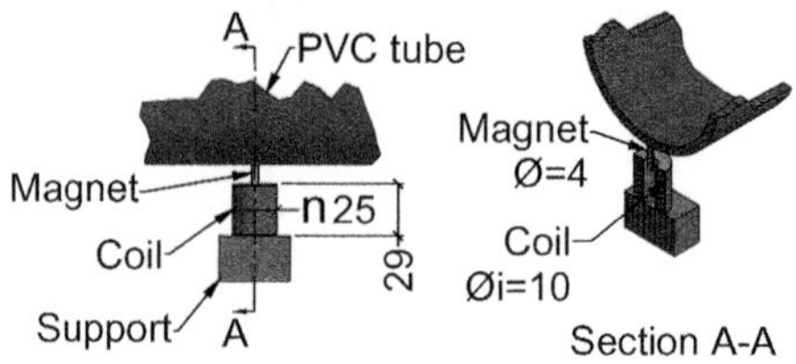

Fig. 3.5 Custom made non-contact electromagnetic actuator attached to an EMM-ARM beam (units: mm)

the current polarity and intensity. The coil's frame was built in a non-ferromagnetic material (plastic) to prevent interaction between the magnet attached to the beam and the body of the actuator, which could again introduce an effect similar to that of an elastic spring at mid span of the beam.

From the work of Granja and Azenha (2017) it possible to observe that the amplitude of the excitation force ranged ±8 mN and is constant throughout the frequency range normal for a EMM-ARM test. Despite the very low excitation force range, it is sufficient to create a response with enough acceleration amplitude to be detectable by the accelerometer. The maximum response of the beam, as reported by Granja and Azenha (2017), was in a range of ±6 mg, value that according to the study of Hong and Park (2015) is small enough to avoid detrimental impacts in the mechanical properties.

3.3.2.2 Signal Processing

The use of forced vibration techniques for testing requires a different data processing from the one shown in Fig. 3.4. Apart from the necessity to generate the excitation it is necessary to measure both the excitation and the response of the beam, and the signal processing needs to be changed to an input-output technique. Yet, despite different, such techniques are much simpler than the stochastic methods, since the excitation of the beam is known. In these techniques, the modal parameters (natural frequencies, mode shapes and damping coefficients) can be calculated by estimating the frequency response functions (FRF) or the Impulse Response Functions (IRF), either in frequency or time domain (Caetano 2000; Granja 2016). The modal parameters are estimated through the FRF in the frequency domain (defined here as H) that describes the relationship between the response and the excitation imposed to the system and can be defined by the ratio of the cross-spectrum excitation-response (S_{FX}) and the auto-spectrum of the excitation (S_{FF}) (Oppenheim et al. 1989), as shown in the Eq. (3.2).

$$H(f) = \frac{S_{FX}(f)}{S_{FF}(f)} \tag{3.2}$$

Figure 3.6 shows an example of a flowchart with a brief overall description of the test procedure and data processing to perform the resonance frequency identification

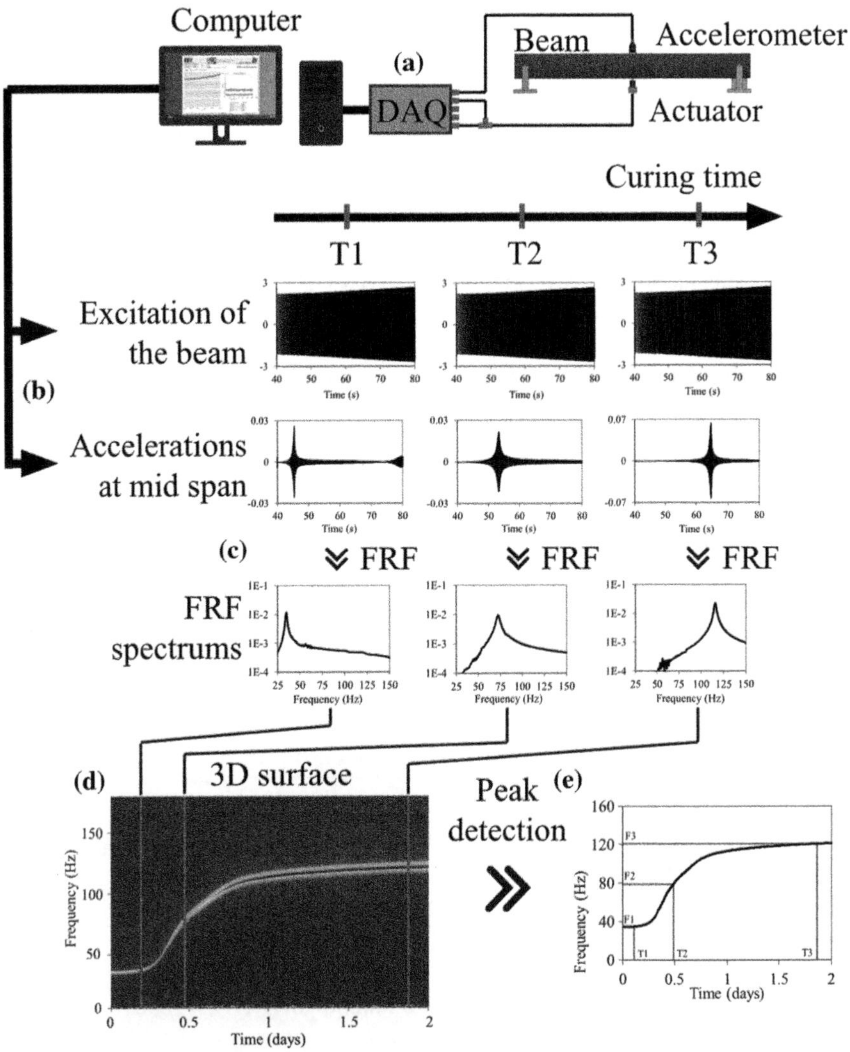

Fig. 3.6 Parametric data processing and stiffness estimation flow chart

along the curing process. First, a sine sweep signal (with linear frequency variation between 10 and 200 Hz) with 40 s duration was continuously generated in a 24-bit NI 4431 dynamic signal analyser during 300 s. The signal is transmitted to the non-contact custom made electromagnetic actuator to perform the excitation of the beam. Simultaneously both the generated signal and the resulting accelerations of the beam were acquired in the same device (Fig. 3.6a). From the recorded 300 s accelerograms and excitation signal (Fig. 3.6b), the FRF of the system was computed and the FRF spectra were obtained (Fig. 3.6c).

After obtaining the FRF spectra, they are cumulatively condensed into a single graph (a surface, as shown in Fig. 3.6d). Then, the resonance frequency of the first mode of vibration is identified through the highest peak in each amplitude spectrum, through a fully automatic algorithm, and the evolution of the frequencies along the curing time is obtained.

In more recent publications (Granja and Azenha 2017; Granja 2016) the EMM-ARM was used with more advanced EMA techniques like the Transfer Function (TF) and the Subspace System Identification (SSI) in time domain. The use of such methods allows substantial reductions in the sensitivity of EMM-ARM to ambient noise disturbances, and increases the accuracy of estimations (even in very unfavourable environments). Also the EMA parametric identification techniques allow the identification of the beam's damping along the material curing that might be used in the future to estimate some viscoelastic properties of the material under test (Granja 2016).

3.4 E-modulus Estimation

The stiffness evolution of the material can be inferred based on the resonance frequency evolution of the composite beam using the corresponding equation of motion. In the case of a simply supported beam with a mass at mid span, one obtains:

$$\bar{E}\bar{I}\frac{\partial^4[\varphi(x)\cdot Y(t)]}{\partial x^4}+\bar{m}\frac{\partial^2[\varphi(x)\cdot Y(t)]}{\partial t^2}=0 \quad (3.3)$$

where $Y(t)$ represents the amplitude of the vertical displacement along the acquisition time, t (in seconds), expressed in relation to the deflection mode $\phi(x)$, x is the coordinate along the span of the beam (m), $\bar{E}$ and $\bar{I}$ are the homogenized elasticity modulus (Pa) and second area moment of inertia (m^4) of the composite cross-section, respectively and $\bar{m}$ is the uniformly distributed mass along the beam (kg/m). Form the Eq. (3.3) it is possible to express $\phi(x)$ as a function of A_1, A_2, A_3 and A_4, as shown below:

$$\varphi(x)=A_1\cdot\cos(a\cdot x)+A_2\cdot\sin(a\cdot x)+A_3\cdot\cos h(a\cdot x)+A_4\cdot\sin h(a\cdot x) \quad (3.4)$$

With

$$a=\sqrt[4]{\frac{\omega^2\cdot\bar{m}}{\bar{E}\,\bar{I}}}$$

where $\omega = 2\pi\cdot f$ is the first angular resonance frequency of the beam (rad^{-1}) and f is the linear frequency (Hz). At this stage, boundary conditions need to be applied to Eq. (3.4) for the present beam, with a double support in one end and a vertical sliding support in the other, resulting in:

$$\begin{array}{ll}\text{At } x=0: \bar{E}\,\bar{I}\cdot\varphi'''(0)=-k\cdot\varphi(0), & \bar{E}\,\bar{I}\cdot\varphi''(0)=0\\ \text{At } x=L: \bar{E}\,\bar{I}\cdot\varphi'''(L)=-\omega^2\cdot\varphi(L)\cdot m_p, & \varphi'(L)=0\end{array} \tag{3.5}$$

where m_p is the concentrated mass located at the mid span (kg), L is half of the span of the beam (m) and k is the vertical stiffness of the supports (N/m). Introducing these boundary conditions in Eq. (3.4) a set of equations is obtained, whose eigenvalues ω may be computed according to:

$$\frac{1}{2k}\begin{bmatrix}\bar{E}\,\bar{I}a^3\omega^2 m_p\cdot\left[1+\cosh(aL)^2-\sinh(aL)^2+2\cos(aL)\cosh(aL)\right]\\ +2\cdot\bar{E}\,\bar{I}\cdot a^6\cdot[\sin(aL)\cdot\cosh(aL)+\cos(aL)\sinh(aL)]\\ +4\cdot\bar{E}\,\bar{I}\cdot a^3\cdot k\cdot\cos(aL)\cdot\cosh(aL)\\ +2\cdot k\cdot\omega^2\cdot m_p\cdot[\cos(aL)\cdot\sinh(aL)+\cosh(aL)\cdot\sin(aL)]\end{bmatrix}=0 \tag{3.6}$$

With the results of modal identification, all variables in Eq. (3.6) are known except for $\overline{EI}$, thus allowing to mathematically obtain this unknown. Bearing in mind the composite tubular section of known internal/external diameters ($\varnothing_i$, $\varnothing_e$), and the known E-modulus of the mould E_m, it is possible to compute the E-modulus of the tested material (concrete) E_c through Eq. (3.7):

$$\bar{E}\,\bar{I}=E_m\frac{\pi\left(\varnothing_e^4-\varnothing_i^4\right)}{64}+E_c\frac{\varnothing_i^4}{64} \tag{3.7}$$

By performing the above procedure for all data packages, the E-modulus *versus* time evolution curve can be obtained.

A similar reasoning can be applied to the version of EMM-ARM aimed to test cement paste, which uses a distinct structural system (cantilever). Despite such fact, the procedure to obtain the resonance frequency of the composite beam is quite similar to the variant for concrete described before. However, after determining the first flexural resonant frequency of the composite beam, Eq. (3.6) is no longer valid due to the different structural system.

Therefore, in order to infer the stiffness of the tested material, it becomes necessary to derive a new formulation for the cantilever structural system (Fig. 3.3). In this case the boundary conditions to introduce in Eq. (3.4), in correspondence to the fixed support in one end, are:

$$\begin{array}{ll}\text{At } x=0: \bar{E}\,\bar{I}\cdot\varphi''(0)=k_\theta\cdot\varphi'(0), & \varphi(0)=0\\ \text{At } x=L: \bar{E}\,\bar{I}\cdot\varphi'''(L)=-\omega^2\cdot\varphi(L)\cdot m_p, & \varphi''(L)=0\end{array} \tag{3.8}$$

where k_θ is the rotational stiffness of the support (N/m). Therefore, the relation between the frequency of the first mode of vibration and the stiffness of the composite beam can be computed according to Eq. (3.9):

$$\begin{aligned} & a^3[\cosh(a\cdot L)\cdot\cos(a\cdot L)+1] \\ & +\frac{\omega^2\cdot m_p}{\bar{E}\,\bar{I}}[\cos(a\cdot L)\cdot\sinh(a\cdot L)-\cosh(a\cdot L)\cdot\sin(a\cdot L)] \\ & +\frac{\bar{E}\,\bar{I}\cdot a^4}{k_\theta}[\cos(a\cdot L)\cdot\sinh(a\cdot L)-\cosh(a\cdot L)\cdot\sin(a\cdot L)] \\ & -2\cdot\frac{a\cdot m_p\cdot\omega^2}{k}\cdot\sin(a\cdot L)\cdot\sinh(a\cdot L)=0 \end{aligned} \quad (3.9)$$

Then the E-modulus of the material inside the mould can be estimated based on the geometry and stiffness of the acrylic tube through the Eq. (3.7).

3.5 Comparison of EMM-ARM with Other Methods

The E-modulus estimations with EMM-ARM have been extensively compared with several other distinct methods to infer the E-modulus or other related properties. In this subchapter two different experimental campaigns will be presented, where the E-modulus estimations through EMM-ARM for two different variants (concrete and cement paste) will be compared with several other methods used in the literature for E-modulus estimation of cement-based materials.

This subchapter presents an extensive comparison between several experimental methodologies capable of quantifying the stiffness of concrete and cement paste, such as: EMM-ARM, cyclic compression/tension tests (CC, BTJASPE, TSTM), ultrasonic wave transmission (UWT), bender-extender elements (BE), and penetration resistance.

3.5.1 *Concrete*

The work that will be presented here was the result of part of an international collaboration between three research centres [University of Minho (UM), *Institut Français des Sciences et Technologies des Transports, de l'Aménagement et des Réseaux* (IFSTTAR) and *Université libre de Bruxelles* (ULB)]. The results have been published in two papers of more general purpose (Boulay et al. 2013a; Delsaute et al. 2016), but are here filtered and analysed from a perspective of EMM-ARM comparison.

In addition to the EMM-ARM technique presented in the previous sections, eight further different techniques have been tested within a round-robin testing program involving three laboratories: University of Minho (UM), IFSTTAR, and University

Table 3.1 Mixture proportions of the concrete (w/c = 0.54)

Components	Mass (kg/m^3)
Cement (CEM I 52.5 N PMES CP2)	340
Sand (0/4)	739
Gravel (8/22)	1072
Total water	184
Density	2335

of Brussels (ULB). They can, a priori, be sorted in two classes, according to their respective ranges of loading rates:

- four techniques of quasi-static loadings [cyclic compression tests: classical cyclic compressive testing CC (all three labs), automated cyclic compressive testing with BTJASPE (Boulay et al. 2012) since fresh state (IFSTTAR), classical automated cyclic compressive testing (ACC) using BTJASPE protocol with initial testing after setting time (IFSTTAR), automated cyclic E-modulus testing with a TSTM (ULB) (Delsaute et al. 2012)];
- and four techniques of high frequency loadings: three classical UPV measurements with PunditLab (UM), FreshCon (ULB) (Carette and Staquet 2015), BTPULS (IFSTTAR) (Boulay et al. 2012); and Smart Aggregates SMAGs (ULB) (Carette et al. 2012).

All of the above-mentioned experimental methods were applied without a rigorous temperature history control history of the specimens due to practical limitations associated to the test setups and the size of specimens themselves. That is why observations are compared after having expressed results at the same concrete maturity using the equivalent time method.

The same raw materials have been used in the three laboratories involved in this comparative programme. The mixture proportions are given in Table 3.1.

The average E-modulus evolutions obtained will all low frequency (or quasi-static) methods are shown in Fig. 3.7. In all the results the overall evolution kinetics is similar, with an initial dormant period until ~5 h after mixing. At this point the E-modulus starts to increase whatever the testing method. After setting, the kinetic of the E-modulus looks very similar, whatever the testing method whereas a limited scattering is observable between the results, especially very early ages. It appears that the difference of protocol of loading (strain rate, stress amplitude) and the type of testing method did not induce any strong effect on the kinetic and the amplitude of the E-modulus. Furthermore, all low frequency testing methods have revealed a good correspondence with classical cyclic compression results.

To illustrate the accuracy and precision of each method the normal distribution of the estimated E-modulus was computed. But first all the values were normalized to a reference value at the age of testing computed from the model (fitting of CC results). The normal distributions of the E-modulus estimations along the whole duration of the tests form the five distinct techniques are shown in Fig. 3.8 and the mean values and standard deviations are summarized in Table 3.2.

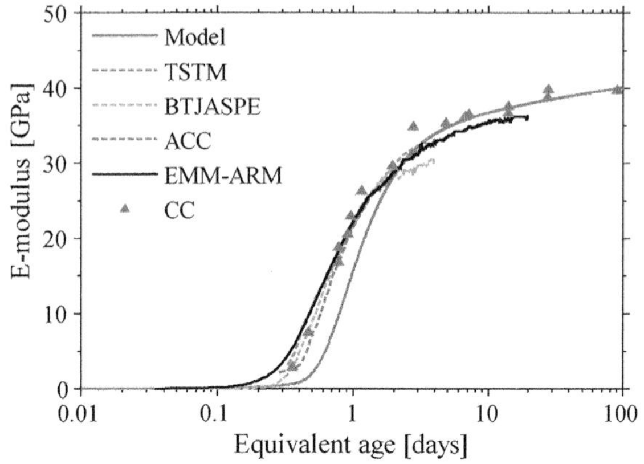

Fig. 3.7 E-modulus obtained with low frequency testing methods

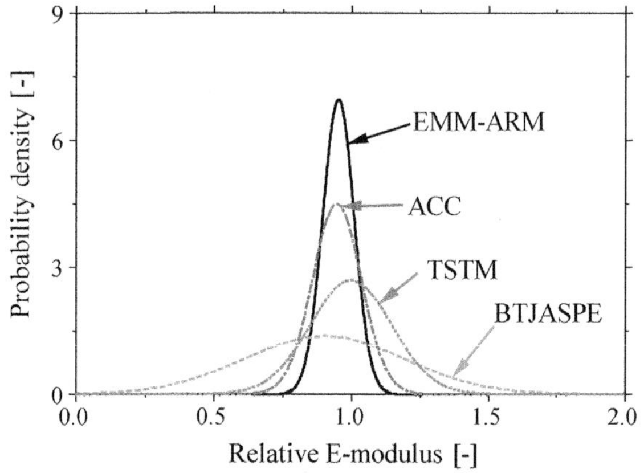

Fig. 3.8 Accuracy and precision of the low frequency experimental methods

Table 3.2 Summary of the mean values (μ) and standard deviations (*SD*) of the E-modulus estimated with the low frequency experimental methods

Method	EMM-ARM	BTJASPE	ACC	TSTM
μ	0.952	0.902	0.946	0.997
SD	0.0618	0.2914	0.0888	0.1490

It can be observed that TSTM was the most accurate method with an average (μ) of 0.997, though the precision was low since the Standard Deviation (*SD*) is 0.1490. Regarding the EMM-ARM results, it is possible to state that the method is able to estimate the so-called quasi-static E-modulus of the concrete with a very high precision and a good accuracy. These results enable the validation of the E-modulus estimations obtained with EMM-ARM since the accuracy and precision of the estimations are within the range of the competing static methods.

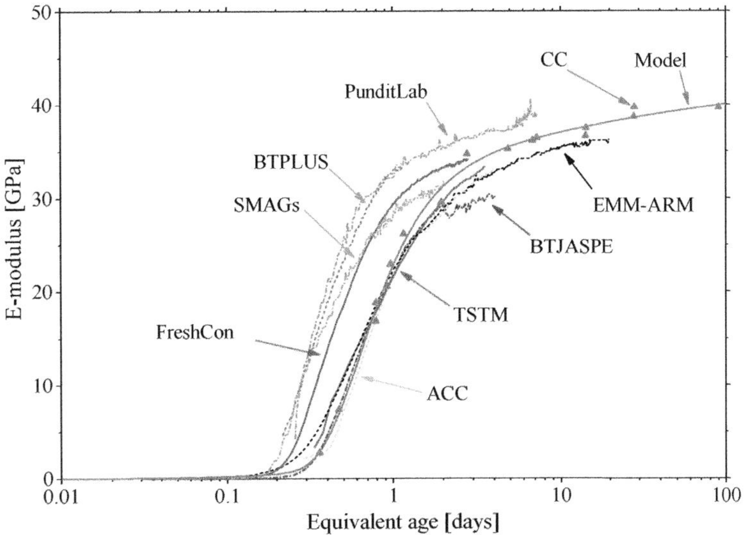

Fig. 3.9 Comparison of E-modulus obtained with static and dynamic testing methods

Figure 3.9 presents the synthesis of all results. Only mean values of each testing method are shown here. As expected, clear differences appear between static and dynamic results. Dynamic modulus is generally higher than static results. A faster evolution of dynamic results is also observable.

3.5.2 Cement Paste

In this subsection, the validation of the EMM-ARM version for cement paste is demonstrated through comparison with a set of directly or indirectly competing methods, such as: classical cyclic compression (CC), ultrasonic pulse velocity (UPV), bender-extender elements (BE) and Vicat needle. For more details about the experimental setups, materials and results see (Granja et al. 2014).

The experiments were conducted on cement pastes containing Type I and Type II Portland cement. In the scope of this experimental program, two cement paste compositions with water/cement ratio w/c = 0.50 were adopted. The mixture proportions of the cement pastes as well as the corresponding nomenclatures are presented in Table 3.3.

The comparison between the elastic modulus results obtained by EMM-ARM and by classic methods (CC and Vicat) for the cement pastes 32.5 and 42.5 is shown in Fig. 3.10. It can be seen that the values obtained through the EMM-ARM are similar to those collected in CC tests in terms of magnitude and evolution kinetics. However, the results for the cement paste 32.5 (Fig. 3.10a) show a non-negligible difference

Table 3.3 Cement pastes adopted for validation of EMM-ARM

Reference	Cement type	w/c ratio	Density (kg/m^3)
32.5	CEM II/B-L 32.5 N	0.50	1787.4 $\pm$ 4.0
42.5	CEM I 45.5 R	0.50	1840.3 $\pm$ 4.1

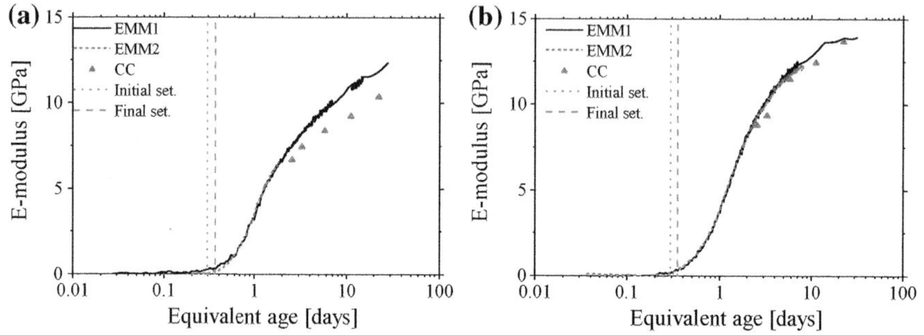

Fig. 3.10 Comparison of the results of EMM-ARM, Vicat, and cyclic compression for cement pastes: **a** 32.5, **b** 42.5

of 1.4 GPa at $t_{eq} = 22.4$ days (538 h). This deviation may possibly be explained by differences in the curing conditions of the samples. In fact, the EMM-ARM samples remained in perfectly sealed conditions during the whole test, while the samples used for the CC tests were exposed to drying during the curing period. This small variation in the curing conditions may have influenced the hydration process at the surface of the CC specimens (Parrott 1990), which may have significant effects in view of the small size of the specimen, thus resulting in lower stiffness. As the expected porosity of 32.5 is higher than that of 42.5, it is plausible that this deficient curing of CC specimens may have affected 32.5 more significantly, as opposed to 42.5.

The results presented in Fig. 3.10 also show good agreement between EMM-ARM and the data collected by the Vicat needle, in the sense that the end of setting determined by Vicat testing is approximately coinciding with the end of the dormant period observed in EMM-ARM, followed by a strong acceleration of the hydration kinetics.

Taking into account that the methodologies based on wave propagation measure dynamic parameters, for the purpose of comparison of all methodologies under study, the results were normalized ('Norm' in Fig. 3.11) by dividing all results of each specimen/methodology by their corresponding values at $t_{eq} = 7$ days. Moreover, in order to simplify the analysis and to compare both methods based on wave propagation (BE and UPV) with the results of quasi-static methods, the velocity values were squared (V^2) prior to normalization, as V^2 is proportional to the elasticity modulus (Granja et al. 2014).

The results of all experimental methods involved in this comparison is given in Fig. 3.11, which demonstrates a quite reasonable reciprocal agreement, thus mutually

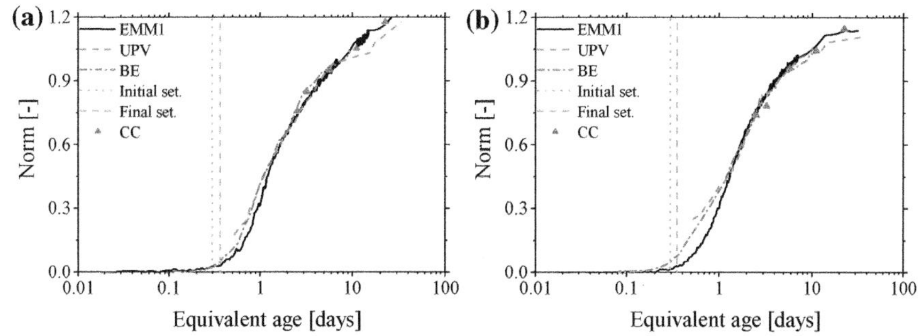

Fig. 3.11 Comparison of the results of all methodologies used in this study for cement pastes **a** 32.5 and **b** 42.5

validating the studied methodologies. The good performance of EMM-ARM in the scope of this comparative study, together with its ability to provide precise, continuous and quantitative estimates of E-modulus confirms the versatility and applicability of this methodology.

In regard to setting times, there is also a good coherence between Vicat, EMM-ARM and BE, as observable in Fig. 3.11. Thus, these results have confirmed the applicability of the wave propagation methods to monitor the stiffness of cement pastes since the fresh state and throughout the entire hardening process, as already mentioned by other authors (Boumiz et al. 1996; Reinhardt and Grosse 2004). Despite this fact, the results obtained by these wave-propagation based methods should be regarded qualitatively, since these refer to dynamic properties, whose conversion to static properties is often arguable, particularly at very early ages. Indeed, it should be noted that these wave velocity methods (UPV and BE) seem to exhibit a slightly more accelerated evolution kinetics than EMM-ARM, which is more evident for the 42.5 paste, as shown in Fig. 3.11b. This fact may be related to the early evolution of Poisson's ratio. Similar findings have been reported in other research works, where the consideration of constant Poisson's ratios led to apparent earlier acceleration of stiffness when estimated through pulse velocity methods (Boulay et al. 2013a).

3.6 Repeatability of E-modulus Estimations

In the previous subchapter it has been shown that EMM-ARM is capable of very consistent results when compared with other methods for direct or indirect assessment of E-modulus. The present section pertains to a repeatability assessment of EMM-ARM (Dotson 2015). According to JCGM (2012) repeatability in the context of a physical measurement is "*the measurement precision under a set of conditions of measurement (...) that includes the same measurement procedure, same operators, same measuring system, same operating conditions and same location, and replicate measurements on the same or similar objects over a short period of time*". In accordance to the previous subchapter, the repeatability study was divided in two parts: the first one dedicated to the version of EMM-ARM for concrete testing and the second on to the cement paste version.

3.6.1 Concrete

To assess the repeatability of EMM-ARM a series of seven tests was performed in five different concrete mixes with at least 2 beams tested simultaneously. The list of all the tests performed with the information of the concrete mix tested, the number of specimens and temperature of testing is presented in Table 3.4. All the tests were performed at room temperature with exception to the test 7 that was performed under imposed variable temperature. The mean temperature history measured in the specimens tested in test 7 is shown in Fig. 3.12. For more information about the experimental setups, procedures and materials see (Delsaute et al. 2016; Azenha et al. 2012b; Granja and Azenha 2017; Granja 2016).

The mixture proportions for the five different concretes are given in Table 3.5.

The E-modulus evolutions obtained in the seven different tests are shown in Fig. 3.13. In this figure only the mean curve (full lines), the mean curve + the standard deviation (dashed lines) and the mean curve − the standard deviation (dash dot lines) are represented. As can be observed in all the tests, the deviation from the mean curve is very small with average error of 1.5% at the end of all the test. In fact, even in tests performed with the same concrete but with different temperature histories (1 and 7), the final values measured at the age of 7 days have a very small discrepancy (a difference of 0.5 GPa or 1.8%). These results enable to state that the E-modulus estimations obtained with EMM-ARM in the version for concrete have a very good repeatability.

Table 3.4 Tests performed in the repeatability study for the EMM-ARM concrete version

Test	Concrete	Number of tests	Testing temperature
1	M1	3	Room temperature 20 ± 2 °C
2	M2	3	Room temperature 20 ± 2 °C
3	M3	3	Room temperature 20 ± 2 °C
4	M4	3	Room temperature 20 ± 2 °C
5	M5	6	Ambient temperature 17 ± 7 °C
6	M6	3	Ambient temperature 23 ± 8 °C
7	M1	2	Variable see Fig. 3.12

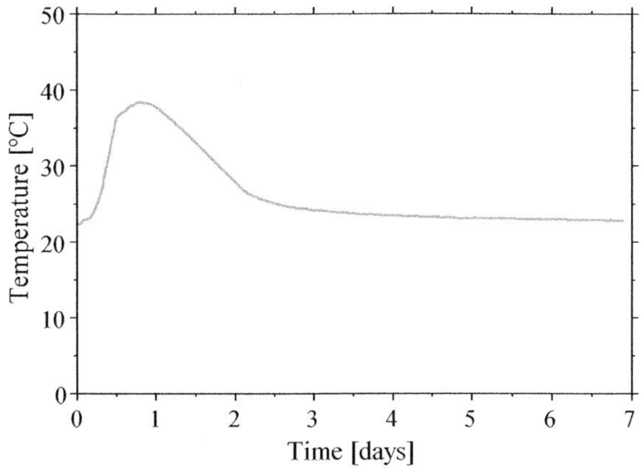

Fig. 3.12 Temperature history of test 7

Table 3.5 Mixture proportions of the different concrete mixes used

Component	Mix M1 (kg/m^3)	Mix M2 (kg/m^3)	Mix M3 (kg/m^3)	Mix M4 (kg/m^3)	Mix M5 (kg/m^3)	Mix M6 (kg/m^3)
Sand	250 (0/2)	395 (0/2)	245 (0/2)	739 (0/4)	300 (0/2)	330 (0/2)
	460 (0/4)	395 (0/6)	786 (0/6)	–	540 (0/6)	560 (0/6)
Gravel	1140 (4/16)	840 (6/22)	417 (6/14)	1072 (8/22)	550 (6/14)	550 (6/14)
	–	–	478 (14/20)	–	420 (11/22)	440 (11/22)
Cement	218 (CEM II/A-L 42.5R)	320 (CEM I 42.5R)	280 (CEM I 42.5R)	340 (CEM I 52.5 N PMES CP2)	320 (CEM II/A-L 42.5R)	220 (CEM II/A-L 42.5R)
Fly ash	112	–	40	–	100	130
Filler	–	260	–	–	–	–
Water (L/m^3)	155	160	143	184	165	160
Super plasticiser	3.30 (Sikament 400+)	7.64 (Viscocrete 3006)	6.25 (Rheobuild 1000)	–	3.36 (BasfGlenium Sky 548)	2.45 (BasfGlenium Sky 548)
Plasticiser	–	–	–	–	0.84 (BasfPozzolith 540)	2.10 (BasfPozzolith 540)

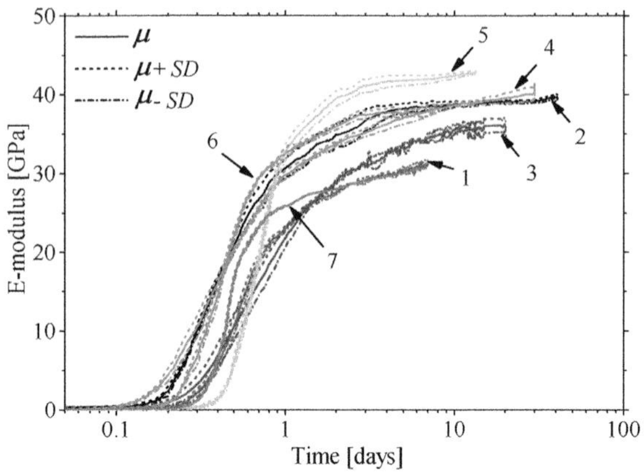

Fig. 3.13 Concrete E-modulus evolutions obtained in seven EMM-ARM tests

3.6.2 Cement Paste

The repeatability of EMM-ARM for cement pastes was evaluated by performing seven tests with seven different cement pastes with at least 2 specimens. The list of all the tests performed with the information of the cement paste composition, the number of specimens and temperature of testing is presented in Table 3.6. For more information about the experimental setups, procedures and materials see (Granja et al. 2014; Granja 2016).

The composition and properties of the different cements used in the cement pastes are shown in Table 3.7.

The E-modulus evolutions obtained in the seven different tests are shown in Fig. 3.14. In this figure only the mean curve (full lines), the 'mean curve + the standard deviation' (dashed lines) and the 'mean curve − the standard deviation' (dash-dotted lines). As can be observed in all the test results, the deviation from the mean curve is very small with average error of 1.1% at the end of all the tests. These results enable to state that the E-modulus estimations obtained with EMM-ARM in the version for cement pastes have a very good repeatability.

Table 3.6 Tests performed in the repeatability study for the EMM-ARM cement paste version

Cement paste	Cement	w/c	Number of tests	Testing temperature (°C)
1	Grey type 1	0.50	7	20
2	Grey type 2	0.50	2	20
3	Grey type 3	0.35	2	20
4	White	0.32	2	20
5	White	0.40	2	20
6	White	0.48	2	20
7	White	0.40	2	10

Table 3.7 Composition of the different cements used in the cement pastes

Cement	Grey type 2	Grey type 1	Grey type 3	White
C_3S (%)	42.47	51.63	70.80	66.89
C_2S (%)	13.58	17.18	10.20	20.00
C_3A (%)	6.82	7.03	6.40	3.51
C_4AF (%)	8.84	9.86	7.30	1.00
Free lime (%)	–	1.05	–	0.31
Gypsum (%)	–	–	5.30	–

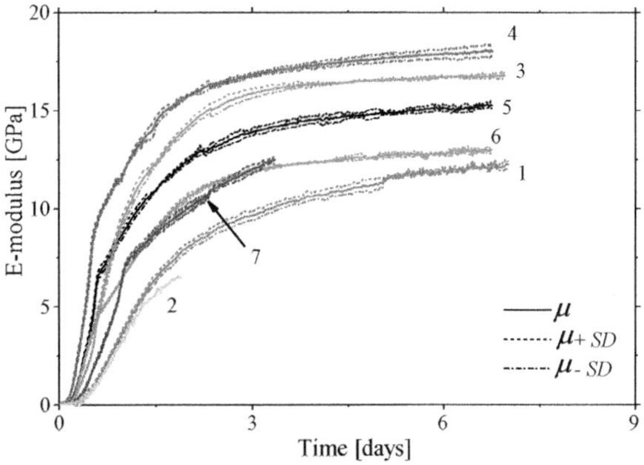

Fig. 3.14 Cement paste E-modulus evolutions obtained in seven EMM-ARM tests

3.7 Accuracy of the E-modulus Estimations

According to JCGM (2012) measurement accuracy is the "*closeness of agreement between a measured quantity value and a true quantity value of a measurand*". Thus, in this section the analysis of the EMM-ARM E-modulus estimations accuracy will be discussed. Once more, the analysis was divided in two parts related to the material: concrete and cement paste.

3.7.1 Concrete

From all the experiments performed during the Ph.D. work of José Granja (Granja 2016) and the results of the first implementation of EMM-ARM (Azenha et al. 2010) the relation between the E-modulus values obtained from the classical cyclic compression tests (CC) and the EMM-ARM estimation at the same age of testing was computed and the results are shown in Fig. 3.15.

As can be seen in the figure, the relation between the two estimations is always very close to 1, and the dispersion of the results is small. Nevertheless, to better

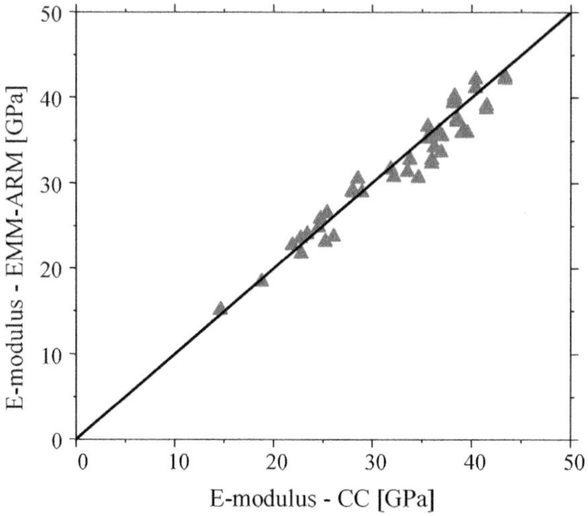

Fig. 3.15 Relation between the concrete E-modulus estimated from CC and EMM-ARM

interpret the results a normal distribution was fitted to the results. A mean value of 0.981 was obtained, along with a standard deviation of 0.0496. The probability density function of this normal distribution is shown in Fig. 3.16. As can be seen the accuracy of the EMM-ARM estimations is very high, with a mean error of 1.9%. Additionally, based on this data, it is possible to observe that there is a 90% possibility that the EMM-ARM E-modulus estimations are between 0.899 and 1.062 of the CC values.

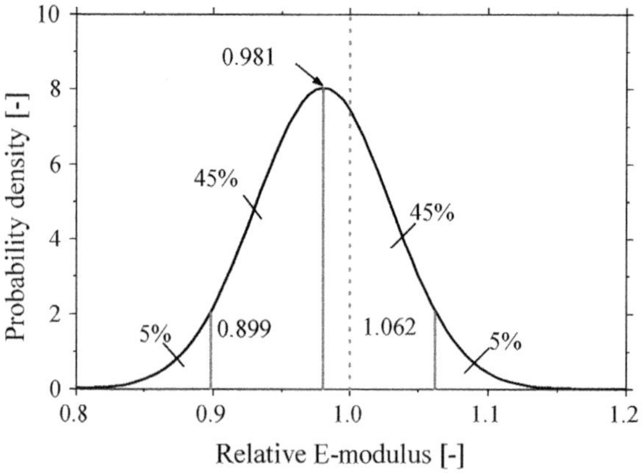

Fig. 3.16 Probability density of the normal distribution of the relation between the concrete E-modulus estimations from EMM-ARM and CC

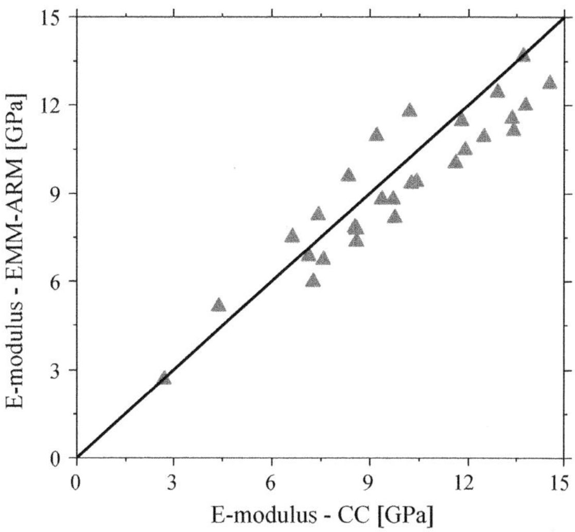

Fig. 3.17 Relation between the cement paste E-modulus estimated from CC and EMM-ARM

3.7.2 Cement Paste

The relation between the E-modulus obtained from the classical cyclic compression tests (CC) and the EMM-ARM estimation at the same age of testing extracted from all the experiments performed until now with EMM-ARM (Maia et al. 2012; Granja 2011, 2016; Granja et al. 2013, 2014; Azenha et al. 2012b) are shown in Fig. 3.17.

As can be observed the values are close to the line that represents the equality between the E-modulus estimated from EMM-ARM and CC. However, the dispersion of the results is slightly higher when compared with the results from the tests on concrete. The mean value and standard deviation of the normal distribution fitted to the results is 0.953 and 0.1121, respectively. Therefore, one can state that the EMM-ARM in the cement paste version have a high accuracy with an average error of 4.7%. By plotting the probability density graph of the normal distribution (Fig. 3.18) one can observe that there is a 90% possibility that the EMM-ARM E-modulus estimations are between 0.769 and 1.137 of the values obtained with the classical cyclic compression tests (CC).

3.8 Conclusions

This chapter gives an overview of all the developments and capabilities that the EMM-ARM method has for continuous monitor the evolution of the Elastic properties of cement-based materials. Several works from different authors were combined to give a clear picture of the method's state of development.

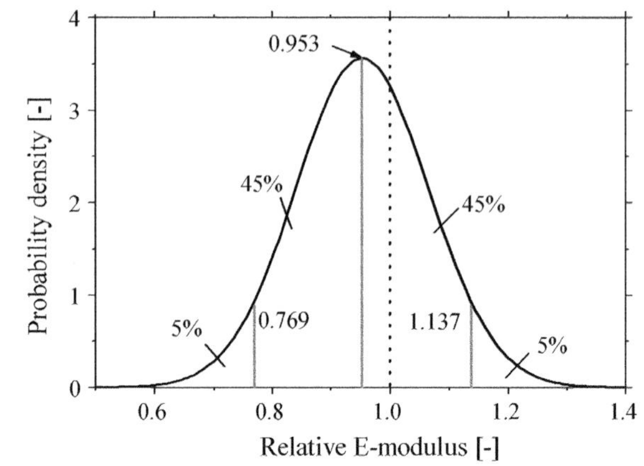

Fig. 3.18 Probability density of the normal distribution of the relation between the cement paste E-modulus estimations from EMM-ARM and CC

The EMM-ARM is under continuous development since its origin in 2009 and now it has more than 10 year of continuous implementation in several laboratories around the world. From these systematic implementations its potentialities are clear and goes from just been able to give in situ real-time information about the elastic modulus of concrete (Granja and Azenha 2015) to support advanced material computation like the one presented in the work of Charpin et al. (2016).

Since the first implementation the robustness and reliability of the method was significantly improved together with an easier way to conduct the experiment through the use of a more user-friendly specimen.

Acknowledgements This work was supported by FEDER funds through the Operational Program for Competitiveness Factors—COMPETE and National Funds through FCT—Portuguese Foundation for Science and Technology under the project IntegraCrete PTDC/ECM-EST/1056/2014, as well to the Research Unit ISISE (POCI-01-0145-FEDER-007633). The first author also acknowledges the Ph.D. grant SFRH/BD/80338/2011 provided by FCT. The authors would also like to acknowledge the benefits of cross-linking research through the network of COST Action TU1404 "Towards the next generation of standards for service life of cement-based materials and structures" (www.tu1404.eu).

References

ASTM. (2002). Standard Test Method for Fundamental Transverse, Longitudinal, and Torsional Resonant Frequencies of Concrete Specimens. *C215*. Annual Book of ASTM Standards, Volume: 04.02: American Society for Testing and Materials.

Azenha, M. (2009). *Numerical Simulation of the structural behaviour of concrete since its early ages*. Ph.D. thesis, Faculty of Engineering of the University of Porto.

Azenha, M., Faria, R., Magalhães, F., Ramos, L., & Cunha, Á. (2012a). Measurement of the E-modulus of cement pastes and mortars since casting, using a vibration based technique. *Materials and Structures, 45*, 81–92.

Azenha, M., Ferreira, C., Silva, J., Correia, A. G., Aguilar, R., & Ramos, L. F. (2011). Continuous stiffness monitoring of cemented sand through resonant frequency. In: D.-H. Chen, J.-R. Chang, & M. Zaman M (Eds.), *Emerging technologies for material, design, rehabilitation, and inspection of roadway pavements* (pp. 174–183). Hunan, China.

Azenha, M., Magalhães, F., Faria, R., & Cunha, Á. (2010). Measurement of concrete E-modulus evolution since casting: A novel method based on ambient vibration. *Cement and Concrete Research, 40,* 1096–1105.

Azenha, M., Ramos, L. F., Aguilar, R., & Granja, J. L. (2012b). Continuous monitoring of concrete E-modulus since casting based on modal identification: A case study for in situ application. *Cement and Concrete Composites, 34*, 881–890.

Bentz, D. P. (2008). A review of early-age properties of cement-based materials. *Cement and Concrete Research, 38,* 196–204.

Boulay, C., & Colson, A. (1979). Un extensomètre à béton éliminant l'influence des déformations transversales sur la mesure des déformations longitudinales. *Materials and Structures, 14,* 35–38.

Boulay, C., Crespini, M., Carette, J., & Staquet, S. (2012). Elastic properties of concrete at early age: Monitoring of the E-modulus and the poisson's ratio with cyclic loadings and ultrasonic measurements. In *Structural Faults and Repair—2012* (p. 11), France.

Boulay, C., Staquet, S., Azenha, M., Deraemaeker, A., Crespini, M., Carette, J., et al. (2013a). Monitoring elastic properties of concrete since very early age by means of cyclic loadings, ultrasonic measurements, natural resonant frequency of composite beam (EMM-ARM) and with smart aggregates. In J. G. M. V. Mier, G. Ruiz, C. Andrade, R. C. Yu, & X. X. Zhang (Eds.) *VIII International Conference on Fracture Mechanics of Concrete and Concrete Structures* (p. 14). FraMCoS-8, Toledo, Spain.

Boulay, C., Staquet, S., Delsaute, B., Carette, J., Crespini, M., Yazoghli-Marzouk, O., et al. (2013b). How to monitor the modulus of elasticity of concrete, automatically since the earliest age? *Materials and Structures, 47*, 141–155.

Boumiz, A., Vernet, C., & Tenoudji, F. C. (1996). Mechanical properties of cement pastes and mortars at early ages: Evolution with time and degree of hydration. *Advanced Cement Based Materials, 3,* 94–106.

Caetano, E. (1992). *Identificação experimental de parâmetros dinâmicos em sistemas estruturais*. MSc: Faculdade de Engenharia da Universidade do Porto.

Caetano, E. (2000). *Dynamic of cable-stayed bridges: Experimental assessment of cable-structure interaction.* Ph.D. thesis, University of Porto.

Carette, J., Dumoulin, C., Karaiskos, G., Staquet, S., & Deraemaeker, A. (2012). Monitoring of the E-modulus in early age concrete since setting time with embedded piezoelectric transducers. In *Conference: Structural Faults & Repair, 14th International Conference and Exhibition* (pp. 1–7), Edinburgh, Scotland, UK.

Carette, J., & Staquet, S. (2015). Monitoring the setting process of mortars by ultrasonic P and S-wave transmission velocity measurement. *Construction and Building Materials, 94,* 196–208.

Charpin, L., Sanahuja, J., Tran, N. C., Petit, L., Bremond, O., Montalvo, J., et al. (2016). Multiscale modeling of hydration, elasticity and creep of VeRCoRs concrete. Focus on creep characteristic times. In *Technological innovations in nuclear civil engineering (TINCE 2016).*

Chen, C.-W., Je-Nan, J., & Lee, G. (1993). Frequency domain state-space system identification. In *American Control Conference* (pp. 3057–3061), 2–4 June 1993.

Chengju, G. (1989). Maturity of concrete: method for predicting early-stage strength. *ACI Materials Journal, 86,* 341–353.

Delsaute, B., Boulay, C., Granja, J., Carette, J., Azenha, M., Dumoulin, C., et al. (2016). Testing concrete E-modulus at very early ages through several techniques: an inter-laboratory comparison. *Strain, 52,* 91–109.

Delsaute, B., Staquet, S., & Boulay, C. (2012). Monitoring of the creep and the relaxation behaviour of concrete since setting time. In *SSCS 2012 (Numerical modeling—Strategies for sustainable concrete structures).*

Dotson, C. (2015). *Fundamentals of dimensional metrology*. Boston, USA: Cengage Learning.

Dunham, M. R., Rush, A. S., & Hanson, J. H. (2007). Effects of induced vibrations on early age concrete. *Journal of Performance of Constructed Facilities, 21,* 179–184.

Ewins, D. J. (2000). *Modal testing: Theory, practice, and application.* Research Studies Press.

Fernandes, J. F., Bittencourt, T. N., & Helene, P. (2011). Concrete subjected to vibrations in early-ages. *Ibracon Structures and Materials, 4,* 592–609.

Fernandes, P., Granja, J. L., Benedetti, A., Sena-Cruz, J., & Azenha, M. (2015). Quality control and monitoring of NSM CFRP systems: E-modulus evolution of epoxy adhesive and its relation to the pull-out force. *Composites Part B Engineering, 75,* 95–103.

Fernandes, P., Granja, J., Sena-Cruz, J., Azenha, M., & Benedetti, A. (2014). A new methodology for assisting quality control of NSM-CFRP systems since very early ages. In R. El-Hacha, A. Rteil, A. Fam, D. Chen, F. Oudah, J.-F. Chen, K. Harris, K. Abdelrahman, & T. Tannert (Eds) *The 7th International Conference on FRP Composites in Civil Engineering.* Canada: Vancouver.

Garijo, L., Azenha, M., Ramesh, M., Lourenço, P. B., & Ruiz, G. (2019). Stiffness evolution of natural hydraulic lime mortars at early ages measured through EMM-ARM. *Construction and Building Materials, 216,* 405–415.

Giner, V. T., Ivorra, S., Baeza, F. J., Zornoza, E., & Ferrer, B. (2011). Silica fume admixture effect on the dynamic properties of concrete. *Construction and Building Materials, 25,* 3272–3277.

Granja, J. L. (2011). *Experimental evaluation of the elastic modulus of cementitious materials at early ages (in Portuguese).* M.Sc. thesis, University of Minho.

Granja, J. (2016). *Continuous characterization of stiffness of cement-based materials: experimental analysis and micro-mechanics modeling.* Ph.D. thesis, University of Minho.

Granja, J., & Azenha, M. (2015). Continuous monitoring of concrete mechanical properties since early age to support construction phasing. In *Mechanics and physics of creep, shrinkage, and durability of concrete and concrete structures*, Vienna, Austria.

Granja, J., & Azenha, M. (2017). Towards a robust and versatile method for monitoring E-modulus of concrete since casting: Enhancements and extensions of EMM-ARM. *Strain, 53,* 1–19.

Granja, J. L., Azenha, M., & Ferreira, C. (2013). Evaluation of the E-modulus of cement paste since fresh state based on resonant frequency: comparison with other methodologies and maturity studies. In *1st International RILEM Conference on Rheology and Processing of Construction Materials and the 7th RILEM Conference on Self-Compacting Concrete.*

Granja, J., Azenha, M., Sousa, C. D., & Ferreira, C. (2014). Comparison between different experimental techniques for stiffness monitoring of cement pastes. *Journal of Advanced Concrete Technology, 12,* 46–61.

Granja, J. L., Fernandes, P., Benedetti, A., Azenha, M., & Sena-Cruz, J. (2015). Monitoring the early stiffness development in epoxy adhesives for structural strengthening. *International Journal of Adhesion and Adhesives, 59,* 77–85.

Hassan, A. M. T., & Jones, S. W. (2012). Non-destructive testing of ultra high performance fibre reinforced concrete (UHPFRC): A feasibility study for using ultrasonic and resonant frequency testing techniques. *Construction and Building Materials, 35,* 361–367.

He, J., & Fu, Z. F. (2001). *Modal analysis.* Oxford, UK: Elsevier Science.

Hong, S., & Park, S.-K. (2015). Effect of vehicle-induced vibrations on early-age concrete during bridge widening. *Construction and Building Materials, 77,* 179–186.

JCGM. (2012). *International vocabulary of metrology—Basic and general concepts and associated terms* (VIM, 3rd ed., p. 200). JCGM.

Jin, X., & Li, Z. (2001). Dynamic Property determination for early-age concrete. *ACI Materials Journal, 98,* 365–370.

Jones, R. (1949). A non-destructive method of testing concrete during hardening. *Concrete and Constructional Engineering, 44,* 127–128.

Juang, J.-N. (1994). *Applied system identification.* Englewood Cliffs, New Jersey: Prentice-Hall Inc.

Juang, J. N., & Pappa, R. S. (1985). An eigensystem realization algorithm for modal parameter identification and model reduction. *Journal of Guidance, Control and Dynamics, 8,* 620–627.

Juang, J. N., & Suzuki, H. (1988). An eigensystem realization algorithm in frequency domain for modal parameter identification. *Journal of Vibration, Acoustics, Stress, and Reliability in Design, 110,* 24–29.
Kolluru, S. V., Popovics, J. S., & Shah, S. P. (2000). Determining elastic properties of concrete using vibrational resonance frequencies of standard test cylinders. *Cement Concrete and Aggregates, 22,* 81–89.
Lee, K.-M., Kim, D.-S., & Kim, J.-S. (1997). Determination of dynamic Young's modulus of concrete at early ages by impact resonance test. *KSCE Journal of Civil Engineering, 1,* 11–18.
Maia, L., Azenha, M., Faria, R. & Figueiras, J. (2011a). Influence of the cementitious paste composition on the E-modulus and heat of hydration evolutions. *Cement and Concrete Research, 41,* 799–807.
Maia, L., Azenha, M., Geiker, M., & Figueiras, J. (2011b). Describing paste E-modulus evolution of commercial cements and the relationship with volume solids formation.
Maia, L., Azenha, M., Geiker, M., & Figueiras, J. (2012a). E-modulus evolution and its relation to solids formation of pastes from commercial cements. *Cement and Concrete Research, 42,* 928–936.
Maia, L., Figueiras, H., Nunes, S., Azenha, M., & Figueiras, J. (2012b). Influence of shrinkage reducing admixtures on distinct SCC mix compositions. *Construction and Building Materials, 35,* 304–312.
McKelvey, T., Akcay, H., & Ljung, L. (1996). Subspace-based multivariable system identification from frequency response data. *Automatic Control, IEEE Transactions on, 41,* 960–979.
Oppenheim, A. V., Schafer, R. W., & Buck, J. R. (1989). *Discrete-time signal processing*. New Jersey, Englewood Cliffs, USA: Prentice-Hall.
Overschee, P. V., & Moor, B. D. (1996). *Subspace identification for linear systems: Theory, implementation, applications*. London: Kluwer Academic Publishers.
Parrott, L. J. (1990). A review os methods to determine the moisture conditions in concrete. British Cement Association Publication C/7.
Peeters, B., & de Roeck, G. (2001). Stochastic system identification for operational modal analysis: A review. *Journal of Dynamic Systems, Measurement, and Control, 123,* 659–667.
Pintelon, R., Guillaume, P., Rolain, Y., Schoukens, J., & van Hamme, H. (1994). Parametric identification of transfer functions in the frequency domain-a survey. *IEEE Transactions on Automatic Control, 39,* 2245–2260.
Powers, T. C. (1938). Measuring Young's modulus of elasticity by means of sonic vibrations. *ASTM Proceedings, 38,* 460–467.
Rainieri, C., & Fabbrocino, G. (2014). *Operational modal analysis of civil engineering structures: An introduction and guide for applications*. New York: Springer.
Reinhardt, H. W., & Grosse, C. U. (2004). Continuous monitoring of setting and hardening of mortar and concrete. *Construction and Building Materials, 18,* 145–154.
Ren, W.-X., & Zong, Z.-H. (2004). Output-only modal parameter identification of civil engineering structures. *Structural Engineering and Mechanics, 17,* 1–16.
Reynders, E. (2012). System identification methods for (operational) modal analysis: Review and comparison. *Archives of Computational Methods in Engineering, 19,* 51–124.
Rilem. (1975). Modulus of elasticity of concrete in compression. CPC 8. RILEM TC14-CPC.
Rodrigues, J. (2004). *Stochastic modal identification, analysis methods and applications in civil engineering structures (in Portuguese)*. Ph.D. thesis, Engineering Faculty of University of Porto.
Silva, J. (2010). *Contribuição para o estudo do tratamento de solos: Avaliação da Deformabilidade*. Dissertação de mestrado: Universidade do Minho.
Silva, J., Azenha, M., Correia, A. G., & Ferreira, C. (2013). Continuous stiffness assessment of cement-stabilised soils from early age. *Géotechnique, 63,* 1419–1432.
Silva, J., Azenha, M., Correia, A. G., & Granja, J. (2014). Continuous monitoring of sand-cement stiffness since layer compaction with a resonant frequency based method: issues on mould geometry and sampling. *Soils and Foundations, 54,* 56–66.

Wang, X., & Subramaniam, K. V. (2011). Ultrasonic monitoring of capillary porosity and elastic properties in hydrating cement paste. *Cement and Concrete Composites, 33,* 389–401.
Welch, P. D. (1967). The use of fast Fourier transforms for the estimation of power spectra: A method based on time averaging over short modified periodograms. *IEEE Transactions on Audio and Electroacoustics, 15,* 70–73.
Zhao, H., Huang, D., Wang, X., & Chen, X. (2014). Dynamic elastic modulus of cement paste at early age based on nondestructive test and multiscale prediction model. *Journal of Wuhan University of Technology-Material Science Edition, 29,* 321–328.

Chapter 4
Monitoring the Viscoelastic Behaviour of Cement Based Materials by Means of Repeated Minute-Scale-Duration Loadings

Brice Delsaute and Stéphanie Staquet

Abstract This chapter presents a new methodology developed to monitor the E-modulus and the ageing of the short term creep simultaneously. The methodology is based on a repeated minute-scale-duration loadings test for which periodically a loading is applied and kept constant during few minutes and finally removed. The repeated minute-scale-duration loadings test is used to quantify the E-modulus when the load is applied or removed and the aging creep when the load is kept constant or till the moment that the load is applied. This methodology is coupled to a permanent loading which allows characterizing creep for long duration. Different test setups and protocols of loading were developed in three laboratories to monitor the ageing of the E-modulus and the creep function on concrete or cement paste since setting or one day after casting. This chapter reports on the differences between the devices used, between the protocols of loading and between the data analysis. Different applications of the repeated minute-scale-duration loadings test are also presented.

Keywords Early age · Restrained shrinkage · Repeated minute-scale-duration loadings · Ageing creep · E-modulus

4.1 Introduction

Through the materials used in construction, concrete has a major importance and is the second most used material in the world after water. Currently, ordinary concretes are composed by sand, aggregate, water and Portland cement. Due to their extensive use, the cement production has a significant carbon footprint with its contribution to the CO_2 emission. It is estimated that between 5 and 8% of the worldwide CO_2 emission comes from the cement industry (Andrew 2017). In addition, the aggregates, which constitute the inert skeleton of concrete, are unequally shared and are hardly inexhaustible resources increasing the CO_2 emission of concrete materials. During the last decades, a lot of research has been performed in order to reduce the environment impact of the concrete industry by replacing partially or totally components

B. Delsaute (✉) · S. Staquet
Service BATir, LGC, Université Libre de Bruxelles (ULB), Brussels, Belgium
e-mail: bdelsaut@ulb.ac.be

M. Serdar et al. (eds.), *Advanced Techniques for Testing of Cement-Based Materials*, Springer Tracts in Civil Engineering, https://doi.org/10.1007/978-3-030-39738-8_4

of concrete. In consequence, there is always the need to get more detailed insight into the concrete behavior. Even nowadays, many areas need to be explored using innovative experimental techniques and new modeling approaches. Considering the new raw materials (e.g. recycled aggregate, mineral addition, admixture) changing the concrete characteristics, new testing methodology should be developed in order to monitor as fast and easy as possible the concrete properties.

Concrete has also the particularity to be a complex material for which its properties continuously change. Cementitious materials are obtained by hydration of binders, and the hydration lasts months to decades. It evolves from a nearly liquid state to a viscoplastic material within a few hours, followed by the setting of the concrete. Then the mechanical properties start to develop and the material exhibits viscoelastic behaviour. During the first days after mixing, the evolution of the concrete properties is very intense. This period is called the early age. Finally, the concrete properties continue to evolve on a period counted in years. Therefore, the microstructure of cementitious materials evolves with time and concrete properties depend on the maturity of the material.

Concrete exhibits both elastic and viscous characteristics, i.e. the material deforms instantaneously and progressively under, and induced by, a constant sustained load. It is thus generally considered that cementitious materials display a viscoelastic behavior. Such assumption is however limited to stress level corresponding to 30–50% of the strength under compression (Mazzotti and Savoia 2001) and 60% of the strength under tension (Domone 1974). Above these levels of stress, the behavior of concrete is non-linear because of the intensive micro-cracks formation.

For the design of concrete structure, it is needed to know in depth the early age viscoelastic properties of concrete for mainly three structural reasons: the early removal of the formwork, the loss of prestress and the deflection of precast concrete elements and the restraint of the autogenous, thermal and/or drying strains. In structural elements with thick sections, the internal temperature can reach a significant level. When the formwork is removed from the concrete, the internal temperature drops slowly while the surface, in direct contact with the environment, cools down rapidly. It results in thermal gradient across the section of concrete elements and therefore this induces stresses (Klemczak and Knoppik-Wrobel 2011). In the prestress/precast industries, the prestress is applied as soon as possible in order to decrease the time of each production cycle. The reduction is possible by using new self-compacting concrete which develops high strength at very early age. However creep and shrinkage are very intense during this period and cause significant prestress loss and deflection (Gardner and Fu 1987; Rosowsky et al. 2000; Kim et al. 2011; Maia and Figueiras 2012; Butler et al. 2016). For general concrete structures built in several phases, free strain of cement based materials are restrained externally (boundary conditions of the structure) and internally (interface between concrete and reinforcement, aggregate and mortar, sand and cement paste or even at the scale of the hydrate). These restrictions of the free displacement induce stresses inside concrete. Consequently, it is important to model accurately the early age behavior of concrete for their design, which influences the whole service life. Even though the mechanical behavior of hardened concrete can usually be accurately estimated, it is not always the case for the early age behavior of concrete, when the mechanical properties change rapidly in function of the advancement of the hydration reactions. Several devices and test methodology were already developed for the monitoring of several concrete properties since the casting or the setting such as the autogenous deformation (Boulay 2012), the coefficient of thermal expansion (Delsaute and Staquet 2017), the compressive strength by coupling electrical resistivity and crushed tests (Milenkovic et al. no date), the dynamic elastic properties (Delsaute et al. 2016; Saleh et al. 2017) and the heat release (Instruments 2017). Even if experimental tests consisting in applying a permanent loading at different ages were previously reported in (Laplante 1993a; Le Roy 1995; Gutsch 2002; Atrushi 2003; De Schutter 2004; Tamtsia et al. 2004; Briffaut et al. 2012; Forth 2014; Jiang et al. 2014), no specific testing methodology aimed at monitoring early age viscoelastic properties has already been developed for concrete.

Several models were developed in the past for the characterization of the evolution of the properties of concrete. These models are based on a large data base of experimental results. A large part of these results are related to concrete in a hardened state. During the hardening process, models are not yet able to predict correctly the evolution of properties such as the E-modulus, the Poisson's ratio and basic creep properties. Models were more developed on the basis of fitting than in an understanding of the physical mechanisms which govern the evolution of these properties. It is therefore questionable if such models are able to consider correctly the evolution of properties of new concrete coming into the market particularly at early age. A more physical approach is needed for the creation of new standards and recommendations. Finally, the experimental work should focus on the determination of model parameter related to the physical mechanisms. Experimental data should also be provided for the calibration or validation of new models.

In this chapter, after an initial overview of the physical mechanisms related to the basic creep, the development of a new test protocol for the monitoring of the viscoelastic properties of cement based materials is introduced. The method consists to apply repeated creep test of minute-scale-duration on a same sample. Different test setups, test protocols and data treatment developed in three laboratories are presented and compared. Finally, applications of the repeated minute-scale-duration loadings are presented.

4.2 Physical Mechanisms of the Basic Creep

Some studies related to the compressive creep have been carried out and are reported in (Laplante 1993a; Le Roy 1995; Gutsch 2002; Atrushi 2003; De Schutter 2004; Tamtsia et al. 2004; Briffaut et al. 2012; Forth 2014; Jiang et al. 2014) for concrete with an age at loading of one day and more. These studies have shown that creep is strongly influenced by the age of concrete at loading inducing changes in terms of amplitude and kinetic. Among many previous studies like those reported in

(Laplante 1993a; Le Roy 1995; Gutsch 2002; Atrushi 2003; De Schutter 2004; Tamtsia et al. 2004; Briffaut et al. 2012; Forth 2014; Jiang et al. 2014), several theories were developed to clarify mechanisms related to this behavior. However each theory alone does not allow explaining all experimental observations. Globally each theory can be linked to two mechanisms: direct mechanisms linked to the cement paste and responsible of the highest part of the creep amplitude and indirect mechanisms linked to the heterogeneity of the concrete (Wittmann 1982; Guénot-Delahaie 1997). Direct mechanisms are related to the water mobility and can be separated in short and long terms phenomena (Ulm et al. 1999; Acker and Ulm 2001; Bernard et al. 2003b). The short term phenomenon is reversible with a small characteristic time of about 10 days and linked to a stress-induced water movement towards the largest diameter pores and also occurs under increasing volume for uniaxial compression. The long term phenomenon is irreversible with a high characteristic time and related to viscous flow in the hydrates and occurs under almost constant volume. The creep rate of this long term phenomenon evolves as a power function t^n (Wittmann 1982; Bažant 1984; Bernard et al. 2003a) with an exponent n between -1 and -0.9 according to (Bažant 1995), between -0.72 and -0.69 according to results of (Wittmann 2015) on concrete and an exponent n between -0.86 and -0.6 on cement paste according to results of (Tamtsia et al. 2004). Nanoindentation tests were carried out on C–S–H by Vandamme and Ulm (2009). It was shown that C–S–H exhibits a logarithmic creep which is in agreement with results obtained on concrete. The indirect mechanisms are due to micro-cracks which occur progressively in the cement paste and at the interface between cement paste and inclusions. Their presence can cause a redistribution of stress in the material. Rossi et al. (1994, 2012, 2013) proposed an approach of the creep mechanisms by means of a micro-cracking process which occurs during loading. Over the time, an increase of the density of micro-cracks occurs.

4.3 Development of a Test Protocol

The viscoelastic behaviour of cement based materials in sealed condition is composed of two parameters: the elastic modulus and the basic creep coefficient. These properties depend on several material parameters. Moreover the creep properties are very sensitive to the conditions of preservation of the specimen during the test, to many factors related to the test rig and to the stress level applied. For those reasons several recommendations where developed in the past. For both properties, recommendations (ISO 1920-9:2009 2009; ASTM C469/C469M-14 2014; ASTM C512/C512M-15 2015) were developed for hardened concrete. The main common requirements for all the apparatus for testing elastic modulus and basic creep in sealed condition are their ability to:

- ensure uniform stress distribution over the cross section of the specimen;
- apply a load very quickly;
- operate in a room with controlled temperature and humidity;
- keep a constant known load during the whole testing period.

In addition, the measurement of the strain must be carried out in the central part of the specimen in order to eliminate the edge effects. The specimen must be perfectly sealed to avoid any external drying or water uptake. Finally, a dummy specimen should be tested at the same time to separate free and mechanical strain. The typical duration of a macroscopic creep experiment ranges from hours to several days (Bažant et al. 1976; Tamtsia et al. 2004; Rossi et al. 2012), weeks (Laplante 1993b; Tamtsia and Beaudoin 2000; Atrushi 2003; Briffaut et al. 2012), months (Rossi et al. 1994; Zhang et al. 2014), or even years (Bažant et al. 2011, 2012; Zhang 2014). Many researchers reported challenges in performing and standardizing creep experiments, e.g. creep compliances found in different databases exhibit significant dispersion, underlining that the problem is not yet solved in a satisfactory fashion. Additionally, in order to gain reliable insight into long-term creep, creep testing on concrete is recommended to be carried out over hundreds of days (Le Roy et al. 1996; Havlásek and Jirásek 2012), and this requires not only the availability of testing facilities over long periods of time, but it is also difficult to control experimental parameters as temperature, applied load, and moisture distribution within the sample.

Even if experimental tests consisting in applying a permanent loading at different ages were previously reported in (Laplante 1993a; Le Roy 1995; Gutsch 2002; Atrushi 2003; De Schutter 2004; Tamtsia et al. 2004; Briffaut et al. 2012; Forth 2014; Jiang et al. 2014), no specific recommendation or standard testing methodology aimed at monitoring early age creep has already been developed for concrete. For that purpose, new testing equipment and new testing protocol were recently designed to monitor the viscoelastic properties of cement based materials since setting time. The first part of this section introduces the development of a new testing methodology based on observations performed on permanent creep tests at early age, the second part describes the new testing equipment developed in three laboratories and the third section described and compared the test protocols developed in the three laboratories.

4.3.1 Preliminary Observations Performed on Creep Test of Long Duration

Several studies were performed in the past for the determination of the viscoelastic properties of concrete at early age [see e.g. (Laplante 1993a; Le Roy 1995; Gutsch 2002; Atrushi 2003; De Schutter 2004; Tamtsia et al. 2004; Briffaut et al. 2012; Forth 2014; Jiang et al. 2014)]. It was generally observed that the amplitude and the kinetic of the basic creep are strongly influenced by the age at loading. Consequently, models are often composed of two terms: an amplitude term and a kinetic term. For consideration of ageing both parameters can be expressed in function of the age at loading (Straub 1930; Shank 1935; Ross 1937; Lorman 1940; Bažant and Osman 1976; Thomas Thelford 1993; Atrushi 2003), the advancement degree of reaction

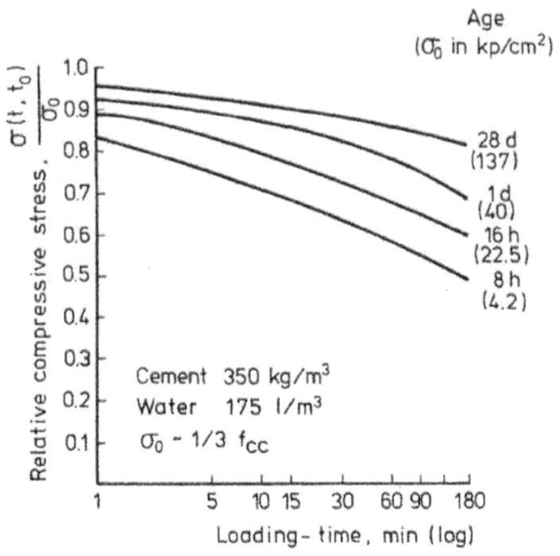

Fig. 4.1 Compressive relaxation tests of short duration time of 180 min performed since very early age till 28 days (Wierig 1971)

at loading (De Schutter 2004; Jiang et al. 2014) or the equivalent age at loading (Laplante 1993b; Guenot et al. 1996; Gutsch 2002).

In 1971, Weirig (1971) performed several compressive relaxation tests with a fairly short loading duration time of 180 min since very early age (8 h) till an age of 28 days (Fig. 4.1). For very early age loading, 50% of the stresses are relaxed in just 3 h. While at an age of 28 days, 17% of the stresses are relaxed for a same duration. In 1936, Jones and Richart (1936) [quoted from Neville et al. (1983)] performed compressive creep tests on five years old concrete at several stress/strength ratio. They have shown that the creep strains occurring after one minute under load are significantly influenced by the stress/strength ratio, especially for stress/strength ratio higher than 40% as shown in Fig. 4.2. It was therefore observed that short relaxation test of 180 min and short creep test of 1 min shows a considerable age and stress level dependency respectively.

Recently, Delsaute et al. (2014, 2016a) have reported results on permanent compressive creep test for ages at loading of 15–20–24–40 and 72 h on an ordinary concrete (Fig. 4.3a) for duration of loading of one week. For such duration of loading, it has been highlighted that the effect of the age at loading can be separated mainly in two parts:

- The amplitude of the basic creep coefficient is strongly influenced by ageing during the first hours after loading. Earlier is the application of the loading and higher is the basic creep coefficient. On the contrary, no ageing effect is detected for the kinetic of the basic creep coefficient (Fig. 4.3b).
- After the first hours of loading, no significant effect of the ageing is observed on the kinetic and the amplitude of the basic creep coefficient. However a general trend shows that, after few hours of loading, ageing increases the amplitude of the basic creep coefficient as it is shown in Fig. 4.3c. All basic creep coefficients are set to zero after two hours of loading to highlight this observation.

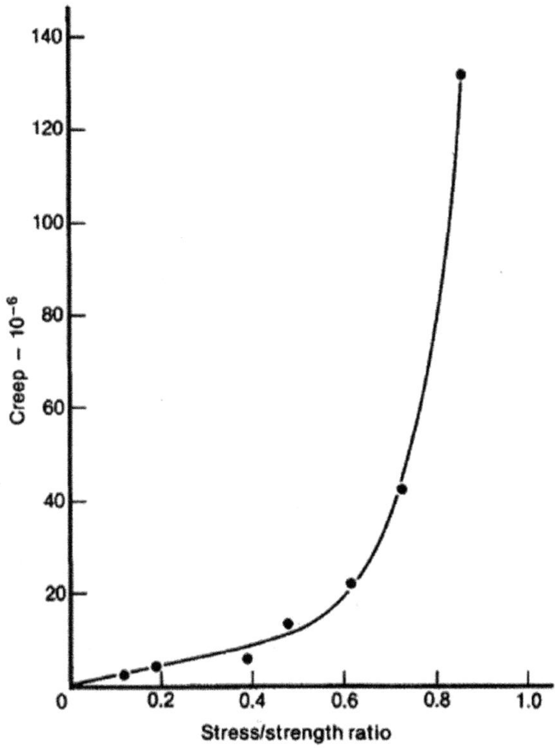

Fig. 4.2 Compressive creep strain after one minute under load according to the stress/strength ratio (Jones and Richart 1936)

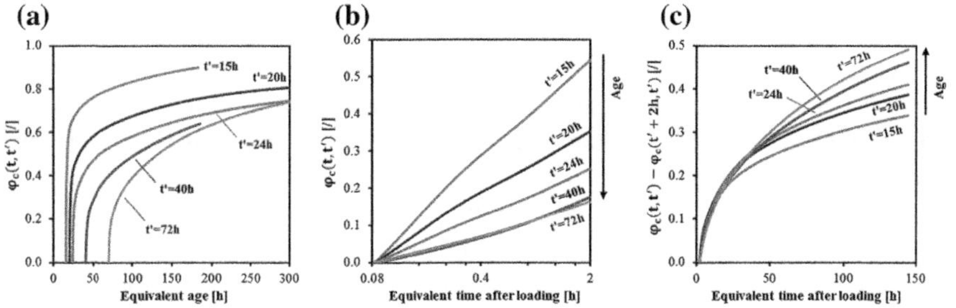

Fig. 4.3 **a** Creep coefficient for several ages at loading. **b** First two hours after loading of the creep coefficient. **c** Creep coefficient set to zero at an age after loading of 2 h (Delsaute et al. 2016a)

These observations were confirmed with experimental results of several authors and with several types of concrete mixtures [Ordinary concrete (Briffaut et al. 2012), high performance concrete (Atrushi 2003; Torrenti and Le Roy 2017) and ternary blended concrete (Jiang et al. 2014)].

From these experimental results, it was observed that the ageing of the basic creep can be mainly assessed by several creep tests of short duration. However such experimental methodology implies the completion of a large number of tests which is time consuming. To overcome this limitation, a new experimental strategy has been proposed by means of a repeated minute-scale-duration loadings test.

4.3.2 Testing Devices

During the last years, several devices were developed to monitor at the same time the elastic properties and the ageing of the short term creep since very early age. As for the monitoring of the autogenous deformation (see Chap. 5), requirements for the apparatus for very early age testing in sealed condition are their ability to:

- perfectly seal the specimen in order to avoid any external drying or water uptake;
- keep the temperature constant, which requires external control because the hydration of cement paste releases heat;
- limit the friction with the specimen in case of confined mold.

Moreover, for the study of the E-modulus and the basic creep at very early age, special attention was taken on the following criteria during their development:

- the need of high accuracy sensor (force and displacement) due to the very low amplitude of the displacement and force applied at very early age;
- the consideration of the free strain with a dummy specimen to separate creep strain from the total strain (by using the superposition principle of Boltzmann);
- the adaptation of the measurement system for very early age measurement.

In order to apply the new testing methodology consisting to apply repeated creep test of short duration, several devices were developed in three laboratories, namely the Université Libre de Bruxelles (ULB), the Institut Français des Sciences et Technologies des Transports, de l'Aménagement et des Réseaux (Ifsttar) and the Technische Universität Wien (TUWIEN). Horizontal and vertical test setups have been designed. These devices can mainly be divided in two categories according to the conditions of the sample: confined or unconfined sample. To perform mechanical test since casting or setting, it is needed to test the sample directly in the mould. The sample can also be unconfined but then the sample can be tested only few hours after setting for practical reasons (minimum concrete age required before mould release, grinding and protecting the sample from external humidity exchange and fixing the measurement system).

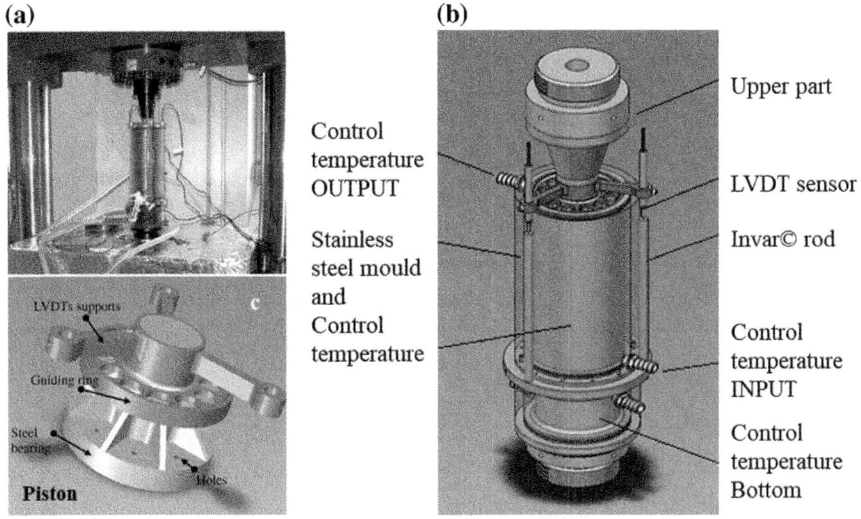

Fig. 4.4 BTJASPE, vertical testing device developed at Ifsttar (Delsaute et al. 2016a)

4.3.2.1 Confined Sample

BTJASPE

The BTJASPE (Boulay et al. 2014) (Fig. 4.4), french acronym of "BéTon au Jeune Age, Suivi de la Prise et du module d'Elasticité", is a device which has been designed to monitor the setting and, in compression, the modulus of elasticity, the creep, the relaxation and the coefficient of thermal expansion soon after the casting. Loadings in compression can be applied on a sample confined in a mould. The servo-controlled testing machine has a capacity of 1000 kN. The machine is fully programmable and controlled (force and displacement of each sensor) by computer. A cylindrical concrete sample (100 mm in diameter, 200 mm in height) is cast inside a stainless steel mould (254 mm depth) whose thickness wall is 1 mm (Fig. 4.3b). Before casting, the mould inner wall is greased and a thermocouple is placed in the center of the mould. The ambient temperature is recorded with a thermocouple. A circulation of water around the mould allows a perfect control of the sample temperature by means of two external circulator baths (one for the bottom bearing and one along the height of the sample). Displacements are measured with three invar rods placed at 120° which are fixed in the bottom part of the mould and the piston which support the 3 LVDT's. The piston has the particularity to have small holes which allow the bleed water to pass through so the upper bearing is always in contact with the solid phase of the concrete (Fig. 4.3c). Through these holes, the amount of drying is negligible so that the sealed conditions are obtained.

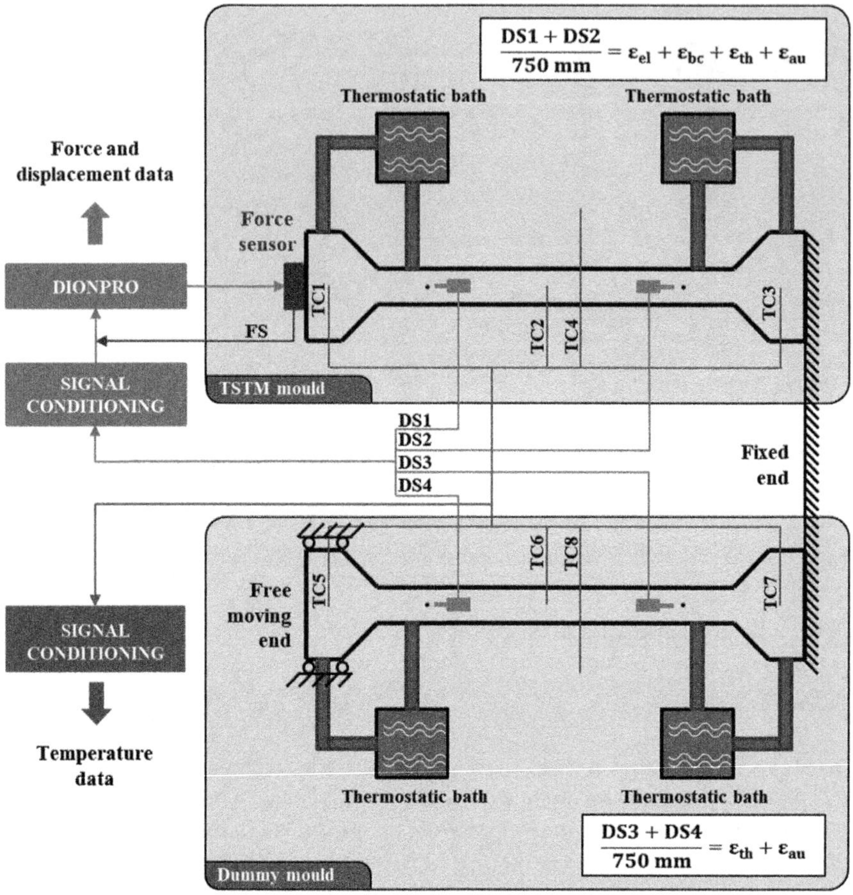

Fig. 4.5 TSTM, horizontal testing device developed at ULB (Delsaute et al. 2016a)

TSTM (Temperature Stress Testing Machine)

The TSTM, or Temperature Stress Testing Machine, is a device which has been developed for the monitoring of restrained shrinkage (Fig. 4.5). A revisited design has been developed at ULB for extending its use to other applications such as creep and relaxation (Staquet et al. 2012). Loadings in tension and compression can be applied on a sample confined in a mould. The TSTM is an electro-mechanical testing machine with a capacity of ±400 kN. The machine is fully programmable and controlled by computer (force and displacement of each sensor).

A prism-shaped mould device (100 × 100 × 1000 mm^3) is used and placed between the two crossheads of the machine which are composed of a fixed part and a mobile part. The crossheads make a dog bone-shaped. Several thermocouples are positioned, during the casting, at each end and at mid-length of the central section of the mould. The temperature of the sample is controlled by a flow of water all around the mould. A watertight plastic film in contact with PTFE (poly-tetra-fluoroethylene) sheets covering the mould and the crossheads, allows reducing the friction between the concrete and the mould. The mould is surrounded by a thermal regulation and a thermal insulation which assure the control of the temperature inside the sample. All equipment is in an air-conditioned room. Displacements are measured by Foucault current's sensors without contact placed on invar supports which are fixed on steel bars. The distance between both sensors is 750 mm (where deformations are uniforms in the sample). The tip of each sensor is placed very close to invar rods anchored in the concrete at a depth of 50 mm. In addition to the elastic and viscoelastic deformations, the thermal and free shrinkage deformations must also be known. For this purpose, a dummy mould was designed. This mould has exactly the same geometry as the first one. The only difference is the free movement of one of the ends. Complete details about the TSTM are presented in (Staquet et al. 2012; Delsaute 2016) and the Chap. 6.

4.3.2.2 Unconfined Sample

Vertical Testing Device Developed at TUWIEN

A vertical testing device has been designed at TUWIEN to monitor the E-modulus (Karte et al. 2015) and the short term of the basic creep on cement pastes (Irfan-ul-Hassan et al. 2016; Göbel et al. 2018a, b), mortar and concrete (Irfan-ul-Hassan et al. 2017; Ausweger et al. 2019). Loadings in compression can be applied on the sample since an age of 20 h. The device is an electromechanical universal testing machine with a capacity of 150 kN. The machine is fully programmable and controlled (force and displacement) by computer. A cylindrical concrete sample (70 mm in diameter, 300 mm in height) is cast inside a specific hollow cylindrical plastic mold. After casting, the specimens are stored in a climate chamber at 20 °C and surrounded by several layers of food preservation foil in order to avoid water exchange with the environment. Before testing, the specimens are demolded, and both circular end faces are shaved with a Stanley knife, in order to minimize possible inhomogeneities resulting from production and/or storage, and to achieve coplanarity of the two faces. During testing, for achieving a central load application, a serial arrangement of the specimen with two metal cylinders was used and is equipped by so-called cylindrical bottlenecks, see Fig. 4.6. The bottlenecks exhibit diameters of 30 mm. During testing, the test setup is placed inside an insulated temperature chamber, equipped with a temperature control unit. Deformations of the samples are quantified by means of five Linear Variable Differential Transducers (LVDTs) placed at 72° and giving access to the relative displacements between two aluminum rings, which are fixed to the specimens by means of three screws each. Thereby, the rings are positioned at a mutual distance of 164 mm (Fig. 4.6). Before the actual testing, the position of the specimen is iteratively improved in order to come close to a central load application.

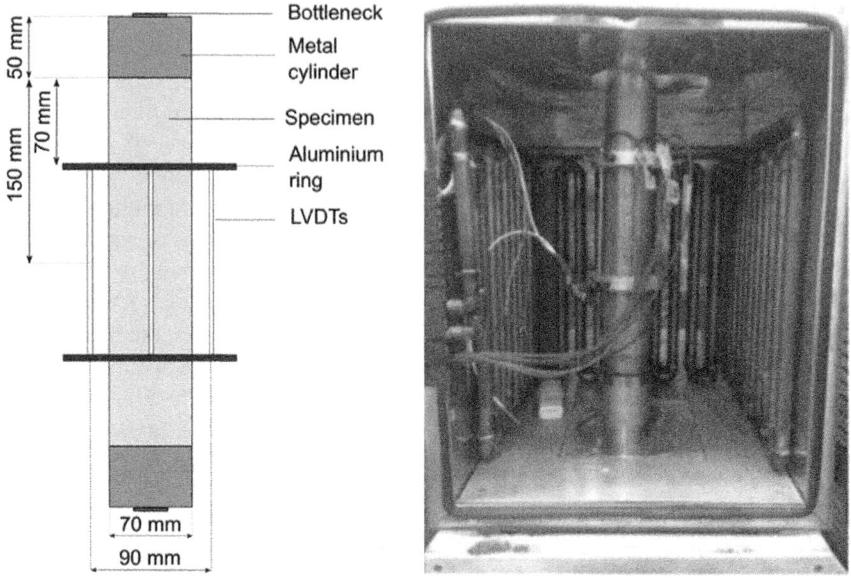

Fig. 4.6 Vertical testing device developed at TUWIEN (Irfan-ul-Hassan et al. 2016)

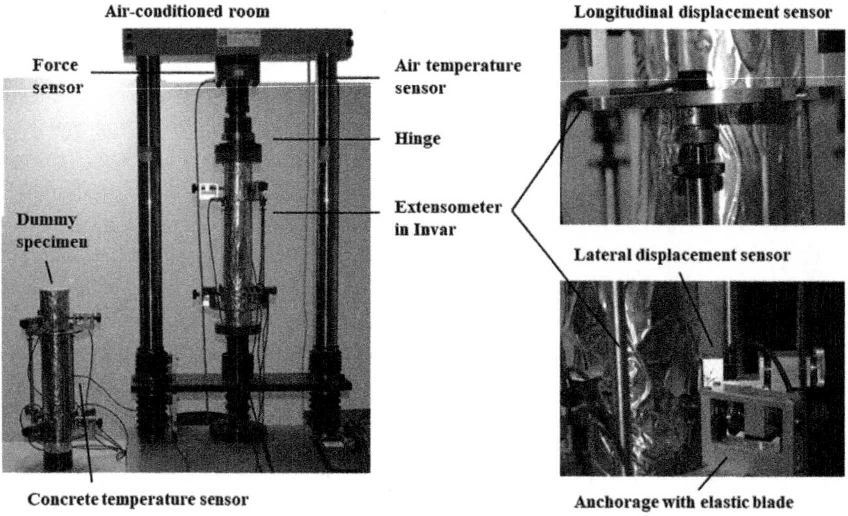

Fig. 4.7 Vertical testing device developed at ULB (Delsaute and Staquet 2019)

Vertical Testing Device Developed at ULB

A second vertical testing device has been designed at ULB (Fig. 4.7) to monitor the elastic and basic creep properties (Delsaute 2016). Compressive or tensile loadings can be applied few hours after the final setting. An electromechanical testing setup with a capacity of ±100 kN is used. The machine is fully programmable and controlled by computer (force and displacement of each sensor). A cylindrical concrete sample (97 mm in diameter, 550 mm in height) is cast inside a Polyvinyl chloride (PVC) mould. For each test, a dummy specimen with exactly the same dimension is also produced. Before casting, the mould inner wall is greased to facilitate the removal of the mold and a thermocouple is placed in the center of the mould of the dummy specimen. After casting, all specimens are placed in an air-conditioning chamber at 20 °C and a relative humidity of 90%. Few hours after the final setting samples are demolded, grinded on both circular end faces and surrounded by 2 self-adhesive aluminum sheets in order to keep the sample in sealed conditions. The sealing has been checked by weighting the sample before and after the test with a high precision balance. As samples are grinded few hours after setting, a small layer of sieved concrete (around 1 cm) is placed on the top of the sample in order to avoid any problem of pullout of the aggregate during the grinding. For test in compression, two trays are connected to the test rig (on the bottom and the top). The tray in the top part of the test rig is composed of a hinge. All equipment is in an air-conditioned room. Displacements are measured with an extensometer in INVAR© composed by two rings spaced of 350 mm and three rods which support the 3 longitudinal displacement sensors placed at 120°. Three lateral displacement sensors are placed on the lower ring. Both rings are fixed to the specimens by means of three elastic anchorages (Boulay and Colson 1981) allowing the measurement of the longitudinal and lateral displacement.

4.3.2.3 Comparison Between the Test Devices

Although these equipments bear some similarities (existence of a thermal regulation, sealing of the sample …), it appears that their design differs by several points:

- The type of testing device (Servo-controlled testing machine or Electro-mechanical testing machine), their load capacity and the position of the sample (horizontal or vertical).
- Starting time of the test: casting, setting or several hours after setting.
- The situation of the sample (type of thermal regulation, confined or unconfined and the type of sealing).
- Size of the sample: length of the straight part (between 250 and 1000 mm) and the cross section (between 3848 and 10,000 mm^2).
- Type of measurement system (anchorated rod, fixation of the displacement sensor on the top part of the sample, extensometer), the length of measurement (between 164 and 750 mm) and the number of displacement sensor (between 2 and 5).
- The type of thermal regulation (liquid regulation around the mould or air-conditioned system mold).
- The presence of a dummy specimen.
- The measurement of the lateral displacement.

A comparison of the testing devices is presented in Table 4.1.

Table 4.1 Comparison between testing devices for repeated minute-scale-duration loadings test developed in three laboratories

Device	BTJASPE	TSTM	Vertical testing device (TUWIEN)	Vertical testing device (ULB)
Type of device	Servo-controlled testing machine	Electro mechanical testing machine		
Test start	Casting	Setting	Age of 20 h	Few hours after setting
Situation of the sample	In the mould since casting		The mould is removed before the test starts	
Direction of the measurement	Vertical	Horizontal	Vertical	Vertical
Load capacity (kN)	1000	±400	150	±100
Thermal regulation	Circulation of water around the mould and all equipment is in an air-conditioned room		Before testing, the sample is in a climatic chamber. During testing, the sample is in an air-conditioned chamber (TUWIEN) or room (ULB)	
Record of the temperature	Yes	Yes	Only before testing	Yes (dummy specimen)
Shrinkage specimen	No	Yes	No	Yes
Sample size	Cylinder—ϕ = 100 mm and h = 200 mm	Prism—1000 × 100 × 100 mm	Cylinder—ϕ = 70 mm and h = 300 mm	Cylinder—ϕ = 97 mm and h = 550 mm
Gauge length (mm)	200	750	164	350
System of measurement	Fixation of the displacement sensors on the piston	2 anchorated rod inside the sample	Extensometer with longitudinal measurement	Extensometer with longitudinal and lateral measurement.
Sealing	Mold and piston	Plastic foil, Teflon and mould	Several layers of food preservation foil	2 layers of aluminium sheet
Displacement sensor	3	2	5	3 (longitudinal) + 3 (lateral)
Type of loading	Compression	Compression and tension	Compression	Compression or tension

4.3.3 Parameters of the Protocol

A new protocol of loading was developed for the monitoring of the short term basic creep. This methodology consists in applying repeated minute-scale-duration loadings on a sample. The scheme of the protocol of loading is presented in Fig. 4.8. The loading has to be applied as fast as possible and automatically to monitor the creep properties. For this purpose, several parameters must be defined:

- Time of application of the first repeated minute-scale-duration loading;
- Protocol of loading and unloading (control in force/displacement, amplitude of the load/displacement);
- Duration of loading;
- The period between two loadings;
- The state of the sample between two loadings (partially or completely unloaded).

Studies have been performed at ULB, Ifsttar and TUWIEN to define a test protocol. A comparison of the different test protocols is presented in Table 4.2. According to the test setup, the test can start since setting or after. It was observed (Delsaute et al. 2016a) that the test can start since the final setting without inducing any macroscopic damage within the sample. This was verified by comparing the evolution of the E-modulus obtained with the repeated minute-scale-duration loadings and classical extensometry tests on cylinders (Boulay et al. 2014; Delsaute et al. 2016a, c). At Ifsttar, the loading was controlled in displacement. Every 30 min, the load is applied in 7 min and corresponds to a deformation of 350 μ strain (displacement velocity of 50 μstrain/min). The load is kept constant during 5 min. At ULB, the loading was controlled in force and corresponds to 20% of the compressive strength. Every 30 min or 1 h, the load and unload are applied in 10 s. The load is kept constant during 5 min. At TUWIEN, the loading was controlled in force and corresponds to 15% of the compressive strength. A continuum micromechanics model was used to quantify the evolution of the compressive strength evolution (Pichler et al. 2013). Validation of this strategy was performed on cement pastes, mortars and concretes with several water-to-cement ratios (Königsberger et al. 2017). Each hour, the load and unload were carried out with a stress-rate of 2 and 1 MPa/s respectively. The

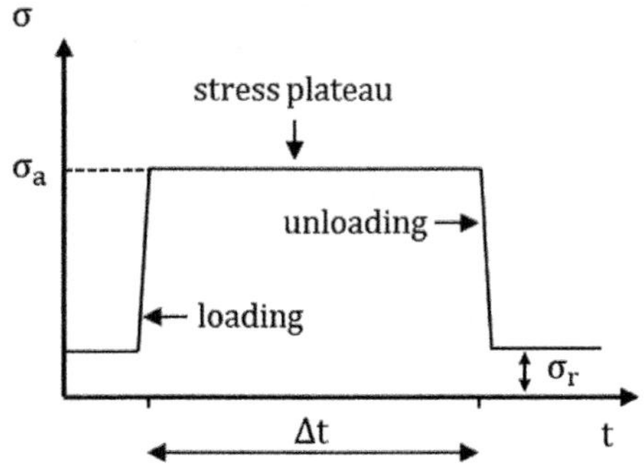

Fig. 4.8 Scheme of the test protocol of the repeated minute-scale-duration loadings

Table 4.2 Differences between test protocol of ULB, Ifsttar and TUWIEN

	ULB	IFSTTAR	TUWIEN
Start of the test	Setting (Delsaute et al. 2016a) or few hours after setting (Delsaute et al. 2016b)	Casting (E-modulus) or setting (creep)	20h00
Control during loading	Force	Displacement	Force
Stress plateau	20% of f_c (Delsaute et al. 2016a) and 40% of f_t (Delsaute et al. 2012)	350 μstrain (Boulay et al. 2012)	15% of f_c (Irfan-ul-Hassan et al. 2015, 2016)
Time to apply the load/stress rate	10 s	7 min	2 MPa/s
Duration of the stress Plateau (min)	5	5	3
Remaining force after unloading	0 kN (Delsaute et al. 2016a) or 0.3 kN (Delsaute et al. 2016b)	0 kN	0.2 kN
Periodicity of the loading	30 min or 1 h	30 min	1 h

load is kept constant during 3 min. According to the test setup a residual force is maintained after the unloading to ensure that the whole test setup stays in an upright position without tipping over. With the TSTM and the BTJASPE, no force remains after the unloading. While, with the vertical testing device developed at ULB and TUWIEN, a remaining force of 0.3 and 0.2 kN respectively is applied on the sample.

Therefore, one main strategy was developed in the three laboratories: a continuous monitoring of the viscoelastic properties by means of repeated loading with a constant period between each loading since very early age till one week or more. However this strategy has the drawback to hog the device for only one composition during the whole testing period. To resolve this issue, a second strategy was used at ULB (Delsaute 2016). A continuous monitoring is performed during the first days after casting, when the hardening process is very intense. Then, when the concrete properties do not evolve anymore strongly, the sample is tested only at specific ages (e.g. 2, 3, 7 and 28 days). This strategy allows carrying out several tests on various mixtures during the same period. This strategy is possible with vertical testing device developed at ULB and TUWIEN for which the sample can be removed from the device and later handed over. With the TSTM and the BTJASPE, the removal of the sample from the mould induces the final stop of the test.

4.4 Results and Data Treatment

Generally authors use the creep strain ε_c, the creep compliance $J(t, t')$, the specific creep $C(t, t')$ or the creep coefficient φ_c to define creep. The link between these three last parameters is given as:

$$C(t, t') = J(t, t') - \frac{1}{E(t')} = \frac{\varepsilon_{tot}(t, t') - \varepsilon_{sh}(t)}{\sigma} - \frac{\varepsilon_{el}(t')}{\sigma} = \frac{\varepsilon_c(t, t')}{\sigma} = \frac{\varphi_c}{E(t')} \tag{4.1}$$

where t is the age of concrete, t' is the age at loading, ε_{tot} is the total strain of the loaded specimen, ε_{sh} is the strain of the dummy specimen, ε_{el} is the elastic strain, σ is the stress on the loaded specimen and E is the elastic modulus. A typical minute-scale-duration loading test is presented in the Fig. 4.9.

The stress and the strain were very well controlled continuously during the loading, the stress plateau and the unloading for the determination of the elastic modulus and the basic creep.

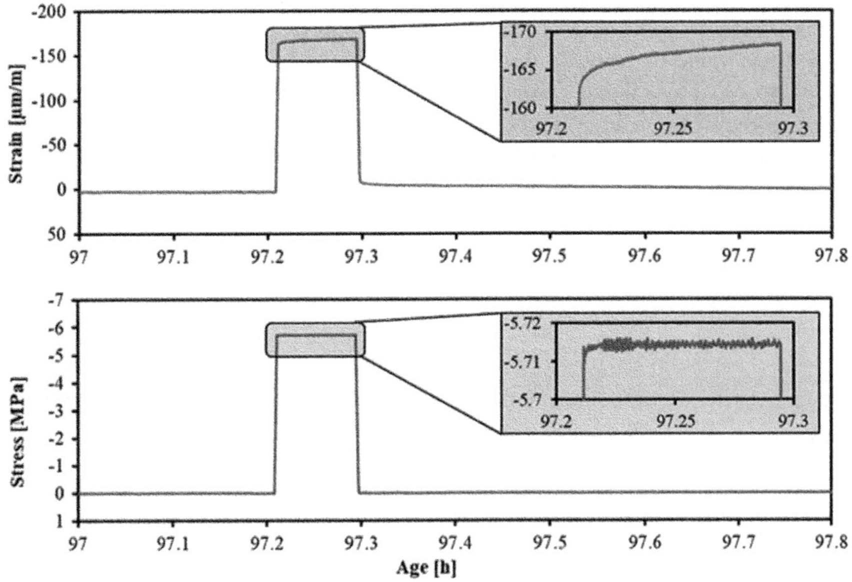

Fig. 4.9 Recording of stress and strain during a creep repeated minute-scale-duration loadings with the TSTM

4.4.1 Data Treatment

The consideration of the set of data is different between the three laboratories. At Ifsttar and ULB, the elastic modulus is defined with the record of the displacement and the force during the loading and the unloading. At ULB, only data between 20 and 80% of the load applied are considered for the computation of the E-modulus while at Ifsttar the whole loading is considered. During each stress plateau, the creep strain is recorded as soon as the stress/strain threshold is reached. Then the creep is defined with the specific creep function $C(t, t')$ (Eq. 4.1). At TUWIEN, both elastic modulus and specific creep function are defined simultaneously with consideration of the whole loading and the stress plateau.

For the mathematical treatment of the data, differences also occur in the three laboratories. At ULB and Ifsttar, the mathematical treatment is based on a first pre-treatment of results on an ordinary concrete. All specific creep functions were superimposed in the same graph to analyse the eventual similarities of the curves and the effect of aging (Fig. 4.10a). As for the creep test presented in Sect. 4.3.1, it was observed that the amplitude of the specific creep depends strongly on the age of the concrete at loading. Two parameters were studied: the evolution of the kinetic and the amplitude according to the age at loading. In Fig. 4.10b, each creep curve was normalized by its value obtained after 5 min of loading. Thus each normalized creep curve has a value of 1 after a duration of loading of 5 min. All normalized creep curves are very close and no ageing effect is detected. In Fig. 4.10c, the amplitude of the creep coefficient after 5 min of loading (coming from the repeated minute-scale-duration loadings test) is compared to the amplitude of the creep coefficient after 2 h of loading (coming from the permanent loading test). Both amplitudes follow the same evolution according to the age at loading. Thus it was concluded that the first term of the short term creep can be defined with repeated minute-scale-duration

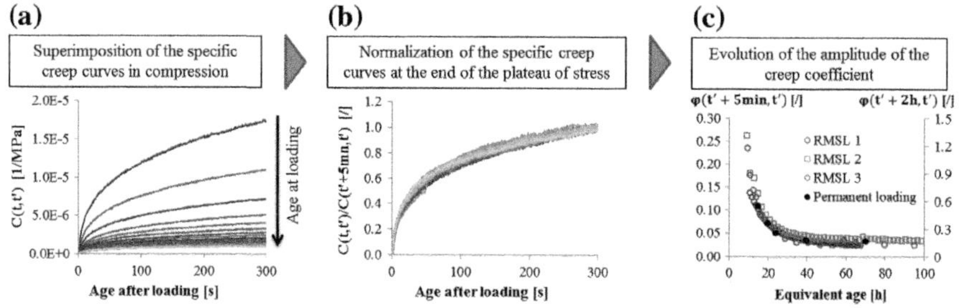

Fig. 4.10 Repeated minute-scale-duration loadings. **a** For each minute-scale-duration loading, creep curves are superimposed and set to zero when the load is kept constant. **b** For each minute-scale-duration loading, creep curves are normalized by their value obtained after 5 min of loading. **c** Evolution of the amplitude of the creep coefficient after 5 min of loading (coming from the repeated minute-scale-duration loadings tests) and 2 h of loading (coming from the permanent loading tests) according to the equivalent time (Delsaute et al. 2016a)

loadings test and is divided in two terms: a dimensionless kinetic term which is constant and an amplitude term which is function of the age of concrete at loading.

Through these observations, different models were developed for the minute-scale-duration loadings test. Boulay et al. (2012) used a Dirichlet model composed of one spring and two Kelvin-Voigt units in series to model the short term creep. As it was observed that short term creep is composed of two terms, one related to the amplitude and another one related to the kinetic, the model was expressed as follows:

$$C(t', t - t') = A_c(t') * \Omega(t - t') \tag{4.2}$$

$$\Omega(t - t') = C_1\left(1 - \exp\left(-\frac{t - t'}{\tau_1}\right)\right) + C_2\left(1 - \exp\left(-\frac{t - t'}{\tau_2}\right)\right) \tag{4.3}$$

where $A_c(t')$ is the amplitude creep term which corresponds to the specific creep function after a duration of loading of 5 min at an age of loading t′. $\Omega(t - t')$ is the kinetic creep term for which the parameter C_1, C_2, τ_1 and τ_2 are fitted to the experimental results (least square method).

Delsaute et al. have developed two models to best fit the creep coefficient. A first model was introduced in (Delsaute et al. 2012, 2013). The model is given in Eq. 4.4. The model is a power law and has the advantage to consider only two parameters: a parameter A_1 linked to the amplitude of the creep coefficient after 5 min of loading and a parameter K_1 linked to the kinetic of the creep coefficient.

$$\varphi_c(t, t') = A_1(t').\left(\frac{t - t'}{t_{A1}}\right)^{K_1} \tag{4.4}$$

where $t_{A1} = 0.083\text{h} = 300$ s. The second model (Delsaute et al. 2017a) developed is a logarithmic law which is also based on an amplitude parameter A_2 linked to the amplitude of the creep coefficient after 5 min of loading and a parameter K_2 linked to the kinetic evolution of the creep coefficient.

$$\varphi_c(t, t') = A_2(v).\ln\left(1 + (e - 1).\left(\frac{(t - t')}{t_{A2}}\right)^{K_2}\right) \tag{4.5}$$

where $t_{A2} = 0.083h = 300$ s. The calculation of both parameters was carried out in two steps. Firstly, each creep curve was fitted with three Kelvin-Voigt chains in order to remove noise from the measurement. The value of the fitted curve after 5 min of loading corresponds to the amplitude parameter for each age of loading. Secondly, the value of the kinetic coefficient is computed with least and square method. A comparison of both models has been performed (Delsaute et al. 2017b) and shows that the best fit is obtained with the logarithmic expression. The evolution of the amplitude parameter A_2 and the kinetic parameter K_2 of an ordinary concrete are presented in Fig. 4.11. A comparison of the creep coefficient of the experimental data and the modeled curves is presented in Fig. 4.12.

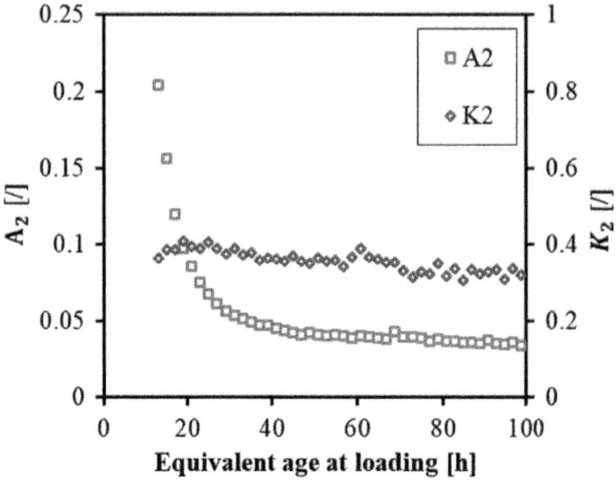

Fig. 4.11 Evolution of the amplitude and kinetic parameter of Eq. 4.9 according to the equivalent age at loading

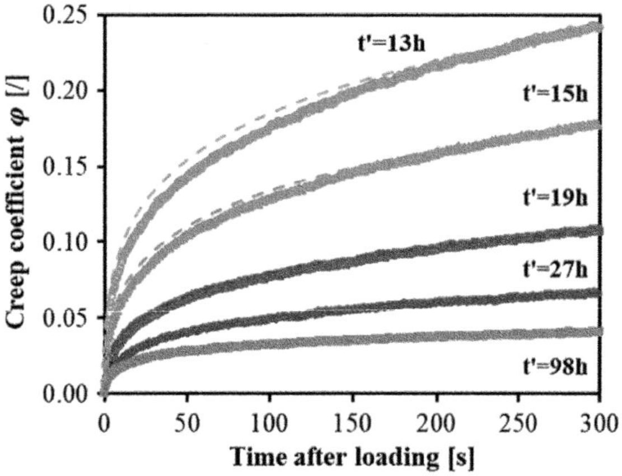

Fig. 4.12 Experimental data of the creep coefficient and modeled curves for several ages at loading from Eq. 4.5

As creep takes place since the moment that a load is applied on the sample, the determination of the elastic modulus should theoretically consider the creep strain occurring during the loading and the unloading. For that reason, at TUWIEN (Irfan-ul-Hassan et al. 2016), the quantification of the elastic and creep properties is based on a "tensile undershooting" criterion. First the determination of the E-modulus is carried out on the unloading phase. Creep strains are defined with the removal of the elastic strain (based on the value of the unloading modulus). After this operation, unexpected tensile creep strain were observed during the loading phase as shown in Fig. 4.13a. This results from an overestimation of the elastic strain with the use of the unloading modulus. Therefore, the E-modulus was increased iteratively till the moment that no tensile creep is observed (Fig. 4.13b). This procedure increases around 3 and 4% the elastic modulus defined on the unloading phase. Finally, basic creep strains are defined since the load is applied (Fig. 4.13c).

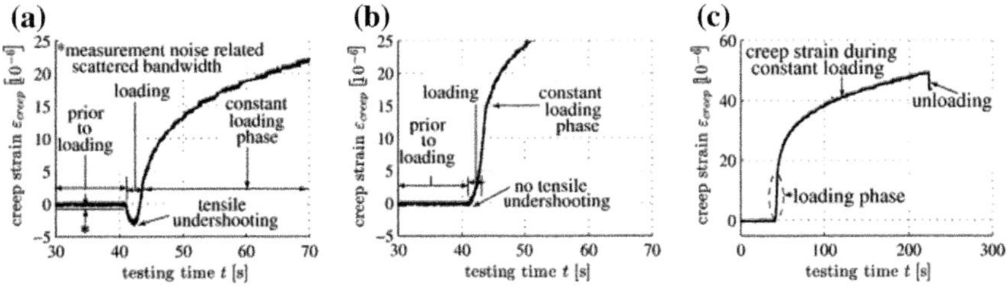

Fig. 4.13 Procedure developed at TUWIEN for the decoupling of the elastic and creep strains (Irfan-ul-Hassan et al. 2016)

Once the E-modulus is defined, Irfan-ul-Hassan et al. (2016) have used a power law expression to fit the basic creep strain during the minute-scale-duration loading. In order to consider the loading phase, the Boltzmann's superposition principle is used:

$$\varepsilon_{tot}(t,t') = \sum_{i=1}^{n} \frac{F(t_i) - F(t_{i-1})}{A} \cdot \frac{1}{E_c}\left(\frac{t - t_i}{t_{ref}}\right)^{\beta} \quad t_n \leq t \leq t_{n-1} \tag{4.6}$$

where F is the load applied, A is the cross section of the sample, $t_{ref} = 1d$, E_c is the creep modulus, β is a dimensionless power-law exponent. The basic creep parameter E_c and β are defined iteratively by optimization. The values of the creep modulus and the power-law exponent of each repeated minute-scale-duration loading of three cement pastes with 3 W/C ratios (0.42, 0.45 and 0.5) are presented in Figs. 4.14 and 4.15 respectively. The creep modulus increases with the age of cement paste at loading. Higher is the W/C, lower is the creep modulus. The power-law exponent decreases slightly with increasing the age at loading. These results are coherent with results obtained at ULB and Ifsttar. On concrete and cement paste, the amplitude of

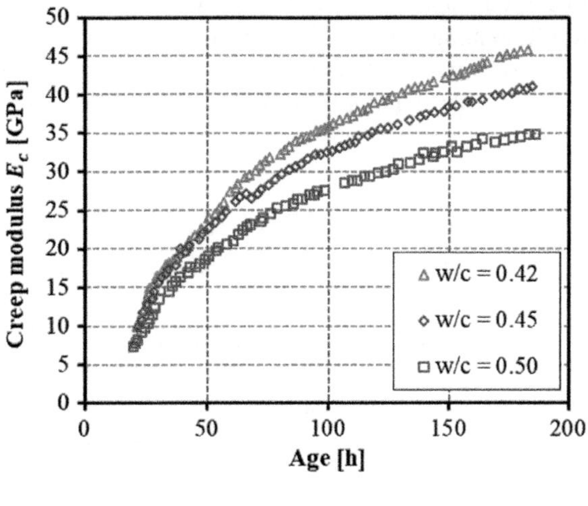

Fig. 4.14 Evolution of the creep modulus E_c for cement pastes with W/C of 0.42, 0.45 and 0.5 according to results of (Irfan-ul-Hassan et al. 2016)

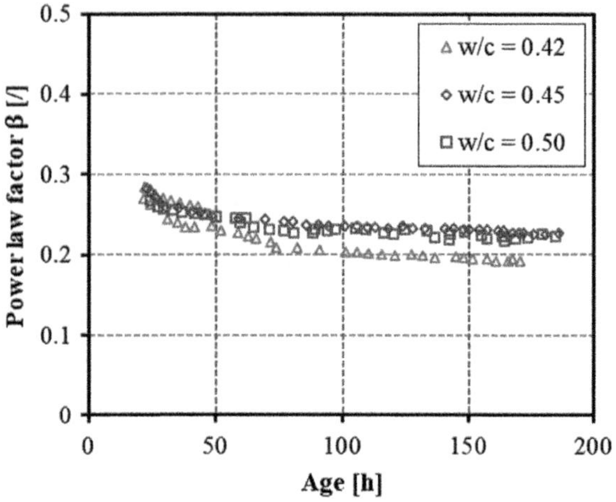

Fig. 4.15 Evolution of the power-law exponent β for cement pastes with W/C of 0.42, 0.45 and 0.5 according to results of (Irfan-ul-Hassan et al. 2016)

the basic creep strain is strongly influenced by the age at loading while the kinetic parameter is slightly or not influenced by the age of the material. The difference of evolution of the kinetic parameter on concrete and cement pastes can be due to the diluting and restraining effects of aggregates.

With the vertical device developed at ULB, for each minute-scale-duration loading, the Poisson's ratio $\boldsymbol{v}_{el}$ is calculated from the set of recordings (longitudinal and radial displacement in the central section) on the whole loading and unloading:

$$v_{el} = \frac{\varepsilon_{el}^{r}}{\varepsilon_{el}^{L}} \tag{4.7}$$

where ε_{el}^{r} and ε_{el}^{L} are respectively the radial and longitudinal strain during the loading or unloading phase. Experimental results (Delsaute et al. 2016b) obtained on a concrete of nuclear power plant (EDF, no date; Staquet et al. 2017) are given according to the equivalent time in Fig. 4.16. A very good correspondence is observed between results coming from the loading and the unloading phase. Moreover, the accuracy of the measurement allows observing a clear decrease of the Poisson's ratio just after setting till a minimum value and then an increase till a threshold value (here 0.17).

4.4.2 *Impact of the Devices on the Determination of the Elastic and Creep Properties*

4.4.2.1 Autogenous and Thermal Strain

For high strength concrete, the development of the autogenous strain is very intense during the first 24 h after setting. Even in 5 min, an influence of the autogenous strain

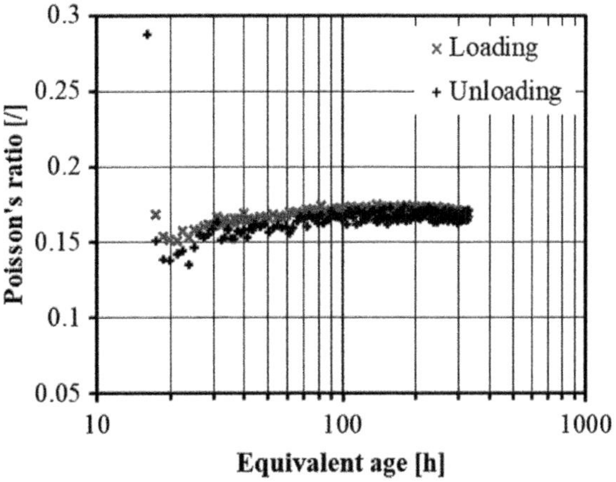

Fig. 4.16 Evolution of the Poisson's ratio of the Vercors concrete

could be observed on the measurement of the creep strain. To illustrate this fact, free strain and the amplitude creep term A_2 (Eq. 4.5) of an ultra-high performance self-compacting concrete studied at ULB with the vertical testing setup are shown in Figs. 4.17 and 4.18 respectively. The test starts at an age of 14 h which corresponds to two hours after the final setting (12h00). In Fig. 4.17, the free strain was measured on the dummy specimen (strain are set to zero at the start of the test). For comparison, the cumulated strain at the end of each repeated minute-scale-duration loading is also given. A very good correspondence is obtained between both evolutions. A same observation was already performed in (Delsaute et al. 2016a) on an ordinary concrete. In Fig. 4.18, the influence of the free strain on the determination of the amplitude creep term is evaluated. The amplitude creep term is first defined with the removal of the free strain coming from the dummy specimen (case a) and then evaluated without the removal of the free strain (case b). During the first hours after the start of the test, the difference between both evolutions of the amplitude creep term

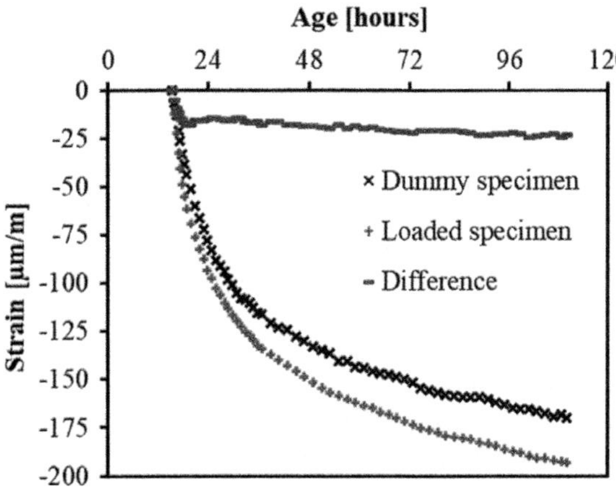

Fig. 4.17 Free strain of the dummy mould and the cumulated one at the end of each repeated minute-scale-duration loading. Data are set to zero at the start of the test

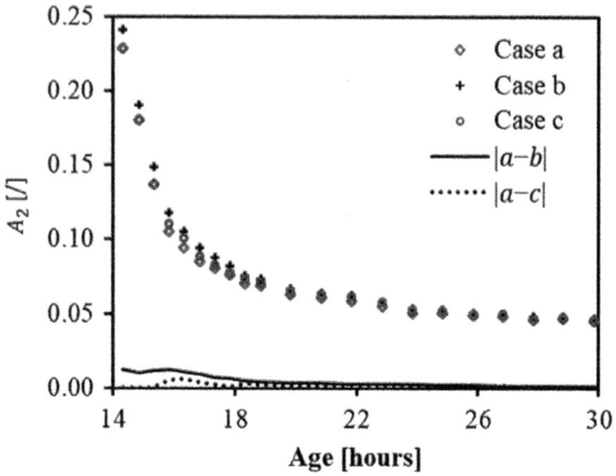

Fig. 4.18 Evolution of the amplitude creep term A_2 (Eq. 4.5) with consideration of the free strain of the dummy specimen (case a), without consideration of the free strain (case b) and with consideration of the strain fitted just before the loading (case c)

reaches a value of 11.4%. Consequently, the influence of the free strain is significant for the determination of the creep strain at very early age even for very short duration of loading. According to the observation carried out on Fig. 4.17, the evolution of the free strain can be defined quite accurately by considering the evolution of the strain taking place before each minute-scale-duration loading. Free strain taking place during the loading are extrapolated with the evolution of the strain taking place during the 10 min before the associated loading. A polynomial expression of the second degree is used to fit the strain. Then the amplitude creep term is defined with the removal of the fitted free strain coming from the loaded specimen (case c). The maximal difference between case a and case c is 6.8%. More generally, it is observed that this procedure reduces the total error by 35% during the whole testing period (since an age of 14 h till 108 h) and by 50% during the 12 firsts hours after the start of the test.

4.4.2.2 Confined Versus Unconfined Sample

Delsaute et al. (2016a, c) have compared results coming from the BTJASPE and the TSTM to results coming from unmolded samples. For the BTJASPE, the longitudinal displacement measured includes the displacement of the bearings. Moreover, the radial displacement of the sample is restrained by the stainless steel mould. Therefore the measurement includes globally the concrete deformation, the deformation of the bearings and the effects of the mould reinforcement. For the determination of the elastic modulus, both artefacts of measurement (deformation of the bearings and the mould reinforcement) were defined by performing a finite element calculation for several E-modulus values and height of the sample (Boulay et al. 2014). An adjustment of the data can then be operated to define the elastic modulus of the sample. For creep, no data treatment was operated because the load remains constant during the stress plateau. Therefore the confinement of the mould is not so involved in the process. For the TSTM, a possible confinement of the mould could appear in the lateral direction. However, the fixation systems of the mould are not 100% rigid and therefore only a very small confinement effect is expected with the TSTM. To confirm these hypotheses, E-modulus results from the BTJASPE and the TSTM were compared to unconfined specimens (Staquet et al. 2014; Delsaute et al. 2016a, c). A very good agreement of the E-modulus results was observed. It was therefore concluded that no or very limited confinement occurs in the TSTM and that the effect of the confinement can be considered when using the BTJASPE.

4.4.2.3 Bleeding

In case of concrete with high water content or with a lack of fines particles in the mix, bleeding can occur. The water-to-cement ratio is higher on the top of the sample and consequently a lower stiffness is expected on the top of the sample. The effect of bleeding on the measurement system depends on the situation of the sample. With the BTJASPE, the sample is tested on the vertical direction. The piston has the particularity to have holes which allows the bleed water to pass through so the upper bearing is always in contact with the solid phase of the concrete. Through these holes, the amount of drying is negligible so that the sealed conditions are obtained. Thus it is assumed, if there is bleed water, that this water is removed from the sample through the hole of the upper bearing when the load is applied. Then the water is still available and can be reabsorbed by the cement paste during the whole test. It is also assumed that if a small weak layer of concrete exists during the test on the top of the sample, this layer is very small and has no real impact on the measurement of the elastic and creep strain. With the TSTM, the sample is tested on the horizontal direction. If there is bleeding at a moment, bleed water would be firstly on the top of the sample and will have no effect on the horizontal stiffness of the concrete and no effect will appear on the measurement. During the test, bleed water could be reabsorbed by the concrete and create a swelling of the concrete on the upper part of the sample and will lead to a bending of the sample. However, it is assumed that the potential swelling will represent a very small layer, will not create a significant bending effect and therefore will not influence the measurement of the elastic and creep strain. With the vertical testing device developed at ULB and TUWIEN, the sample is tested few hours after setting and the upper part of the sample is removed in order to assure the flatness of the sample. In consequence, no effect of bleeding is expected with these devices. In order to assure that no damage, irreversibility or effect of bleeding occurs during the test, classical extensometry can also be performed in parallel to the repeated minute-scale-duration loadings test.

4.4.3 Microstructural Interpretation of the Results

Experimental results coming from the repeated minute-scale-duration loadings test and compressive creep tests of long duration were compared to the microstructure of the cement paste in (Delsaute et al. 2016a). Results from the literature reports that two main microstructural parameters are associated to the aging of the basic creep: the capillary pores and the CSH. According to Powers (1965), the short term creep is related to the migration of water between the layer of absorbed water to the capillary pores where the pressure is lower which leads to a global and progressive contraction of the cement paste. In addition, Acker and Ulm (2001) have suggested that there is a stress-induced water movement towards the largest diameter pores. Based on these observations, the volume fraction porosity of higher diameter was compared to the evolution of the amplitude creep term A_2 (Eq. 4.5). It has been observed that the volume fraction of porosity of diameter higher than 5 μm follows a same trend than the amplitude creep term (Fig. 4.19). During hydration, the volume fraction of the CSH increases and materials properties such as the E-modulus increases also. When a load is applied, the CSH will form in the capillarity pores, will bear a part of the load and will reduce the load carried by the other hydrates. When concrete is loaded at early age, a high reduction of the load occurs on the hydrates and in consequence the amplitude of the basic creep is reduced (Bažant and Prasannan 1989a; Bažant et al. 1997a, b). Based on this observation, the increase of the creep coefficient between an age corresponding to 2 and 145 h after loading was compared to the increase of the volume fraction of CSH during the same period for several ages at loading (Fig. 4.20). For loading applied at very early age, a high increase of the CSH occurs while a low increase of the creep coefficient is observed. Inversely, for higher age at loading, a low increase of the volume fraction of CSH is noticed while a high increase of the creep coefficient takes place.

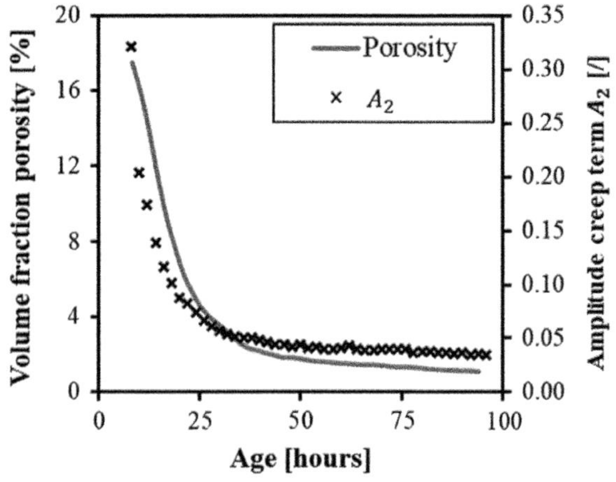

Fig. 4.19 Evolution of the porosity (pores diameter size from 5 to 9 μm) and the amplitude factor

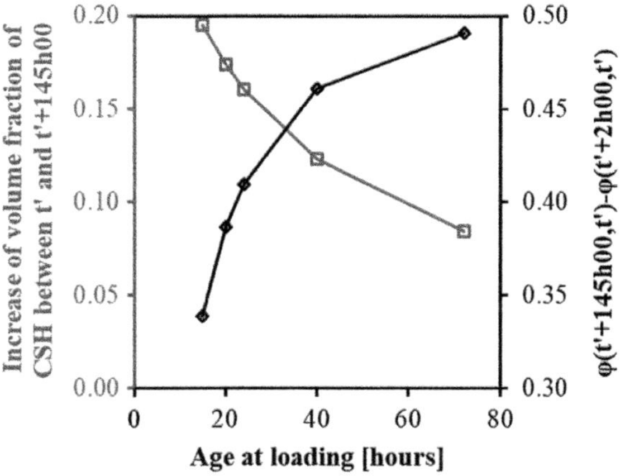

Fig. 4.20 Increase of volume fraction of CSH and creep coefficient for different age at loading

Irfan-ul-Hassan et al. (2015) have compared the inverse of the creep modulus (Eq. 4.6) to the gel space ratio for three cement pastes with different water-to-cement ratio of 0.42, 0.45 and 0.5 (Fig. 4.21). The gel space ratio corresponds to the ratio between the volume of hydration products and the volume of both hydration products and capillary pores. Thus the use of the gel-space ratio as an aging scale allows considering simultaneously the bearing microstructure formed by the hydrate and the voids formed by self-desiccation phenomenon. The relation linking the creep modulus and the gel-space ratio was found independent of the water-to-cement ratio (Fig. 4.21).

A similar observation was previously performed by Pichler, et al. (Pichler et al. 2013) between the compressive strength and the gel-space ratio. It was observed that, for different cement pastes with a same gel-space ratio and a same loading applied, the stress concentrations at the hydrate scale are equivalent. Therefore, results of the creep modulus highlights that the stress concentrations at the hydrate scale trigger the basic creep of young cement pastes.

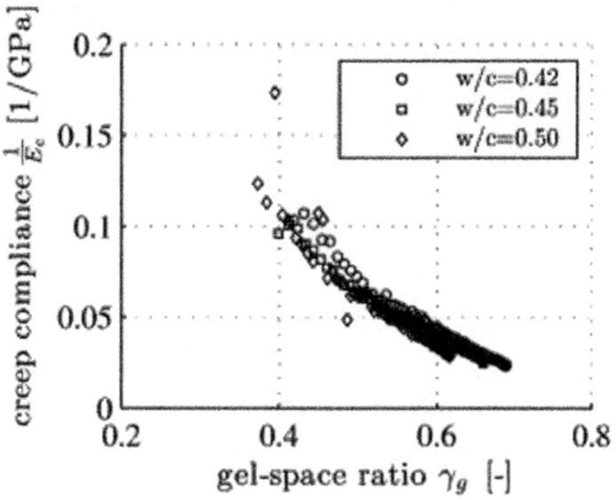

Fig. 4.21 Inverse of the creep modulus according to the gel-space ratio of cement pastes with 3 different w/c ratio (0.42, 0.45 and 0.5) (Irfan-ul-Hassan et al. 2015)

4.5 Application of the Repeated Minute-Scale-Duration Loadings

Repeated minute-scale-duration loadings tests were used for different applications. At ULB and Ifsttar, the main application of this test was to model accurately the viscoelastic behavior since the setting time. At TUWIEN, the new testing methodology was used for the identification of creep properties at the microstructural scale and also to study the water transfer between aggregate and cement paste during hydration.

4.5.1 Modelling Basic Creep Since Setting Time

Delsaute et al. (2017b) reports on a new strategy for the modeling of the basic creep since setting time. This strategy uses as a basis the results coming from two types of test: classical creep test with permanent load applied during one week or more and repeated minute-scale-duration loadings. Three models were proposed. The first model used 3 (Delsaute and Staquet 2014) or 4 (Delsaute et al. 2017b) Kelvin-Voigt units in series for which only units with a characteristic time lower than 2 h have an ageing factor which corresponds to the value of the creep coefficient after 5 min of loading. This model has the advantage to be easily implemented in finite element software which is convenient for the study of the restrained shrinkage. However, this model cannot accurately predict the basic creep for long duration and cannot distinguish the reversible and irreversible part of creep.

The second model is based on the physical mechanisms occurring on concrete under loading during the hydration process. The creep is divided in three terms (Eq. 4.8). The first term C_{IST} is related to the state of the material (capillary pores, CSH) when the load is applied and it corresponds to a logarithmic law composed of a kinetic term k and the amplitude term A_2 (Eq. 4.5). The kinetic is constant and the amplitude depends on the amplitude creep term A_2 (Eq. 4.9). The second term C_{ST} is related to the solidification of the material under loading (Eq. 4.10). The kinetic term K_{ST} is constant and is expressed with two Kelvin-Voigt units in series (Eq. 4.12). The amplitude term $A_{ST}(t')$ depends linearly on the time derivative of the elastic modulus at the age of loading (Eq. 4.13). The third term C_{LT} corresponds to the long term creep and is expressed with a power-law expression (Eq. 4.11).

$$C(t,t') = C_{IST}(t,t') + C_{ST}(t,t') + C_{LT}(t,t') \qquad (4.8)$$

$$C_{IST}(t,t') = \frac{h.A_2(t')}{E(t')}.\ln\left(1 + (t-t')^k\right) \qquad (4.9)$$

$$C_{ST}(t,t') = \frac{A_{ST}(t')}{E(t')}.K_{ST}(t-t') \qquad (4.10)$$

$$C_{LT}(t,t') = \frac{f}{g+1}.(t^{g+1} - t'^{g+1}) \tag{4.11}$$

$$K_{ST}(t-t') = \sum_{i=1}^{2} \frac{1}{E_i}.\left(1 - \exp\left(-\frac{t-t'}{\tau_i}\right)\right) \tag{4.12}$$

$$A_{ST}(t') = l.\frac{dE}{dt} + m \tag{4.13}$$

The third model (Eq. 4.14) is a logarithmic law adapted from the Model Code 2010. The model is composed of a parameter C_{MC} related to the global amplitude of the basic creep and is independent of the age at loading. The effect of the age at loading is considered with a second parameter τ_{MC} which is function of the concrete composition [binder (Delsaute et al. 2017b) and aggregate (Delsaute and Staquet 2019)] and is a power-law function of the inverse of the time derivative of the elastic modulus during very early age and is linearly function of the inverse of the time derivative of the elastic modulus at later ages (Eq. 4.15).

$$\mathrm{C(t,t')} = \frac{1}{\mathrm{C_{MC}}}.\ln\left(1 + \frac{\mathrm{t-t'}}{\tau_{\mathrm{MC}}}\right) \tag{4.14}$$

$$\tau_{\mathrm{MC}} = \begin{cases} q \cdot \left(\frac{dE}{dt} \cdot t_E\right)^p, t' < 40h \\ n \cdot \left(\frac{dE}{dt}\right)^{-1}, t' \geq 40h \end{cases} \tag{4.15}$$

According to the basic creep results of Zhang et al. (2014) on mature cement pastes, Irfan-ul-Hassan et al. (2016) shows that the specific creep can be also accurately fitted with a power law expression during the first days after loading while the long term creep can be represented with a logarithmic law. However, the direct use of both mathematical expressions cannot fit the experimental results.

4.5.2 Identification of Concrete Properties by Means of Multiscale Modelling

4.5.2.1 Downscaling Based Identification

A mentioned in Sect. 4.2, a part of the creep strain exhibited by cementitious materials is related to the creep properties of the hydrates. The hydration process and the subsequent formation of hydrates make concrete an aging material. Hydrates themself show non-aging creep properties (Bažant and Prasannan 1989a, b; Scheiner and Hellmich 2009). Königsberger et al. (2016) have investigated the creep properties of hydrates by using minute-scale-duration creep tests performed on cement pastes with different water-to-cement ratios.

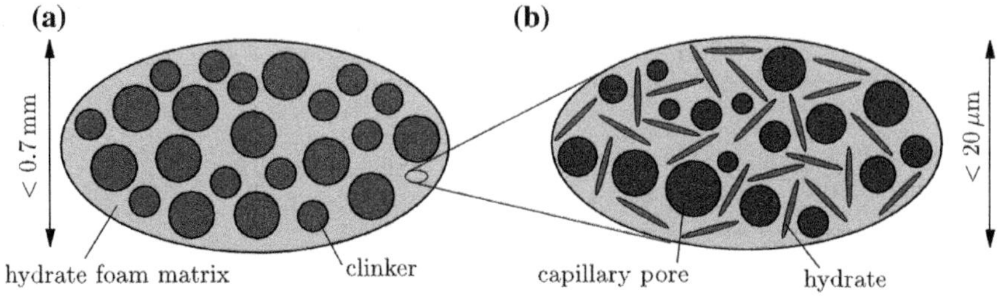

Fig. 4.22 Micromechanical representation of the cement pastes (**a**) and the hydrate foam matrix (**b**) (Königsberger et al. 2016)

A microstructural representation of the cement pastes was used to identify the viscoelastic properties of the hydrates (Fig. 4.22). At a scale of several hundred of microns, the cement paste is defined as a representative volume element composed of cement clinker inclusion in a continuous hydrate foam matrix (Fig. 4.22a). At a lower scale of 20 μm, the hydrate foam matrix is represented by water, capillary pores and thick hydrate needles oriented uniformly in all space directions (Fig. 4.22b). Hydrates creep properties were identified by back-analysing experimental results obtained with the repeated minute-scale-duration loading. For that purpose, homogenization at both scales (cement paste and hydrate foam matrix) was carried out. It was observed that the creep of the hydrates follows a single power-law for which the exponent is close to the one of the cement paste.

4.5.2.2 Water Transfer Between Aggregate and Cement Paste

Porous aggregates are likely to absorb or release water from cement paste (Alhozaimy 2009). If aggregates are not fully saturated before mixing, water is absorbed by the aggregate during mixing and few hours after. When self-desiccation starts, aggregate acts as a water tank and releases water to the cement paste. Therefore, porous aggregate can be considered as an internal curing agent. This is especially true for cement based materials with low water-to-cement ratio. The porosity of the aggregate affects the development of capillary pressure (Bentz and Snyder 1999; Henkensiefken et al. 2009) and the microstructure of interfacial transition zone (ITZ) between the cement paste and the aggregate (Ollivier et al. 1995; Scrivener et al. 2004). Irfan-ul-Hassan et al. have investigated the influence of the water transfer between aggregate and cement paste on the creep properties (Irfan-ul-Hassan et al. 2017). For that purpose, several repeated minute-scale-duration loadings were applied on cement pastes, mortars and concretes with different water-to-cement ratios and volume fractions of aggregate. The influence of the internal curing was modelled by using micromechanics upscaling (from cement paste to mortar and concrete). An extension of the Powers hydration model was performed by considering the water uptake capacity of the aggregate and the cement paste void-filling extent. It was therefore shown that it is possible to model the creep properties of concrete based on cement paste data by using micromechanics approach.

4.6 Conclusion and Outlook

A new experimental methodology is presented for the characterization of the viscoelastic behavior of cement based materials since very early age. This new methodology consists in applying repeated creep test of minute-scale-duration on a same sample. Testing devices, test protocols and data treatment methods were developed recently in three laboratories. A comparison of the testing and modelling strategy was presented. It was shown that basic creep can be tested since final setting time by means of repeated minute-scale-duration loadings. Three main aging parameters are identified to define the viscoelastic properties of cement based material for short duration of loading: the elastic modulus, an amplitude creep term and a kinetic creep term which can be linked to a power law or a logarithmic law.

The new testing method based on repeated minute-scale-duration loading is a field that can offer much to the characterization of cement-based materials especially for the characterization of concrete composed of new raw materials (e.g. recycled aggregate, fibres and geopolymer). A lot of meaningful research efforts have emerged recently showing the capability of this new testing methodology to define with a limited number of tests and a limited time duration the elastic and creep properties of cement-based materials at the macroscopic and microscopic scale. However these works were limited to medium stress levels (up to 40%), laboratory temperature condition (20 °C), binder composed of ordinary Portland cement, compressive loading, sealed condition and uniaxial testing. Therefore suggestions for future research include:

- more sophisticated tests with complex histories of loading and temperature for computational purposes,
- the study of the transversal creep at early age,
- the influence of mineral additions (e.g. slag and fly ash) and admixtures (e.g. superabsorbent polymer and superplasticizer) on the basic creep at very early age,
- the influence of external drying,
- the study of the tensile creep,
- the coupling between damage and creep at early age.

References

Acker, P., & Ulm, F. J. (2001). Creep and shrinkage of concrete: Physical origins and practical measurements. *Nuclear Engineering and Design, 203*(2–3), 143–158. https://doi.org/10.1016/S0029-5493(00)00304-6.

Alhozaimy, A. M. (2009). Effect of absorption of limestone aggregates on strength and slump loss of concrete. *Cement and Concrete Composites, 31*(7), 470–473. https://doi.org/10.1016/j.cemconcomp.2009.04.010.

Andrew, R. M. (2017). Global CO_2 emissions from cement production. In *Earth system science data discussions* (pp. 1–52). https://doi.org/10.5194/essd-2017-77.

ASTM C469/C469M-14. (2014). Standard test method for static modulus of elasticity and Poisson's ratio of concrete in compression. *ASTM International*. https://doi.org/10.1520/c0469_c0469m.

ASTM C512/C512M-15. (2015). Standard test method for creep of concrete in compression. *ASTM International*, 3–7. https://doi.org/10.1520/c0512_c0512m-15.

Atrushi, D. S. (2003). *Tensile and compressive creep of early age concrete: Testing and modelling*. Ph.D. thesis, The Norwegian University of Sciences and Technology, Trondheim, Norway.

Ausweger, M., et al. (2019). Early-age evolution of strength, stiffness, and non-aging creep of concretes: Experimental characterization and correlation analysis. *Materials, 12*(2), 207. https://doi.org/10.3390/ma12020207.

Bažant, Z. P. (1984). Double-power logarithmic law for concrete creep. *Cement and Concrete Research, 14*, 793–806.

Bažant, Z. P. (1995). Creep and shrinkage prediction model for analysis and design of concrete structures—Model B-3. *Materials and Structures, 28*, 357–365.

Bažant, Z. P., & Osman, E. (1976). Double power law for basic creep of concrete. *Matériaux et Constructions, 9*(1), 3–11. https://doi.org/10.1007/BF02478522.

Bažant, Z. P., & Prasannan, S. (1989a). Solidification theory for aging creep. I: Formulation. *Journal of Engineering Mechanics, 115*(8), 1691–1703. https://doi.org/10.1016/0008-8846(88)90028-2.

Bažant, Z. P., & Prasannan, S. (1989b). Solidification theory for concrete creep. II: Verification and application. *Journal of Engineering Mechanics, 115*(8), 1704–1725. https://doi.org/10.1061/(ASCE)0733-9399(1989)115:8(1691).

Bažant, Z. P., Asghari, A. A., & Schmidt, J. (1976). Experimental study of creep of hardened Portland cement paste at variable water content. *Materials and Structures/Materiaux et Constructions, 9*(4), 279–290. https://doi.org/10.1007/BF02478648.

Bažant, Z., Hauggaard, A. B., & Baweja, S. (1997a). Microprestress-solidification theory for concrete creep. II: Algorithm and verification. *Journal of Engineering Mechanics, 123*, 1195–1201.

Bažant, Z. P., et al. (1997b). Microprestress-solidification theory for concrete creep. I aging and drying effects. *Journal of Engineering Mechanics, 123*(11), 1188–1194.

Bažant, Z. P., Hubler, M. H., & Yu, Q. (2011). Pervasiveness of excessive segmental bridge deflections: Wake-up call for creep. *ACI Structural Journal, 108*(6), 766–774. https://doi.org/10.1017/CBO9781107415324.004.

Bažant, Z. P., Yu, Q., & Li, G.-H. (2012). Excessive Long-time deflections of prestressed box girders. I: Record-span bridge in palau and other paradigms. *Journal of Structural Engineering, 138*(June), 676–686. https://doi.org/10.1061/(ASCE)ST.1943-541X.

Bentz, D. P., & Snyder, K. A. (1999). Protected paste volume in concrete: Extension to internal curing using saturated lightweight fine aggregate. *Cement and Concrete Research, 29*(11), 1863–1867. https://doi.org/10.1016/S0008-8846(99)00178-7.

Bernard, O., Ulm, F.-J., & Germaine, J. T. (2003a). Volume and deviator creep of calcium-leached cement-based materials. *Cement and Concrete Research, 33*(8), 1127–1136.

Bernard, O., Ulm, F.-J., & Lemarchand, E. (2003b). A multiscale micromechanics-hydration model for the early-age elastic properties of cement-based materials. *Cement and Concrete Research, 33*(9), 1293–1309. https://doi.org/10.1016/S0008-8846(03)00039-5.

Boulay, C. (2012). Test rig for early age measurements of the autogenous shrinkage of a concrete. *Proceedings of the RILEM-JCI International Workshop ConCrack, 3*, 111–122.

Boulay, C., & Colson, A. (1981). Un extensomètre à béton éliminant l'influence des déformations transversales sur la mesure des déformations longitudinales. *Materials and Structures, 14*(79), 35–38.

Boulay, C., et al. (2012). Monitoring of the creep and the relaxation behaviour of concrete since setting time, part 1: Compression. In *Strategies for sustainable concrete structures* (p. 10). Aix-en-Provence.

Boulay, C., et al. (2014). How to monitor the modulus of elasticity of concrete, automatically since the earliest age? *Materials and Structures, 47*(1–2), 141–155. https://doi.org/10.1617/s11527-013-0051-3.

Briffaut, M., et al. (2012). Concrete early age basic creep: Experiments and test of rheological modelling approaches. *Construction and Building Materials, 36*, 373–380. https://doi.org/10.1016/j.conbuildmat.2012.04.101.

Butler, L. J., et al. (2016). Evaluating the early-age behaviour of full-scale prestressed concrete beams using distributed and discrete fibre optic sensors. *Construction and Building Materials, 126*, 894–912. https://doi.org/10.1016/j.conbuildmat.2016.09.086.

De Schutter, G. (2004). Applicability of degree of hydration concept and maturity method for thermo-visco-elastic behaviour of early age concrete. *Cement and Concrete Composites, 26*(5), 437–443. https://doi.org/10.1016/S0958-9465(03)00067-2.

Delsaute, B. (2016). *New approach for monitoring and modelling of the creep and shrinkage behaviour of cement pastes, mortars and concretes since setting time*. Université Libre de Bruxelles (BATir) and Université Paris Est (IFSTTAR).

Delsaute, B., & Staquet, S. (2014). Early age creep and relaxation modelling of concrete under tension and compression. In *CONMOD 2014: Proceedings of the RILEM International Symposium on Concrete Modelling* (pp. 12–14).

Delsaute, B., & Staquet, S. (2017). Decoupling thermal and autogenous strain of concretes with different water/cement ratios during the hardening process. *Advances in Civil Engineering Materials, 6*(2), 22. https://doi.org/10.1520/ACEM20160063.

Delsaute, B., & Staquet, S. (2019). Development of strain-induced stresses in early age concrete composed of recycled gravel or sand. *Journal of Advanced Concrete Technology, 17*(6), 319–334. https://doi.org/10.3151/jact.17.319.

Delsaute, B., Staquet, S., & Boulay, C. (2012). Monitoring of the creep and the relaxation behaviour of concrete since setting time, part 2: traction. In *Strategies for Sustainable Concrete Structures* (p. 10). Aix-en-Provence.

Delsaute, B., Carette, J., & Staquet, S. (2013). Monitoring of the creep and the relaxation at very early age: complementary results on the CEOS concrete. In *VIII International Conference on Fracture Mechanics of Concrete and Concrete Structures (IA-FramCOS-8)* (pp. 453–458).

Delsaute, B., Staquet, S., & Boulay, C. (2014). Early age creep and relaxation behavior of concrete under tension and compression. In *Concrete Innovation Conference CIC 2014*, Oslo (p. 10).

Delsaute, B., Boulay, C., & Staquet, S. (2016a). Creep testing of concrete since setting time by means of permanent and repeated minute-long loadings. *Cement and Concrete Composites, 73*, 75–88. https://doi.org/10.1016/j.cemconcomp.2016.07.005.

Delsaute, B., Torrenti, J.-M., & Staquet, S. (2016b). Monitoring and modeling of the early age properties of the Vercors Concrete. In *TINCE 2016* (p. 12). Paris.

Delsaute, B., et al. (2016c). Testing concrete E-modulus at very early ages through several techniques: An inter-laboratory comparison. *Strain, 52*(2), 91–109. https://doi.org/10.1111/str.12172.

Delsaute, B., Torrenti, J.-M., & Staquet, S. (2017a). Modeling basic creep of concrete since setting time. *Cement and Concrete Composites, 83*, 239–250. https://doi.org/10.1016/j.cemconcomp.2017.07.023.

Delsaute, B., Torrenti, J.-M., & Staquet, S. (2017b). Modelling of basic creep of concrete with high substitution rate of Portland cement by mineral additions. In S. Staquet & D. G. Aggelis (Eds) *Second international RILEM/COST conference on early age cracking and serviceability in cement-based materials and structures* (pp. 343–348). Brussel.

Domone, P. L. (1974). Uniaxial tensile creep and failure of concrete. *Magazine of Concrete Research, 26*(88), 144–152. https://doi.org/10.1680/macr.1974.26.88.144.

EDF. (no date). http://researchers.edf.com/research-activities/generation/vercors-an-experimental-mock-up-of-a-reactor-containment-building-290900.html.

Forth, J. P. (2014). Predicting the tensile creep of concrete. *Cement and Concrete Composites, 55*(July), 70–80. https://doi.org/10.1016/j.cemconcomp.2014.07.010.

Gardner, N. J., & Fu, H. C. (1987). Effects of high construction loads on the long-term deflections of flat slabs. *ACI Structural Journal, 84*(4), 349–360.

Göbel, L., Osburg, A., & Pichler, B. (2018a). The mechanical performance of polymer-modified cement pastes at early ages: Ultra-short non-aging compression tests and multiscale homogenization. *Construction and Building Materials, 173,* 495–507. https://doi.org/10.1016/j.conbuildmat.2018.04.033.

Göbel, L., et al. (2018b). Viscoelastic behavior of polymer-modified cement pastes: Insight from downscaling short-term macroscopic creep tests by means of multiscale modeling. *Applied Sciences, 8*(4), 487. https://doi.org/10.3390/app8040487.

Guenot, I., Torrenti, J.-M., & Laplante, P. (1996). Stresses in early-age concrete: Comparison of different creep models. *ACI Materials Journal, 93*(3), 254–259.

Guénot-Delahaie, I. (1997). *Contribution à l'analyse physique et à la modélisation du fluage propre du béton.* Ph.D. thesis, Laboratoire Central des Ponts et Chaussées, Paris, France.

Gutsch, A.-W. (2002). Properties of early age concrete-Experiments and modelling. *Materials and Structures, 35,* 76–79. Available at: http://link.springer.com/article/10.1007/BF02482104. Accessed December 17, 2013.

Havlásek, P., & Jirásek, M. (2012). Modeling of concrete creep based on microprestress-solidification theory. *Acta Polytechnica, 52*(2), 34–42.

Henkensiefken, R., et al. (2009). Volume change and cracking in internally cured mixtures made with saturated lightweight aggregate under sealed and unsealed conditions. *Cement and Concrete Composites, 31*(7), 427–437. https://doi.org/10.1016/j.cemconcomp.2009.04.003.

Instruments, T. (2017). *Tam air isothermal calorimetry.* Available at: http://www.tainstruments.com/pdf/brochure/TAMAIRbrochure.pdf. Accessed February 14, 2017.

Irfan-ul-Hassan, M., et al. (2015). Minutes-long creep tests on young cement pastes provide access to creep properties relevant for ageing creep with a duration of 2 Days. In *CONCREEP 10* (pp. 348–357). Reston, VA: American Society of Civil Engineers. https://doi.org/10.1061/9780784479346.041.

Irfan-ul-Hassan, M., et al. (2016). Elastic and creep properties of young cement paste, as determined from hourly repeated minute-long quasi-static tests. *Cement and Concrete Research, 82,* 36–49. https://doi.org/10.1016/j.cemconres.2015.11.007.

Irfan-ul-Hassan, M., et al. (2017). How water-aggregate interactions affect concrete creep: Multiscale analysis. *Journal of Nanomechanics and Micromechanics, 7*(4). https://doi.org/10.1061/(asce)nm.2153-5477.0000135.

ISO 1920-9:2009. (2009). *Testing of concrete—Part 9: Determination of creep of concrete cylinders in compression* (pp. 1–13).

Jiang, W., De Schutter, G., & Yuan, Y. (2014). Degree of hydration based prediction of early age basic creep and creep recovery of blended concrete. *Cement and Concrete Composites, 48,* 83–90. https://doi.org/10.1016/j.cemconcomp.2013.10.012. Elsevier Ltd.

Jones, P. G., & Richart, F. E. (1936). The effect of testing speed on strength and elastic properties of concrete. *ASTM Proceedings, 36,* 380–392.

Karte, P., et al. (2015). Unloading-based stiffness characterisation of cement pastes during the second, third and fourth day after production. *Strain, 51*(2), 156–169. https://doi.org/10.1111/str.12129.

Kim, Y. H., et al. (2011). Experimental study on creep and durability of high-early-strength self-consolidating concrete for precast elements. *ACI Materials Journal, 108*(2), 128–138.

Klemczak, B., & Knoppik-Wrobel, A. (2011). Early age thermal and shrinkage cracks in concrete structures—Influence of curing conditions. *Architecture Civil Engineering Environment, 4,* 47–58.

Königsberger, M., et al. (2016). Downscaling based identification of nonaging power-law creep of cement hydrates. *Journal of Engineering Mechanics, 142*(12), 04016106. https://doi.org/10.1061/(ASCE)EM.1943-7889.0001169.
Königsberger, M., et al. (2017). Hydrate failure in ITZ governs concrete strength: A micro-to-macro validated engineering mechanics model. *Cement and Concrete Research*. https://doi.org/10.1016/j.cemconres.2017.10.002.
Laplante, P. (1993a). *Propriétés mécaniques des bétons durcissants: Analyse comparée des bétons classiques et à très hautes performances*. Ph.D. thesis, Ecole Nationale des Ponts et Chaussées, Paris, France.
Laplante, P. (1993b). *Propriétés mécaniques des bétons durcissants: Analyse comparée des bétons classiques et à très hautes performances*. Ph.D. thesis, Ecole Nationale des Ponts et Chaussées.
Le Roy, R. (1995). *Déformations instantanées et différées des bétons à hautes performances*. Ph.D. thesis, Ecole Nationale des Ponts et Chaussées, Paris, France.
Le Roy, R., de Larrard, F., & Pons, G. (1996). Calcul des déformations instantanées et différées des bétons à hautes performances. *Bulletin des Laboratoires des Ponts et Chauss ees* (special issue XIX).
Lorman, W. R. (1940). The theory of concrete creep. *Proceedings of ASTM, 40,* 1082–1102.
Maia, L., & Figueiras, J. (2012). Early-age creep deformation of a high strength self-compacting concrete. *Construction and Building Materials, 34,* 602–610. https://doi.org/10.1016/j.conbuildmat.2012.02.083.
Mazzotti, C., & Savoia, M. (2001). Experimental study of non-linear creep of concrete at high stress level. In *Creep, shrinkage and durability mechanics of concrete and other Quasi-Brittle materials*.
Milenkovic, N., et al. (no date). Advanced characterisation of the early age behaviour of bulk hydrophobic mortars. *Construction and Building Materials*.
Neville, A. M., Dildger, L. H., & Brooks, J. J. (1983). *Creep of plane and structural concrete*.
Ollivier, J. P., Maso, J. C., & Bourdette, B. (1995). Interfacial transition zone in concrete. *Advanced Cement Based Materials, 2*(1), 30–38. https://doi.org/10.1016/1065-7355(95)90037-3.
Pichler, B., et al. (2013). Effect of gel–space ratio and microstructure on strength of hydrating cementitious materials: An engineering micromechanics approach. *Cement and Concrete Research, 45,* 55–68. https://doi.org/10.1016/j.cemconres.2012.10.019.
Powers, T. C. (1965). Mechanisms of shrinkage and reversible creep of hardened cement paste. In *The structure of concrete and its behaviour under load* (pp. 319–344), London.
Rosowsky, D. V., Stewart, M. G., & Khor, E. H. (2000). Early-age loading and long-term deflections of reinforced concrete beams. *ACI Structural Journal, 97*(3), 517–524. https://doi.org/10.14359/4647.
Ross, A. D. (1937). Concrete creep data. *The structural Engineer, 15*(8).
Rossi, P., et al. (1994). Investigation of the basic creep of concrete by acoustic emission. *Materials and Structures, 27,* 510–514.
Rossi, P., et al. (2012). Basic creep behavior of concretes investigation of the physical mechanisms by using acoustic emission. *Cement and Concrete Research, 42*(1), 61–73. https://doi.org/10.1016/j.cemconres.2011.07.011.
Rossi, P., Tailhan, J.-L., & Le Maou, F. (2013). Comparison of concrete creep in tension and in compression: Influence of concrete age at loading and drying conditions. *Cement and Concrete Research, 51,* 78–84. https://doi.org/10.1016/j.cemconres.2013.04.001.
Saleh, M., et al. (2017). Applicability of ultrasonic measurement on the monitoring of the setting of cement pastes: Effect of water content and mineral additions. *Advances in Civil Engineering*.
Scheiner, S., & Hellmich, C. (2009). Continuum microviscoelasticity model for aging basic creep of early-age concrete. *Journal of Engineering Mechanics, 135*(4), 307–323. https://doi.org/10.1061/(ASCE)0733-9399(2009)135:4(307).
Scrivener, K. L., Crumbie, A. K., & Laugesen, P. (2004). The interfacial transition zone (ITZ) between cement paste and aggregate in concrete. *Interface Science, 12*(4), 411–421. https://doi.org/10.1023/B:INTS.0000042339.92990.4c.

Shank, J. R. (1935). Plastic flow of Portland cement concrete. *Industrial and Engineering Chemistry, 27*(9), 1011–1014.

Staquet, S., et al. (2012). Design of a revisited Tstm system for testing concrete since setting time under free and restraint conditions. In *Concrack 3—RILEM-JCI International Workshop on Crack Control of Mass Concrete and Related Issues Concerning Early-Age of Concrete Structures* (p. 12).

Staquet, S., et al. (2014). Maturity testing through continuous measurement of e-modulus: An inter-laboratory and inter-technique study. In *Proceedings of ECO-CRETE International Symposium on Sustainability* (p. 8).

Staquet, S., et al. (2017). Further works on concrete at early age coordinated within COST TU1404 European project. In *JCI-RILEM International Workshop on Control of Cracking of Mass Concrete and Related Issues concerning Early Age Cracking of Concrete Structures* (p. 10).

Straub, H. (1930). Plastic flow in concrete arches. *Proceedings American Society of Civil Engineers, 56,* 49–114.

Tamtsia, B. T., & Beaudoin, J. J. (2000). Basic creep of hardened cement paste. A re-examination of the role of water. *Cement and Concrete Research, 30*(9), 1465–1475. https://doi.org/10.1016/S0008-8846(00)00279-9.

Tamtsia, B. T., Beaudoin, J. J., & Marchand, J. (2004). The early age short-term creep of hardening cement paste: Load-induced hydration effects. *Cement and Concrete Composites, 26*(5), 481–489. https://doi.org/10.1016/S0958-9465(03)00079-9.

Thomas Thelford. (1993). *Comité Eurointernational du Béton CEB/FIB Model Code 90, Design Code.*

Torrenti, J.-M., & Le Roy, R. (2017). Analysis of some basic creep tests on concrete and their implications for modeling. In *Structural Concrete*. https://doi.org/10.1002/suco.201600197.

Ulm, F.-J., Le Maou, F., & Boulay, C. (1999). Creep and shrinkage coupling: New review of some evidence. In *ACI-RILEM Workshop on Creep and Shrinkage of Concrete Structures* (pp. 21–37).

Vandamme, M., & Ulm, F.-J. (2009). Nanogranular origin of concrete creep. *Proceedings of the National Academy of Sciences of the United States of America, 106*(26), 10552–10557. https://doi.org/10.1073/pnas.0901033106.

Wierig, H. J. (1971). Einige Beziehungen don Eigen-schaften von grünen un jungen Betonen und denen des Festbetons. *Beton, 11+12*, 445–448, 487–490.

Wittmann, F. H. (1982). Creep and shrinkage mechanisms. In Z.P. Bažant & F.H. Wittmann (Eds.) *Creep and shrinkage of concrete* (pp. 129–161).

Wittmann, F. H. (2015). Useful fundamentals of shrinkage and creep of concrete. *Concreep, 10,* 84–93.

Zhang, Q. (2014). *Creep properties of cementitious materials: Effect of water and microstructure: An approach by microindentation*. Ph.D. thesis, Laboratoire Navier, Ecole des Ponts ParisTech, Champs-sur-Marne.

Zhang, Q., et al. (2014). Long-term creep properties of cementitious materials: Comparing microindentation testing with macroscopic uniaxial compressive testing. *Cement and Concrete Research, 58,* 89–98. https://doi.org/10.1016/j.cemconres.2014.01.004.

Chapter 5
Monitoring of the Thermal and Autogenous Strain

Brice Delsaute and Stéphanie Staquet

Abstract During early age, the autogenous and thermal deformations are two of the most important concrete properties that are involved in the cracking risk in cement-based materials and concrete structures (especially for massive structures). For that reason, several test rigs were developed in the past for the monitoring since casting time of these deformations. During early age, the thermal strain depends strongly on the evolution of the coefficient of thermal expansion (CTE), the thermal conductivity, the heat capacity, the heat release of the cement paste and the boundary conditions. During hydration process, both autogenous and thermal strains occur simultaneously and cannot be directly distinguished. In previous work, two kinds of methodology were used to monitor the CTE. The first method uses two distinct tests with two different histories of temperature in order to define autogenous and thermal strain. The second method uses only one test during which repeated thermal variations are applied to a sample. This chapter reports on the most recent advances on the physical mechanisms associated to the development of the autogenous strain and the CTE. The recent devices and test protocols developed for the simultaneous monitoring of the coefficient of thermal expansion and the autogenous strain at three different scales (cement paste, mortar and concrete) are presented. The global testing methodologies bear some similarities but major differences remain in the test set up designs, in the testing processes and also in the data processing. Based on the physical mechanisms, existing test facilities and test protocols, a new test protocol is established and is compared to existing test protocols. Finally, recommendations are proposed for the determination of the autogenous strain and the CTE.

Keywords Early age · Autogenous strain · Coefficient of thermal expansion · Heat release

---

B. Delsaute (✉) · S. Staquet
Service BATir, LGC, Université Libre de Bruxelles (ULB), Brussels, Belgium
e-mail: bdelsaut@ulb.ac.be

M. Serdar et al. (eds.), *Advanced Techniques for Testing of Cement-Based Materials*, Springer Tracts in Civil Engineering, https://doi.org/10.1007/978-3-030-39738-8_5

5.1 Introduction

The free deformations are evolving quite fast during the hardening process and particularly just after the setting of the material. According to the composition, the structural and environmental situation, the amplitude of these deformations can be very significant. Dam, piers of bridge, gas containment are relevant examples for which the free deformations are important. If these deformations are partially or fully restrained, stresses are induced in the material which can lead to cracking. Therefore, it is meaningful to consider correctly the development of the free deformation in the design of concrete structures which are massive, use low water-cement ratio or silica fume, or which need to be waterproof or airtight. When the material is not exposed to drying, the free deformations of the material after setting are composed by the autogenous deformation ε_{au} and the thermal deformation $\varepsilon_{\mathrm{th}}$:

$$\varepsilon_{\mathrm{f}}(\mathrm{t}) = \varepsilon_{au}(\mathrm{t}) + \varepsilon_{\mathrm{th}}(\mathrm{t}) = \varepsilon_{\mathrm{au}}(\mathrm{t}) + \int_0^{\mathrm{t}} \alpha_{\mathrm{T}}(\mathrm{t}) \cdot \dot{\mathrm{T}} \cdot \mathrm{dt} \tag{5.1}$$

The thermal deformations are function of the evolution of the temperature inside the concrete material and the evolution of the coefficient of thermal expansion α_{T} (CTE, expressed in °C^{-1}) of the concrete (Eq. 5.1). The evolution of the temperature depends on the boundary conditions and the dimensions of the structure and the evolution of materials properties of the concrete. The evolution of the temperature caused by the hydration of cement is defined by solving the heat equation (Eq. 5.2) for which the density ρ (expressed in kg m^{-3}), the thermal conductivity λ (expressed in W m^{-1} K^{-1}), the heat capacity $\mathrm{C_p}$ (expressed in J °C^{-1}), and the heat release by the concrete during hydration process Q (expressed in J g^{-1}) must be known.

$$\rho \cdot \mathrm{C_p} \cdot \frac{\partial \mathrm{T}}{\partial \mathrm{t}} = \mathrm{div}(\lambda \cdot \mathrm{gradT}) + \frac{\partial \mathrm{Q}}{\partial \mathrm{t}}. \tag{5.2}$$

The heat release, the thermal conductivity, the heat capacity and the density are material parameters which evolve during the hydration process and are necessary to know in order to define the evolution of the temperature. The heat release can be determined by using adiabatic calorimetry (Darquennes 2009), semi-adiabatic calorimetry (Boulay et al. 2010) or isothermal calorimetry (Instruments 2017). The thermal conductivity and the heat capacity can be determined by generating a thermal gradient inside the sample and measuring the temperature distribution (Campbell-Allen and Thorne 1963; Mounanga et al. 2004; Bentz 2007). The coefficient of thermal expansion can be determined by applying thermal variations on a specimen. Nevertheless, the monitoring of the CTE and the autogenous deformation remains difficult to assess during early age. As a matter of fact, during the hardening process both parameters evolve at the same time. Displacements due to both are superimposed during the measurement.

In the present chapter, a new test protocol is developed for the monitoring of the autogenous strain and the coefficient of thermal expansion since setting. First, an initial overview of the physical mechanisms related to the development of the autogenous strain and the coefficient of thermal expansion is presented. The recent devices developed and their associated test protocol for the characterization of both properties at the three scales (cement paste, mortar and concrete) are then presented and compared. Based on the physical mechanisms, existing test facilities and test protocols, a new test protocol is established. The results obtained on an ordinary concrete with the BTJADE (Boulay 2012) device are presented then. A sensitivity analysis is performed in order to highlight the importance of the data processing in the determination of the CTE. Finally, an extension of this methodology on cement paste and mortar is presented and recommendations are proposed for the determination of the autogenous strain and the CTE.

5.2 Physical Mechanisms

5.2.1 Autogenous Deformation

Le Chatelier (1900) has first shown that the hydration of cement causes shrinkage, because the volume of hydration products is lower than the volume of reactants. This is known as chemical shrinkage. As long as cement paste remains fluid, autogenous deformation is equal to chemical shrinkage (Fig. 5.1a). When the material sets, the stiffness of the skeleton of the cement paste increases and a restriction of the chemical shrinkage takes places (Bouasker et al. 2008). Then, chemical shrinkage is not transformed in volumetric strain due to the formation of a gaseous volume phase in the pore (because of the water consumption by the hydration reaction). The internal voids, or capillarity pores, created in the cement paste cause self-desiccation shrinkage (Fig. 5.1b). Capillarity pores have a size ranging from 10 nm to 10 μm

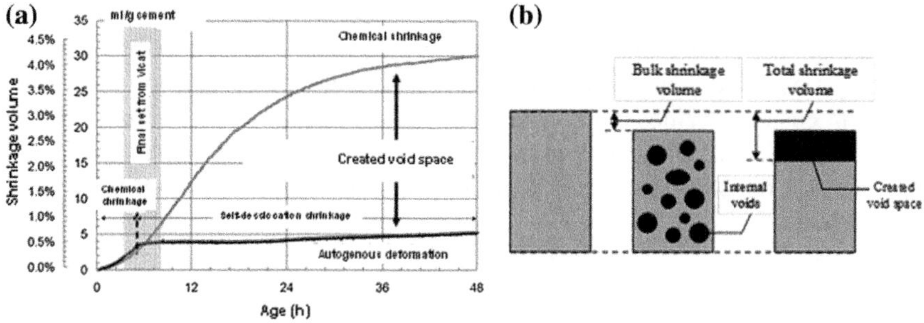

Fig. 5.1 **a** Chemical shrinkage and autogenous deformation (Bouasker 2007), **b** definition of autogenous shrinkage after setting (Garcia Boivin 1999)

(Mindess et al. 2003). Most of studies dealing with autogenous deformation actually refer to self-desiccation, in other words autogenous shrinkage after setting. However, at the very early stages after setting, some cementitious materials exhibit swelling during several hours or days before shrinking (Darquennes et al. 2011; Delsaute and Staquet 2017a).

Despite the many studies carried out in the paste, there is still no general consensus on the mechanisms related to the swelling. Three reasons were observed:

1. **The water absorption of cement paste during hydration** due to bleeding or due to the internal curing coming from porous aggregate (e.g. lightweight aggregate (Kohno et al. 1999; Liu and Hansen 2016) and recycled aggregate (Delsaute and Staquet 2017b, 2019), fillers (Esping 2008; Craeye et al. 2010) or super absorbent polymer (Wyrzykowski and Lura 2013a) which cancels the self-desiccation phenomenon.
2. **The volume of the hydration product**, swelling is also linked to the formation and growth of portlandite crystals (Baroghel-Bouny et al. 2006; Sant et al. 2011; Stefan et al. 2018) and ettringite (Mehta 1973). The formation of both elements will lead to a crystallization stress. These stresses depend on many factors such as the size of the elements and their growth direction (Min and Mingshu 1994; Odler and Colán-Subauste 1999), porosity (size distribution and quantity), the E-modulus and the creep of the cement paste (Chaunsali and Mondal 2016).
3. Another source of swelling comes from **measurement artifacts**. The removal of the thermal strain needs an accurate knowledge of the evolution of the coefficient of thermal expansion during the whole hydration process which is generally not the case (Bjøntegaard et al. 2004). Secondly, round-robin testing (Bjøntegaard et al. 2004) have shown quite important dispersion between results from different laboratory and different test rigs (Bjøntegaard et al. 2004).

The autogenous shrinkage is related to the self-desiccation of the cement paste during the hydration process. This results in a decrease of the relative humidity till a value of around 75% (Le Roy 1995; Barcelo et al. 1999; Hammer 2003; Eppers and Mueller 2008). In this range, the mechanisms involved in self-desiccation shrinkage are capillary tension, disjoining pressure and surface tension (Brooks and Megat Johari 2001).

Capillary pores are initially filled with water. As water is consumed by the hydration of cement, voids appear and liquid-vapour menisci are created in the pores. Thus capillary tension develops in the pore water and compressive stresses are generated in the solid. By considering Young-Laplace and Kelvin laws, the capillary depression σ_{cap} can be expressed as a function of the radius of capillary pores or the internal relative humidity:

$$\sigma_{cap} = P_g - P_l = \frac{R \cdot T \cdot \rho_e}{M} \cdot \ln(RH) = \frac{2 \cdot \sigma}{r} \cdot \cos \alpha_m \tag{5.3}$$

where P_g is the pressure in gaseous phase (air + vapour) [Pa], P_l is the pressure in liquid phase (water) [Pa], R is the universal gas constant [8.314 J mol^{-1} K^{-1}],

T is the temperature [K], ρ_e is the density of water [kg m^{-3}], M is the molar mass of water [kg mol^{-1}], RH is the internal relative humidity, defined as the ratio of vapour pressure to saturated vapour pressure, σ is the surface tension of pore fluid ($\sigma = 72.75 \times 10^{-3}$ N m^{-1} for water) [N m^{-1}], r is the radius of curvature of the meniscus of the pore size [m], and α_m is the liquid-solid contact angle [rad]. From the latter equation, it follows that the finer the capillary pores, the higher the capillary depression, and that large capillarity pores dry first. The hydration products progressively fill the porosity of cement paste and refine the capillary porosity, thus the hydration reactions cause changes in moisture capacity (adsorption isotherm) of cement paste (Xi et al. 1994). The internal surface of hydration products increases which induces a redistribution of pores water and a growing effect of surface forces of solid gel particles. As a consequence effective pore pressure and self-desiccation shrinkage increases as hydration goes on (Bentz et al. 1998; Coussy 2003).

Disjoining pressure develops in the microstructure of cement paste in the areas of hindered adsorption. The formulation of this approach of hindered adsorption was first developed by Powers (1968), then by Feldman and Sereda (1968), Bažant (1972) and Ferraris and Wittmann (1987), Beltzung and Wittmann (2005). At a given temperature, the thickness of the adsorbed water layer depends on the local relative humidity. If the distance between two solid surfaces is lower than two times the thickness of the free adsorbed layer, adsorption is hindered and disjoining pressure develops. It tends to separate the adjacent surfaces. A decrease of the relative humidity results in a reduction of this pressure and causes shrinkage.

The surface tension concerns materials with a large internal surface, such as hydrating cement pastes and cement-based materials (Wittmann 1968). According to Bangham and Fakhoury (1931), a local increase in relative humidity of dry material results in a reduction of surface tension and an expansion. The adsorbed water actually reduces surface tension and lowers the hydrostatic pressure of the gel particles. Therefore, when the thickness of water adsorbed on solid surface is reduced, as it is the case during the hydration process, an increase of the surface tension of the solid occurs and leads to shrinkage. However, the changes of surface tension play a role minor in case of high relative humidity (Powers 1965) as it is the case in sealed condition for cement based materials (Le Roy 1995; Barcelo et al. 1999; Hammer 2003; Eppers and Mueller 2008).

As the interactions between hydration products and the pore solution are not fully understood, research is still needed to understand the mechanisms related to autogenous shrinkage and swelling.

Autogenous shrinkage is influenced by the main composition parameters of concrete, namely: W/C-ratio (Delsaute and Staquet 2017a), cement or binder type (Benboudjema et al. 2019; Staquet et al. 2019), volume of paste and aggregate (Liu and Hansen 2016) and nature of aggregate [porosity (Maruyama and Teramoto 2012) and stiffness (Le Roy 1995)].

5.2.2 Coefficient of Thermal Expansion

The coefficient of thermal expansion (CTE) expresses the variation in a unit length of a material in response to a change in temperature. The CTE is indicated as a linear change (linear expansion coefficient α_T in µm/m/°C) or as a volumetric change (volumetric expansion coefficient β_T in $\mu m^3/m^3/°C$). The volumetric CTE can be approximated as 3 times the linear one. In cement based materials, the coefficient of thermal expansion depends mainly on the internal relative humidity of the cement paste and the nature of the aggregate (fib 2013; Siddiqui and Fowler 2014, 2015). During the hardening process, 4 stages occur (Sellevold and Bjøntegaard 2006) as shown in Fig. 5.2. First the value of the CTE is very high due to the contribution of water when the cement paste is still in a plastic phase. When the cement paste starts to set, a sudden decrease of the CTE is observed till a minimum value corresponding to the CTE of the solid skeleton. During this second stage, the evolution of the CTE is driven by the amount of water which is no yet chemically bound and by the increase of the stiffness of the cement paste. Then an increase of the CTE is observed and is due to the decrease of the internal relative humidity in the cement paste. Finally, no significant evolution of the CTE is observed after a certain age even if the material is still hydrating. This age is function of the composition of the material and the curing temperature. During this last stage, only the boundary conditions and the associated environmental conditions (external relative humidity) of the concrete structure affect the magnitude of the CTE. Therefore the CTE has a minimum value around the setting time. For compositions with low water-cement ratio, a factor 2 can be observed between this minimum and the value of the CTE obtained after several weeks (Sellevold and Bjøntegaard 2006).

During the third and fourth stage, the CTE varies according to the change in the internal relative humidity of the cement paste induced by the hydration process and/or the environmental conditions. Two contributions have to be considered to define the magnitude of the CTE: the thermal expansion of the solid body, which is caused by the molecular movement of the cement paste, and the hygro-thermal mechanism (Bažant 1970; Wyrzykowski and Lura 2013b). Sellevold and Bjøntegaard (2006) have shown that the thermal dilation of the solid body is nearly independent of the

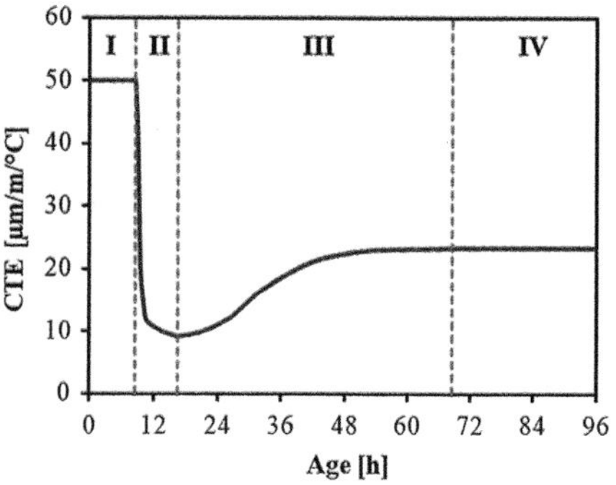

Fig. 5.2 Typical evolution of the CTE of a cement paste

saturation state of the cement paste and that hydration has a very low effect on the dilation of the solid skeleton. In addition, this dilation takes place immediately when the temperature of the solid skeleton changes.

As already mentioned, the magnitude of the CTE is strongly related to the moisture content of the cement paste and, in particular, to the change of internal relative humidity (Grasley and Lange 2006; Sellevold and Bjøntegaard 2006; Yeon et al. 2009; Wyrzykowski and Lura 2013b). A decrease of the internal relative humidity causes an increase of the CTE. The internal relative humidity is function of the temperature. Thus, in a cement based material, an increase of the temperature leads to an increase of the internal relative humidity (Jensen and Hansen 1999) and therefore to a decrease of the CTE. The hygrometric coefficient $\Delta RH/\Delta T$ is used to quantify the change of RH as a result of the unit change of temperature. In Fig. 5.3, based on several results coming from the literature (Bažant 1970; Nilsson 1987; Persson 2002; Radjy et al. 2003; Grasley and Lange 2006), a typical evolution of the hygrometric coefficient is shown as a function of the internal relative humidity of a cement paste. An asymmetrical bell-shaped is observed between both parameters. A maximum is reached around a relative humidity of 60%. Whatever the relative humidity, the hygrometric coefficient is always positive. Between a range of relative humidity between 50 and 100%, the hygrometric coefficient increases when the relative humidity decreases. For relative humidity lower than 50%, a decrease of the hygrometric coefficient is observed. In case of early age concrete in sealed condition, the internal relative humidity of the cement paste is higher than 75% (Le Roy 1995; Barcelo et al. 1999; Hammer 2003; Eppers and Mueller 2008). In such range of relative humidity, the hygrometric coefficient increases always when the relative humidity decreases. This explains why the CTE increases during the hardening process with the self-desiccation of the cement paste. This evolution of the hygrometric coefficient is associated to the redistribution of liquid water inside the cement paste. According to Bazant (1970), water migrates from gel pores to capillary pores when the temperature increases. Recently Wang et al. (2018) have shown

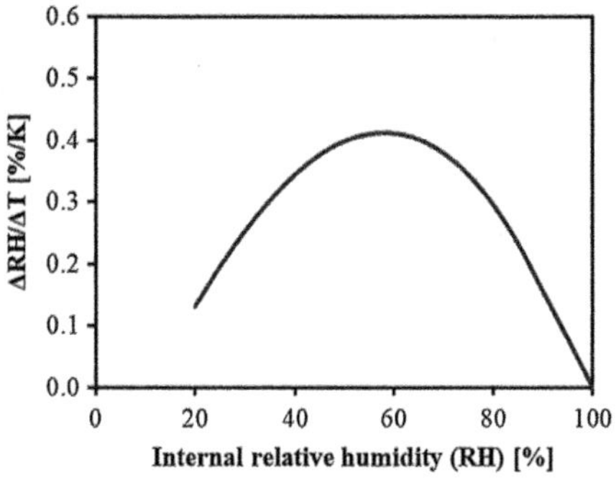

Fig. 5.3 General scheme of evolution of the hygrothermic coefficient of cement paste as a function of the internal relative humidity based on results from the literature (Bažant 1970; Nilsson 1987; Persson 2002; Radjy et al. 2003; Grasley and Lange 2006)

by using micro-poromechanics with a three-scale representation of the cement paste that the hydrometric coefficient is related to the quasi-instantaneous transfer of water between the hydrates and the nanoscopic pores and to the change in pore pressure associated to this water transfer.

Several authors (Bažant 1970; Day 1974; Wittmann and Lukas 1974) reported that the CTE is composed of an **instantaneous and delayed terms**. It was observed that the delayed term is function of the temperature history and the internal relative humidity. The thermal strain induced by the pure thermal dilation of the solid skeleton and the change of internal relative humidity caused by temperature changes develops very fast and consequently are considered to be nearly instantaneous. According to Wyrzykowski and Lura (2013a, b), these thermal strains refer to a time spans ranging from minutes to hours. This time is needed to reach thermal equilibrium and quasi-equilibrium of the corresponding thermal strain in the sample. These mechanisms are responsible of water redistribution between gel and capillarity pores causing delayed thermal shrinkage/swelling. The delayed deformations are also induced by the pressure dissipation in pores which causes a delayed recovery of pure thermal dilation (Scherer 2000, 2003).

Aggregates are inert and have a diluting effect on the evolution of the CTE. After setting, the CTE of a cement paste ranges from 8 to 25 μm/m/°C while the CTE of aggregate is generally smaller with a magnitude ranging from 4 to 14 μm/m/°C according to its mineral composition (Johnson and Parsons 1944). Consequently, for concrete material, an incompatibility of the free thermal strain of the component exists and leads to an internal restraint of the displacement of the cement paste resulting in stress development which may cause micro-cracking and reduce the durability of concrete elements (Cagnon et al. 2016).

The coefficient of thermal expansion is influenced by the main composition parameters of concrete, namely: the nature, the porosity and the content of gravels (Kumar Mehta and Monteiro 2006; Maruyama and Teramoto 2012; Delsaute and Staquet 2017b), the type, the fineness and content of the cement, including type and amount of supplementary cementitious materials (Mitchell 1953; Emanuel and Hulsey 1977; Maruyama et al. 2014; Delsaute and Staquet 2017a), and the adding of other admixture such as super absorbent polymer (Wyrzykowski and Lura 2013a; Staquet et al. 2019). The sand-aggregate ratio, the nature of the sand and the temperature do not affect significantly the magnitude of the CTE of concrete (Ziegeldorf et al. 1978; Alungbe et al. 1992; Maruyama et al. 2014). The water-to-cement ratio changes the evolution of the CTE during the hardening process but does not change significantly the amplitude of the CTE of hardened cement paste (Maruyama et al. 2014) and concrete (Delsaute and Staquet 2017a).

5.2.3 *Correlation Between the Development of the Autogenous Strain and the CTE*

Delsaute and Staquet (2017a) have compared the evolution of the autogenous strain and the CTE for concrete compositions with a water-to-binder ratio of 0.4, 0.5 and 0.6. It was observed that the evolution of both parameters is linked. For each composition, the CTE and the autogenous strain evolve in an opposite direction during the whole test duration and the maximum of the autogenous swelling takes place when the CTE reaches a minimal value. In other words, when concrete exhibits swelling, the CTE decreases while shrinkage is observed when the CTE increases. In addition, results of the CTE were plotted in function of the autogenous strain (Fig. 5.4 for the swelling period and Fig. 5.5 for the shrinkage period). For this comparison, autogenous strain was set to zero when the maximum of swelling is reached. For each composition, a linear relation was observed and is function of the water-to-cement ratio and the nature of the binder during early age. It was therefore concluded that the mechanism related to the expansion of the concrete is also linked to the decrease of the CTE. The increase of the CTE and the autogenous shrinkage are correlated to the mechanism of self-desiccation of the cement paste as explained in previous sections. These observations were then confirmed on mortar with different type of

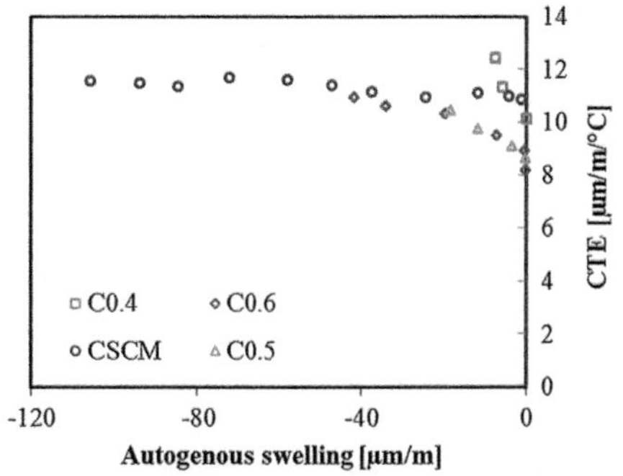

Fig. 5.4 Evolution of the CTE according to the autogenous swelling (Delsaute and Staquet 2017a)

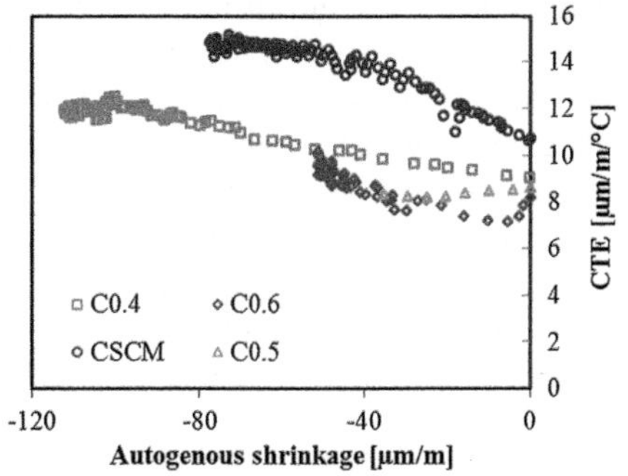

Fig. 5.5 Evolution of the CTE according to the autogenous shrinkage (Delsaute and Staquet 2017a)

cement (Milenkovic et al., no date) and concrete composed of recycled aggregate (Delsaute and Staquet 2017b, 2018a, b, 2020).

5.3 Test Setup

The characterization of the autogenous strain is very sensitive to the conditions of storage of the specimen and to many factors related to the test rig. For that reason several recommendations were developed in the past (Jensen and Hansen 2001; Hammer and Bjøntegaard 2006; ASTM Standard C1698 2014; Bjøntegaard et al. 2015). The main common requirements for all the apparatus for autogenous deformation test are their ability to:

- perfectly seal the specimen in order to avoid any external drying or water uptake;
- keep the temperature constant, which requires external control because the hydration of cement paste releases heat;
- limit the friction with the specimen.

Several devices aimed at monitoring the autogenous deformations were developed in the past. The methods can be in general divided in two categories: linear (Paillere and Serrano 1976; Baroghel-bouny 1994; Kovler 1994; Kronlöf et al. 1995; Mejlhede Jensen and Freiesleben Hansen 1995; Bjontegaard 1999; Bouasker 2007; Boulay 2012) and volumetric (Loukili et al. 2000; Charron 2003; Mitani 2003; Bouasker 2007; Stefan 2009; Loser et al. 2010; Wyrzykowski and Lura 2013a, b). Whereas volumetric methods are mainly used on cement pastes and mortars, the linear methods have been used on three scales including concrete. Horizontal and vertical setups have been designed. Both methods can be divided into two groups: rigid moulds and flexible moulds. Rigid moulds have been reported to underestimate the magnitude of autogenous shrinkage because of friction between the specimen and the mould. It is assumed that volumetric free deformations are isotropic. Thus volumetric free deformation corresponds to three times of the linear one. Zhutovsky and Kovler (2017) have shown that results obtained with both methods are close when considering the maturity of the samples. However, no clear evidence of this relation has been shown in the literature till now. This could be explained by the different artifacts of measurement coming from volumetric and linear methods. In the linear method, elongation and shortening of the sample is monitored with either direct measurement of the sample length change or by displacement measurement with e.g. LVDT, strain gauges (Boulay 2003). For correct consideration of the temperature variation, it is needed to have knowledge in depth of the thermal sensitivity of the test rig. The volume change of the cement paste or mortar is monitored by using hydrostatic weighing method. The sample is immersed in a bath containing liquid of known density and is supported to a scale. The volume change is measured through the apparent mass changes of the immersed sample. This method was significantly improved during the last decade in order to remove artifacts coming from the absorption of the water by the membrane (Mitani 2003), the liquid used for the thermal regulation (Lura and Durand 2006), bleeding of the cement paste (Le Roy 1995).

No standard was developed till now for the simultaneous monitoring of the autogenous strain and the coefficient of thermal expansion since setting. However, as mentioned in (Benboudjema et al. 2019), the ASTM C1698-09 (ASTM Standard C1698 2014) standard exists for the determination of the autogenous strain since setting on cement paste and mortar. No standard has been developed for concrete. For the CTE, the ASTM C531-00 (ASTM C531-00 2012) standard was developed for hardened mortar and concrete. Hammer and Bjøntegaard (2006) have developed a recommendation for the simultaneous characterization of the autogenous strain and the CTE. In addition to the requirements stated above, the test rig should be designed with the consideration of several other performance criteria:

- the side, length or diameter of the sample must be 4 times higher than the nominal maximum diameter of the aggregate;
- Thermal gradient between the core and the surface of the specimen and between the core of the specimen and the target temperature must be never more than 2 °C;
- Thermal dilation of the test rig and/or the measuring devices must be minimized and controlled. This means that the test rig and/or measuring devices must be calibrated;
- During the first 24 h after the final setting of the material, measuring intervals must be less than 15 min. Measurements must continue till an age of 2 weeks minimum;
- The resolution of the measurement devices must be less than 15 μm/m;
- In case of bleeding, arrangement must be taken in order to minimize its effect. Rotation of the specimen or removal of the bleed water on the top surface of the specimen at the end of the bleeding period can prevent effect of bleeding on the development of the autogenous strain and the CTE.

5.3.1 Cement Paste and Mortar

Volumetric and linear methods were developed for the monitoring of the autogenous strain and the coefficient of thermal expansion on cement paste and mortar. Loser, et al. (2010) have developed a volumetric method for which a sample with a mass of around 20–30 g is inserted in an elastic membrane. The sample is hung to a high accuracy scale (sensitivity of 0.1 mg and accuracy of 0.28 mg), then immerged in an oil bath and placed in a permeable box which reduces strongly the impact of oil circulation induced by the thermal regulation on the weight measurement. The temperature around the sample is measured and is considered as the temperature of the cement paste or mortar during the test for the determination of the autogenous strain and the coefficient of thermal expansion. In order to validate the test rig, results of the autogenous strain and the coefficient of thermal expansion were compared to results obtained on prisms (linear method) and a good agreement was observed. The test rig is shown in Fig. 5.6.

Fig. 5.6 Volumetric testing device designed for cement paste and mortar with a sample weight of 25–30 g (Loser et al. 2010)

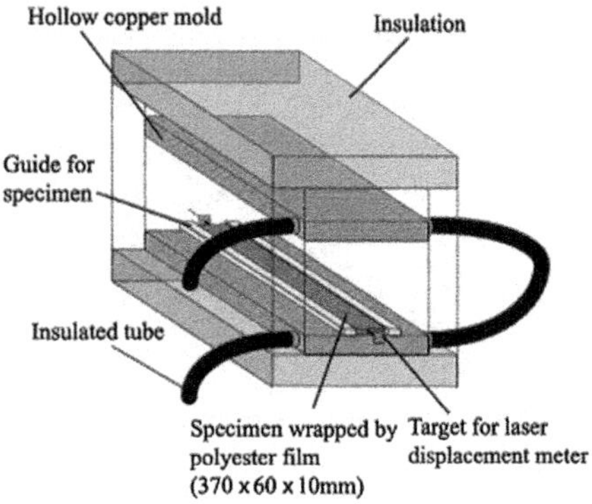

Fig. 5.7 Linear horizontal testing device designed for cement paste and mortar with laser displacement sensor in a 370 × 60 × 10 mm beam (Maruyama and Teramoto 2012)

Maruyama and Teramoto (2011) have developed a horizontal device for which the sample has a cross section of 10 mm × 60 mm and a length of 370 mm (Fig. 5.7). The sample is casted in a hollow copper mold in which water from temperature controlled water tank circulates. A layer of polyester with a thickness of 0.05 mm is placed between the sample and the mold in order to reduce the friction and to keep the material in sealed condition. The whole system is surrounded by an insulation to improve the efficiency of the thermal regulation. The temperature of the sample is recorded with two thermocouples placed in one extremity and in the center of the specimen. At both extremities of the sample, a measurement guide is anchored in the sample in order to monitor the displacement of the sample by means of two laser displacement sensors (resolution of 0.5 μm).

5.3.2 Concrete

Measuring the thermal dilation of concrete at early ages can represent an important challenge due to the fact that: (i) volumetric methods cannot be applied in non-homogeneous 'small' samples; (ii) the size of concrete specimens causes the cooling/heating processes to take a few hours and it is very hard to get more than a few measurements in the first 24 h of age (the CTE is evolving significantly during such measurement). Several test apparatus were developed in different laboratories for the determination of the autogenous strain and the CTE. In Table 5.1, the parameter of several test apparatus coming from the literature are presented. The main variations in the test apparatus come from:

- Dimension of the sample;
- Sample orientation (horizontal (H) or vertical (V));
- The system used to limit the friction between the mold and the specimen and to assure the sealing of the specimen;
- The thermal regulation;
- The measurement length;
- The displacement measurement system (internal or external and the sort of displacement sensor);
- The knowledge of the dilation of the test apparatus and the instrumentation (temperature compensation in the measurement);
- The temperature measurement of the sample;
- The need of sample preparation (for measurement after setting).

The dimensions of the sample are generally based on the nominal maximum diameter of the aggregate of the concrete and the length of measurement necessary to ensure sufficient measurement accuracy. In order to characterize the material since the setting or even before, the orientation of the sample is generally horizontal (ease of casting, no influence of the dead load of the specimen on the measurement). The friction between the sample and the mold is minimized by using specific material like the combination of plastic film and grease. The temperature of the concrete can be regulated by means of thermally controlled insulated mold (Koenders 1997; Lokhorst 1998; Bjontegaard 1999; Gutsch 2000; Hammer et al. 2003) or by placing the test apparatus in a environmental chamber controlled in temperature. Two types of instrumentation are used for the monitoring of the displacement: external sensor (displacement transducer) and embedded extensometer. Embedded extensometer behaves as an inclusion and should be small as possible and with a low rigidity in order to not restrain the displacement of the concrete being tested. Linear variable displacement transducers (LVDT) are the most used sensor for the measurement of concrete displacement (Boulay 2003) because of their accuracy and the fact that they are reusable. The temperature of the sample is generally measured in the center of the specimen by using a thermocouple while the temperature of the testing room is recorded only by few authors. However, when thermal variation are applied on a sample, thermal variation of the test rig occurs and should be considered for the correct determination of the CTE. Several examples of test apparatus designed for the monitoring of the autogenous strain and the coefficient of thermal expansion are shown in Fig. 5.8.

Table 5.1 Test apparatus for the determination of the CTE and the autogenous deformation

References	Sample dimension	Sample orientation	System to reduce friction and sealing	Thermal regulation	Measurement length	Displacement sensor	Temperature measurement
Koenders (1997), Lokhorst (1998)	150 × 150 × 1000 mm	H	Lubricating grease, plastic cover foil and sheet of felt	Thermally controlled insulated mould in a climate-controlled room	750 mm	4 LVDT's	3 thermocouples at mi-heigth of the sample
Gutsch (2000)	100 × 100 × 620 mm	H	2 layers of plastic foil with vaseline	Thermally controlled insulated mould in a climate-controlled room	Unknown	Quartz-glass tube or/and electric strain gauged glued	Unknown
Bjontegaard (1999), Hammer (2003)	100 × 100 × 500 mm	H	2 layers of plastic foil with talcum powder in between and an aluminium foil	Thermally controlled insulated mould in a climate-controlled room	470 mm	2 LVDT's	1 thermocouple in the center of the specimen
Kada et al. (2002)	100 × 100 × 400 mm	H	The sample is wrapped in plastic bags	The sample is immersed in a heat-controlled water bath	Unknown	Vibrating wire extensometer embedded in the sample	Temperature recorded at the center of the sample with the vibrating wire extensometer

(continued)

Table 5.1 (continued)

References	Sample dimension	Sample orientation	System to reduce friction and sealing	Thermal regulation	Measurement length	Displacement sensor	Temperature measurement
Cusson and Hoogeveen (2006, 2007)	75 × 75 × 295 mm	H	A layer of vaseline and a thin plastic film	Environmental chamber	Unknown	2 LVDT's	1 thermocouple in the center of the specimen and 1 thermocouple in the environmental chamber
Yeon et al. (2013)	Ø: 75 mm, h: 237.5 mm	H	A single layer of soft fabric (radial direction) and a styrofoam (longitudinal direction)	/	153 mm	Embedded vibrating wire strain gage	A thermistor is integrated to the displacement sensor
Viviani et al. (2006, 2007)	Ø: 315 mm, h: 1000 mm	V	/	/	800 mm	Embedded fibre-optic sensor	2 thermocouples (at the center and the extremity)

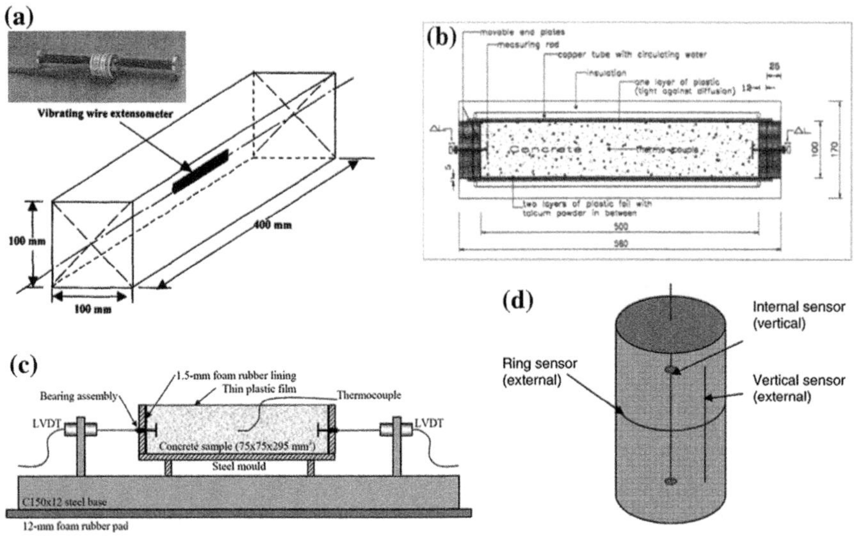

Fig. 5.8 Examples of test apparatus designed for the monitoring of the autogenous strain and the coefficient of thermal expansion, **a** Kada et al. (2002), **b** Bjøntegaard et al. (2015), **c** Cusson and Hoogeveen (2007) and **d** Viviani et al. (2007)

5.4 Test Protocol and Data Treatment

5.4.1 *Review of the Literature*

As explained by Boulay (2003), several authors have developed different methodologies to assess the CTE and the autogenous strain. A first method is based on the removal of the autogenous deformation by differences between two samples subjected to two histories of temperature. The maturity of both samples is considered when autogenous deformations are eliminated (Laplante and Boulay 1994). For consideration of the ageing and the main temperature effect, concrete properties can be expressed in function of the equivalent time t_{eq} (Eq. 5.4). Equivalent time is based on the Arrhenius equation and is function of the age of the material t, the evolution of the temperature T (°C), a reference temperature T_r (here 20 °C), the universal gas constant R (=8.314 J/mol/K) and the apparent activation energy E_a (J/mol).

$$t_{eq}(t,T) = \int_0^t \exp\left(\frac{E_a}{R} \cdot \left(\frac{1}{273 + T(s)} - \frac{1}{273 + T_r}\right)\right) \cdot ds \qquad (5.4)$$

However, the temperature has an effect on the evolution of the autogenous deformation and the CTE. Thus it is needed to have very close temperature histories which lead to a lack of accuracy in the determination of the CTE (Boulay 2003).

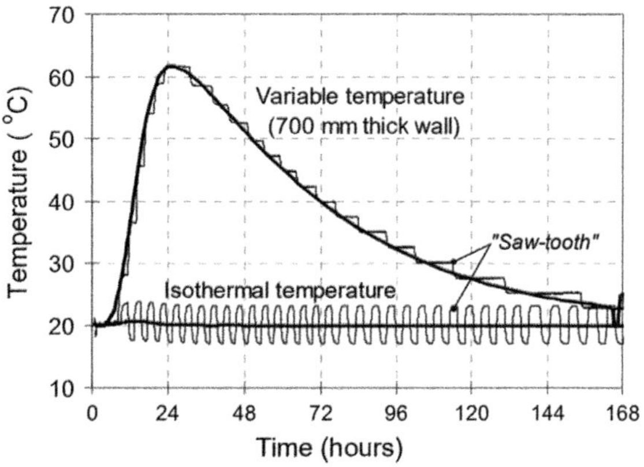

Fig. 5.9 Test procedures proposed by Hammer and Bjøntegaard (2006) for the characterization of the autogenous strain and the CTE

The second method uses only one sample on which thermal variations are applied (Bjøntegaard et al. 2004). The test procedure is based on the repeated application of thermal variation around a constant temperature (isothermal condition) or around a variable temperature (non-isothermal condition). Hammer and Bjøntegaard (2006) recommend imposing thermal variations as fast as possible in order to reduce the amount of autogenous deformation taking place during the thermal variation. The amplitude of the thermal variation should be high enough to insure a good accuracy in the determination of the CTE and small enough in order to reduce the influence of the temperature in the determination of the CTE. However, no specific value was provided for the frequencies and the amplitude of the thermal variation. An example of test procedure in isothermal and non-isothermal condition is shown in Fig. 5.9.

As presented by Delsaute and Staquet in (2017a), several protocols and mathematical treatments were developed for the determination of the CTE. In Table 5.2, the parameter of several test protocols coming from the literature are presented. The main variations in the test protocol come from:

- The amplitude of the thermal variation (ΔT)
- The duration of the isothermal phase (Δt)
- The nature of the material tested [cement paste (CP), mortar (M) or concrete (C)]
- The evolution of the temperature between the different cycles.

No specific recommendation is provided in the literature for the calculation of the CTE. When applying repeated thermal variation on a sample, the CTE was generally computed for hardened concrete by using Eq. 5.5 in which ε_{tot} is the total strain measured, α_{ta} is the CTE of the test apparatus and ΔT is the thermal variation applied. This method has the disadvantage to not consider the autogenous strain development of the sample during the thermal variation. In case of a decrease/increase in temperature of the sample when the specimen is shrinking, an overestimation/underestimation of the CTE is done which is especially significant at very early age.

$$\alpha_{\mathrm{T}}(\mathrm{t}) = \frac{\varepsilon_{tot} - \alpha_{ta} \cdot \Delta T}{\Delta T} \tag{5.5}$$

Table 5.2 Parameters of test protocols for the determination of the CTE and the autogenous deformation by using only one sample (Delsaute and Staquet 2017a)

References	ΔT, °C	Δt, min	Kind of device	Remark
Loser et al. (2010), Wyrzykowski and Lura (2013a, b)	6	90	Volumetric (CP and M)	Trapezoidal variation
Cusson and Hoogeveen (2006, 2007)	5	240	Linear (C)	
Ozawa and Morimoto (2006)	7	N.A.[a]	Linear (C)	Sinusoidal variation
Maruyama and Teramoto (2011, 2012), Maruyama et al. (2014)	5	25	Linear (CP and M)	Isothermal and various temperature histories with alternated triangular variation
Bjøntegård and Sellevold (2001)	6	120	Linear (C)	

[a]N.A. = not available

To overcome this issue, Maruyama and Teramoto (2011) have developed a test protocol presented in Fig. 5.10a for cement paste and mortar. Four thermal variations of 5 °C are applied in 100 min on a sample. The coefficient of thermal expansion is defined as the average slope between the sample strain and the temperature during the four thermal variations (Fig. 5.10b) and corresponds to the midpoint age of the 4 thermal variations. To decrease the fluctuation in the results, only results between a thermal variation of 0.5 and 4.5 °C is considered for the computation of the CTE. Such calculation reduces the influence of the autogenous strain in the determination of the CTE. However, such calculation method makes the assumption that the evolution of

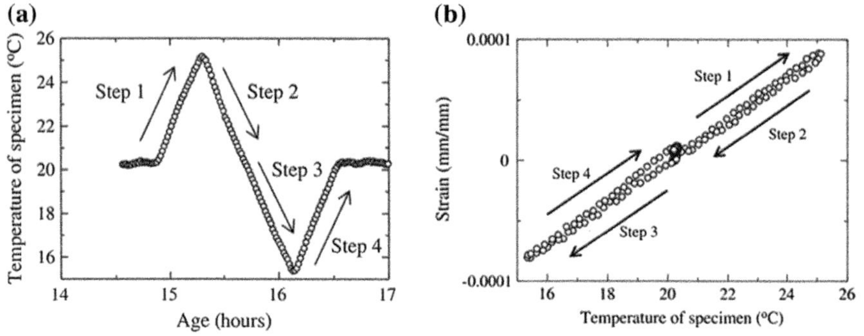

Fig. 5.10 Scheme of determination of the coefficient of thermal expansion of Maruyama and Teramoto (2011), **a** representative evolution of the temperature during thermal variations, **b** evolution of the temperature—strain relationship when thermal variations are applied

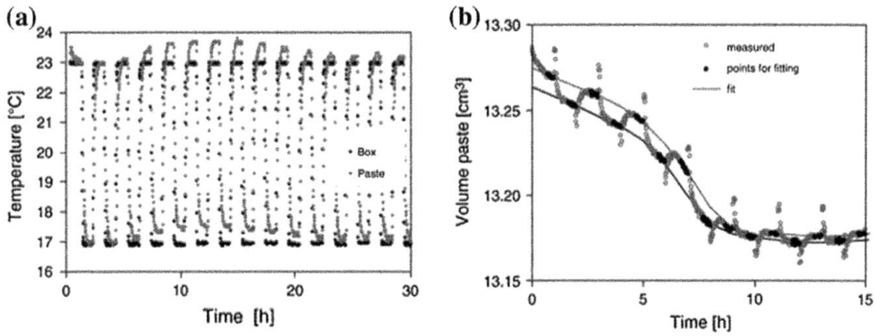

Fig. 5.11 Scheme of determination of the coefficient of thermal expansion of Loser et al. (2010), **a** representative evolution of the temperature, **b** evolution of the volume of the paste when thermal variations are applied

the autogenous strain is linear during the thermal variations which still induces small error in the determination of the CTE.

To solve this problem, the non-linear development of the autogenous strain should be considered in the calculation of the CTE. On cement paste and mortar, Loser et al. (2010) have applied repeated thermal variations of 6 °C (between 17 and 23 °C) since casting (Fig. 5.11a). Only data obtained at the end of each plateau of temperature (when no thermal gradient is assumed in the sample) have been considered (Fig. 5.11b). By means of cubic smoothing splines, two encasing envelopes of the volume of the paste are defined for a curing temperature of 17 and 23 °C. The CTE is then defined by knowing both the difference in volume and temperature between the two envelope curves. An example of such treatment is presented in Fig. 5.11 for volumetric method (Loser et al. 2010). This method has the disadvantage to not consider the maturity of the sample. During thermal variation, an acceleration or deceleration of the hydration process is induced. For a good consideration of the maturity of the sample, strain should be considered according to the equivalent age (Eq. 5.4).

Recently, Zhutovsky and Kovler (2017) have developed a new methodology based on ultrasonic pulse velocity measurement for the assessment of the coefficient of thermal expansion. The calculation of the CTE is based on poromechanics theory. This approach has the advantage of defining the CTE without having to perform a decoupling with the autogenous strain. However, this method does not consider the self-desiccation of the material and thus the increase of the coefficient of thermal expansion which leads to an underestimation of the CTE after the final setting of the cement paste.

5.4.2 Development of a New Test Protocol for Concrete

At the concrete scale, no specific data treatment associated to a test protocol was presented in the literature to properly define the development of the CTE and the autogenous strain since setting. For that reason, the development of a new experimental protocol and a new mathematical analysis strategy were defined to characterize the autogenous deformation and the coefficient of thermal expansion during the same single test at the Université Libre de Bruxelles (Delsaute and Staquet 2017a). This new methodology considers correctly the evolution of the autogenous strain when thermal variations are applied to the sample since the setting.

The free strains of the concrete or mortar are measured from casting using the BTJADE (from the French acronym 'BéTon au Jeune Age, Déformation Endogène') device (Boulay 2012). The test rig (Fig. 5.12) is composed of a vertical flexible corrugated PVC mold to monitor the free strain and fixed metallic parts. The whole frame is placed in a temperature controlled bath and is designed to accommodate 3 tests rigs simultaneously. The sensors for each channel are the central displacement, the temperature of the sample with a thermocouple, the temperature of the surrounding water in the tank with a platinum probe and the temperature in the air above the cover of the tank, near the displacement transducer, with a platinum probe also. For a good consideration of the heterogeneity of the concrete, the diameter of the cylinder has to be greater than the coarse aggregate (a factor 5 is a minimum). For that reason, a diameter of 125 mm is used. The base length of the sample is approximately 225 mm. More details about the BTJADE are given in Boulay (2012).

When measurements are done, free strain of the concrete is defined by removing the influence of the temperature variations itself inside the tank and the influence of the temperature of the room on the displacement transducer located above the cover of the tank. As the test is carried out in sealed condition, the free strain corresponds to the sum of the autogenous strain and the thermal strain. Therefore the total strain of the concrete ε_{tot} (m/m) is defined as:

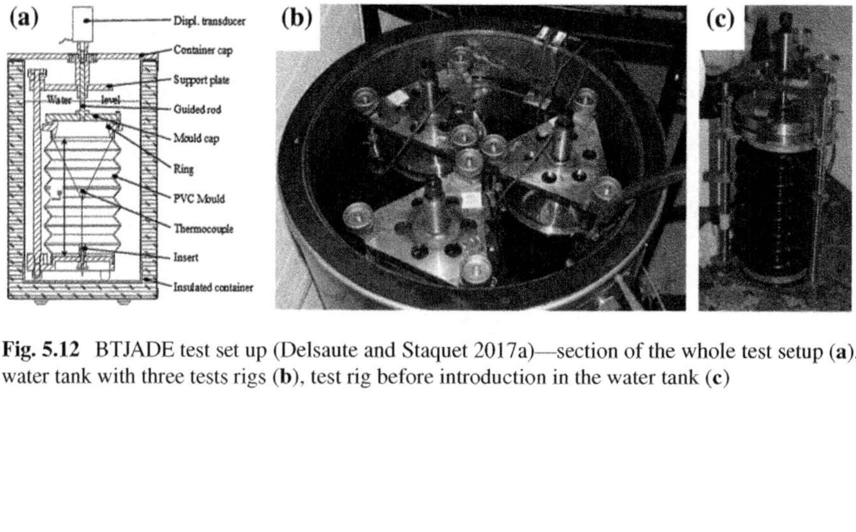

Fig. 5.12 BTJADE test set up (Delsaute and Staquet 2017a)—section of the whole test setup (**a**), water tank with three tests rigs (**b**), test rig before introduction in the water tank (**c**)

$$\varepsilon_{tot} = \varepsilon_{au} + \alpha_c \cdot \Delta T_c = \frac{\Delta L}{L_0} - \alpha_{wt} \cdot \Delta T_{wt} - \alpha_{at} \cdot \Delta T_{at} \tag{5.6}$$

where ΔL is the displacement measured by the displacement sensor (mm), ε_{au} is the autogenous strain (m/m), α_{c} is the coefficient of thermal expansion of the concrete (m/m/°C), ΔT_{c} is the thermal variation of the concrete, the index wt is related to the thermal variation of the water in the tank and the index at is related to the thermal variation of the air above the cover of the tank. The thermal dilation coefficient of the test rig α_{wt} and α_{at} are defined by replacing the concrete sample by a stainless steel rod whose the coefficient of thermal expansion is well known. α_{wt} is determined with a room temperature constant and steps of temperature imposed to the water inside the tank. α_{at} is obtained with a constant temperature of the water inside the tank and steps of temperature imposed to the air of the room (Boulay 2012).

The base length of the sample is approximately 225 mm and the inner diameter 125 mm. When measurements are done, the free strain of the concrete is defined by removing the influence of the temperature variations themselves inside the tank and the influence of the temperature of the room on the displacement transducer located above the cover of the tank. As the test is carried out in sealed condition, the free strain corresponds to the sum of the autogenous strain and the thermal strain.

At the basis the test rig has been design for the monitoring of the autogenous strain. In order to validate the test apparatus, results coming from a horizontal testing device has been compared as presented in (Delsaute et al. 2016). A general very good agreement was observed between both results.

The test protocol is based on repeated thermal variations which are applied since the casting of the sample. The thermal variations are designed by optimizing the three following parameters: the duration of the temperature increase, the thermal variation amplitude and the duration of the isothermal phase in the tank. The duration of the temperature increase in the tank must be as fast as possible in order to minimize concrete autogenous deformation amplitude between two sets of measurements and to maximize the number of thermal variation at early age when the CTE evolve significantly. The thermal variations amplitude must be high enough to allow measuring significant displacement (digital displacement transducer with accuracy of 1 μm over the whole stroke) and small enough in order to reduce the influence of the temperature variation on the development of the autogenous strain (Jensen and Hansen 2001). The duration of the isothermal phase in the tank must be high enough to avoid any thermal gradient inside the concrete sample at the end of the plateau of temperature.

Several preliminary tests were performed on hardened concrete with the BTJADE device in order to optimize these parameters. The test protocol is presented in Fig. 5.13. The recording of data provided by each thermal and displacement sensor is done every minute. The phase of temperature change is limited to 10 min; the magnitude of the temperature variation is ±3 °C and the duration of the isothermal phase is 150 min. The duration of the isothermal phase can be adapted according to the composition. For composition with high thermal conductivity, temperature stabilization in the concrete is already reached after 120 min. At this period, an identical

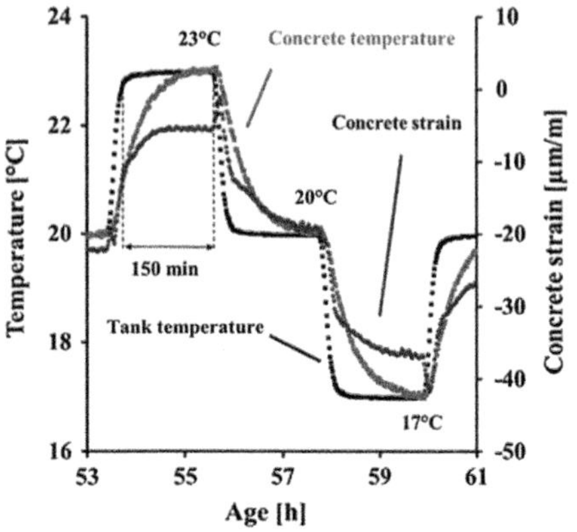

Fig. 5.13 Preliminary test on hardened concrete

temperature is measured in the tank and in the concrete sample for hardened concrete as shown in Fig. 5.13. As the thermal inertia of the test rig is lower than the one of the concrete sample, a faster change in temperature takes place in the test rig. This induces an error in the computation of concrete strain (Eq. 5.6) when the temperature of the water changes in the tank. This is observed in the evolution of the concrete strains in Fig. 5.13 at the beginning of each thermal variation.

Therefore no thermal gradient occurs at the end of the plateau of temperature. 10 cycles were performed on this hardened concrete for which it is assumed that no significant autogenous strain occurs (the age of the sample was higher than one month at the start of the test). Results of the CTE are presented in Fig. 5.14. An average value of 11.6 µm/m/°C is obtained. The minimal value is 11.1 µm/m/°C and the maximal value is 12.0 µm/m/°C. The standard deviation is very low and is equal to 0.3 µm/m/°C which corresponds to 2.3% of the average value of the CTE. Through these results, it is concluded that such protocol is accurate enough for the assessment of the CTE on hardened concrete. As the variation of the thermal conductivity with aging of concrete is limited (De Schutter and Taerwe 1995; Briffaut et al. 2012), it is assumed that no thermal gradient occurs because of the temperature variation in

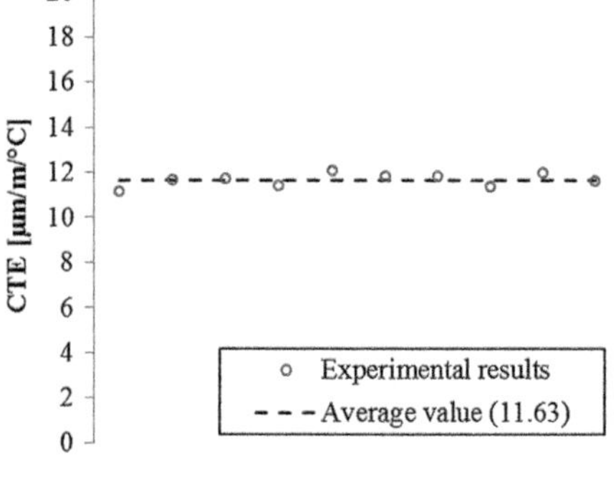

Fig. 5.14 Results of the CTE for a hardened concrete

the tank during the hardening process of the concrete. The thermal gradient due to the hydration is assumed to be unchanged.

This last assumption is verified on an ordinary concrete on which two tests was performed with different temperature histories. The composition of the concrete is presented in the references (Boulay et al. 2014). For the first test, a constant temperature of 20 °C is imposed. The second test uses the protocol of temperature illustrated previously in Fig. 5.13. For each test, two test rigs are used in order to assure the repeatability of the results. For a good comparison, both tests use a concrete coming from the same batch. The scheme of the experiment is presented in Fig. 5.15.

The test starts approximately 1 h after mixing. Results related to the evolution of the temperature at early age are presented in Figs. 5.16 and 5.17. Figure 5.16 shows the evolution of the temperature in the water tank and inside the sample. For each thermal variation, the temperature of the sample reaches the temperature imposed by the thermal regulation. At the end of the plateau of temperature, the temperature is stable in the sample. In Fig. 5.17, the temperature inside samples from both tests is compared. Moreover the comparison is extended to two additional curves which correspond to the temperature of the sample with a constant thermal cure of 20 °C

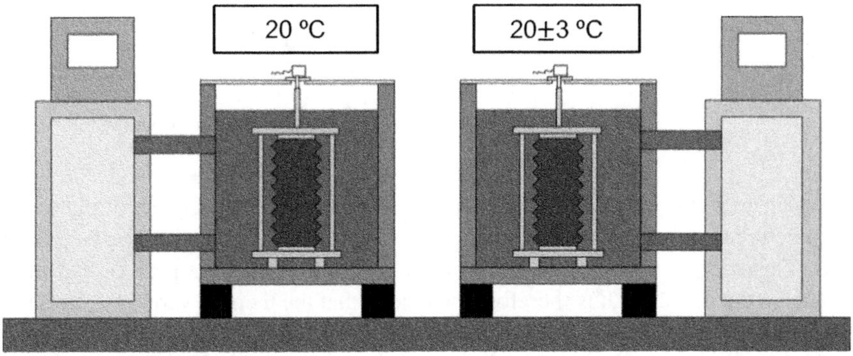

Fig. 5.15 Scheme of the experiment

Fig. 5.16 Thermal variation in the water tank and inside the sample

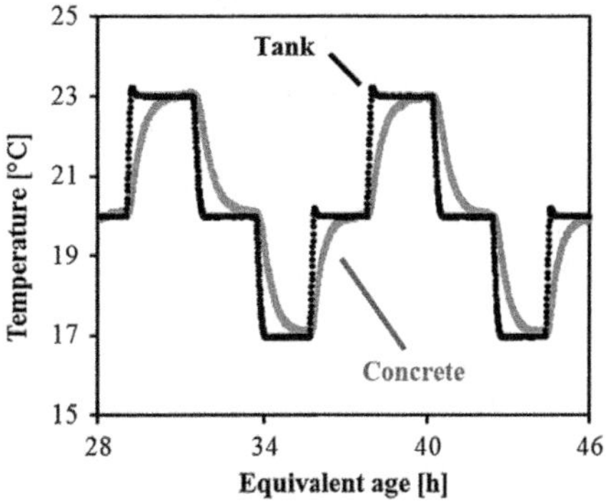

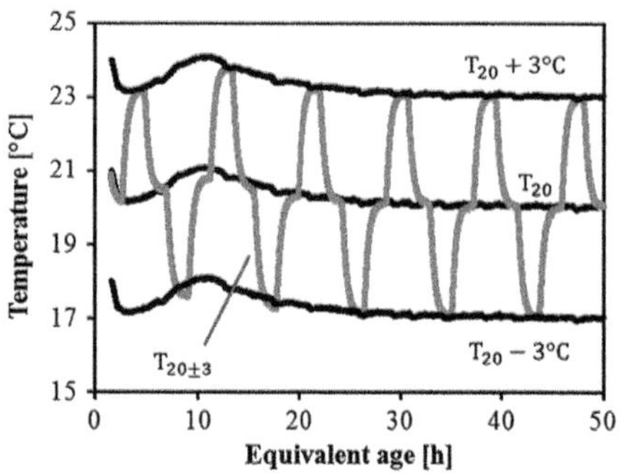

Fig. 5.17 Evolution of the temperature of the sample for both tests

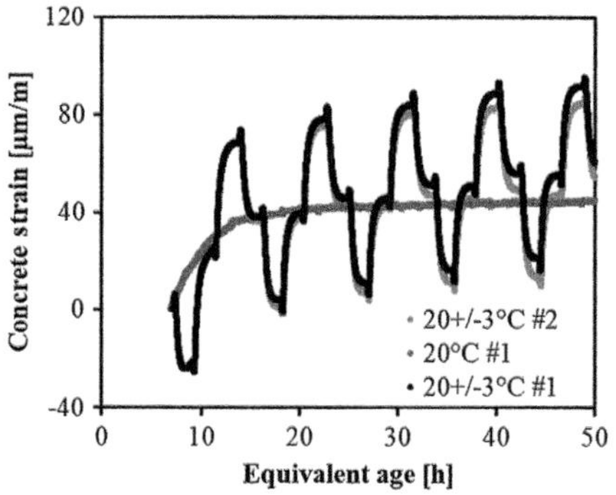

Fig. 5.18 Evolution of the total strain for both temperature histories

on which an addition or a subtraction of 3 °C is operated. For each end of plateau of temperature, a very good correspondence of the temperature is obtained between both tests. This very good correspondence is also observed during the peak of hydration occurring the first 24 h. It is therefore concluded that the thermal variations cause no significant change in the thermal gradient due to hydration.

The total concrete strains from both tests are compared in Fig. 5.18 since final setting time. Results obtained at the end of the plateau of temperature at 20 °C are very close to the results obtained at a constant temperature of 20 °C. It is therefore possible to fit results obtained at the end of the plateau at 20 °C in order to define the total strain obtained for a cure at 20 °C. The same methodology can be used to define the free strain at 17 and 23 °C. In Fig. 5.19, the temperature and the total strain in the sample with repeated thermal variation are compared. A good coherence between both evolutions is observed. At the beginning of each thermal variation, a discontinuity in the evolution of the total strain is observed. This discontinuity comes from the computation of the total concrete strains. When the displacement induced by the dilation/contraction of the test rig is removed from the total measurement, it is assumed that each component of the test rig has an instantaneous thermal response. The thermal inertia of each component is not considered. That is why such variations in the computed total concrete strains are observed. Therefore only results obtained at the end of the plateau of temperature have a physical meaning.

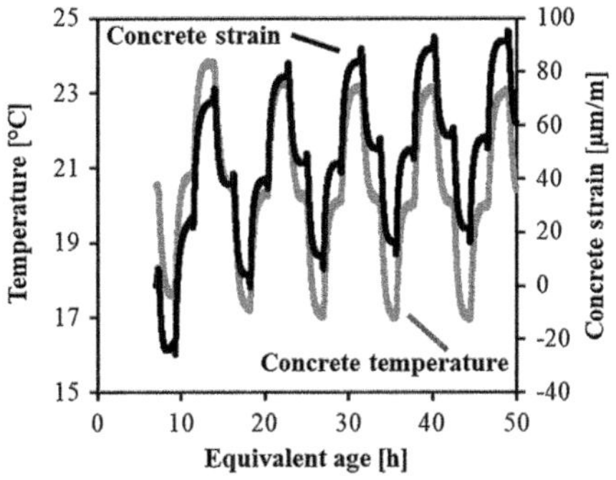

Fig. 5.19 Evolution of the total strain and the temperature for the test with repeated thermal variations

The coefficient of thermal expansion (CTE) and the autogenous strain are computed in several steps. First the concrete strain is defined by removing thermal displacement of the test rig (an acquisition is carried out each minute). Then only physical values of the total strain are considered. This physical value corresponds to the moment when the test rig and the concrete have no thermal gradient and have a constant temperature. This moment corresponds to the end of the plateau of temperature. An average value of the last five data recorded during the plateau in temperature is considered in order to reduce the noise coming from the measurement of the temperature and the displacement sensors. To define the autogenous strain and the CTE, it is needed to have two temperature histories. For that purpose, a fictive temperature history is created by considering only the values obtained during the plateau of temperature at 20 °C. With a spline interpolation, it is possible to create new data (total strain and temperature) which correspond to a constant cure at 20 °C. Therefore a predictive value of the total strain and temperature is obtained for a constant temperature of 20 °C in the tank. For both temperature histories, the variation of the total strain, the autogenous strain and the thermal strain between two plateaus of temperature is given in Eq. 5.7.

$$\Delta\varepsilon_{tot} = \Delta\varepsilon_{au} + \alpha_c \cdot \Delta T \tag{5.7}$$

The effect of the temperature on the evolution of the autogenous deformation and the CTE (Bjøntegaard 2011) was previously demonstrated by several authors. However the difference of evolution was highlighted for a very large difference of temperature. For a very small difference of temperature, as here 3 °C, the effect of the temperature on the evolution of the autogenous deformation and the CTE is very low. In consequence it could be considered that, for both temperature histories, the evolution of the autogenous strain during 150 min and the value of the CTE is the same. Therefore, with value of the thermal variation and total strain variation for both temperature histories, the CTE is defined as expressed in Eq. 5.8. In this equation, the index 1 and 2 are relative to the two different temperature histories.

$$\alpha_c = \frac{\Delta\varepsilon_{tot,1} - \Delta\varepsilon_{tot,2}}{\Delta T_1 - \Delta T_2} \tag{5.8}$$

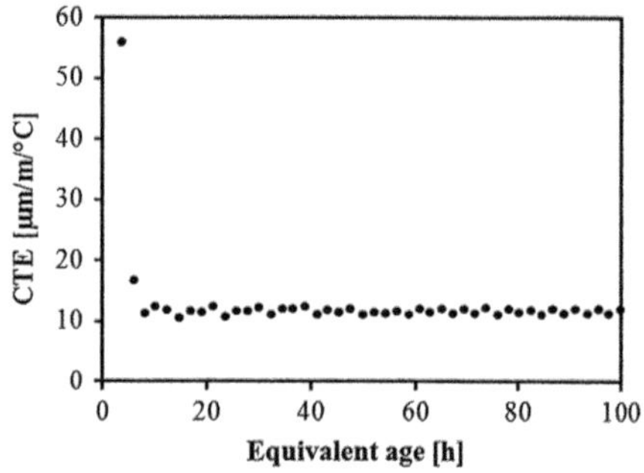

Fig. 5.20 Evolution of the coefficient of thermal expansion according to the equivalent age

Results of the CTE are given in Fig. 5.20. The CTE changes strongly during the few first hours after casting till an age of 8 h. Afterwards, the CTE follows a very constant value. The age when this change takes place corresponds to the setting of the concrete. This observation was already done by Loser et al. (2010) for cement paste and by Delsaute and Staquet (2017a) for concrete. For this composition, the CTE seems very constant after setting. This is in coherence with results obtained in the literature on ordinary concrete. Indeed, for concrete with a high water-cement ratio, no strong variation of the CTE occurs because the value of the relative humidity stays high during the whole hydration process. Figure 5.21 presents only results of the CTE obtained after setting. The scattering in the results is very low. However just after setting a higher scattering is observed. It is due to the strong increase of the autogenous strain during this period. An average value of 11.6 µm/m/°C is obtained for the CTE.

When the CTE is defined, the evolution of the autogenous deformation is defined with the removal of the thermal strain on the total strain (Eq. 5.7). Figure 5.22 presents the evolution of the autogenous strain by considering a constant value of the CTE of 11.6 µm/m/°C. Results are strongly dependent on the time when the autogenous strains are initialized. In order to correctly compare results, each curve is initialized at an equivalent age of 140 h (Fig. 5.23). Results from both tests give very equivalent results and show a very good repeatability.

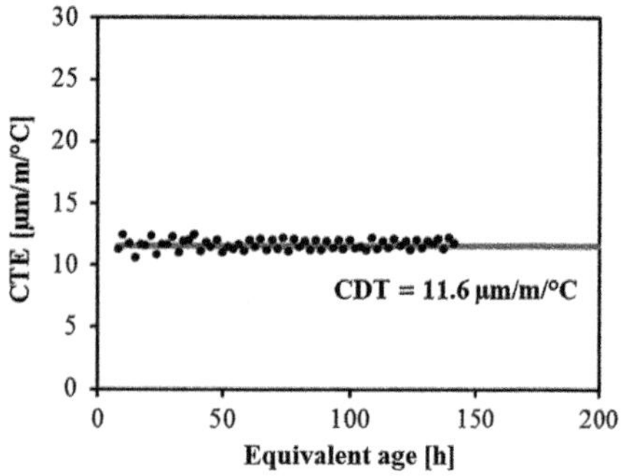

Fig. 5.21 Evolution of the coefficient of thermal expansion after setting

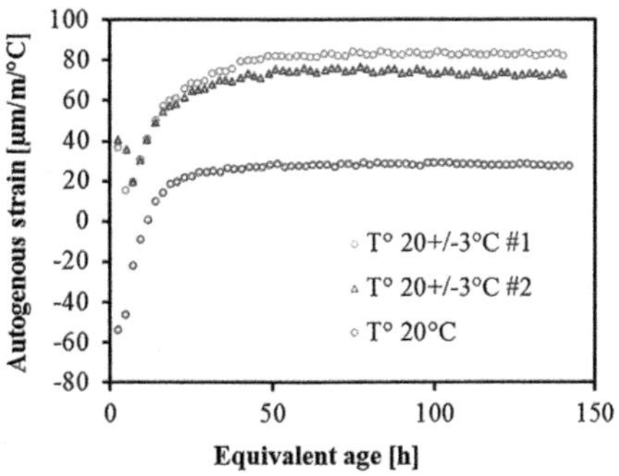

Fig. 5.22 Autogenous strain

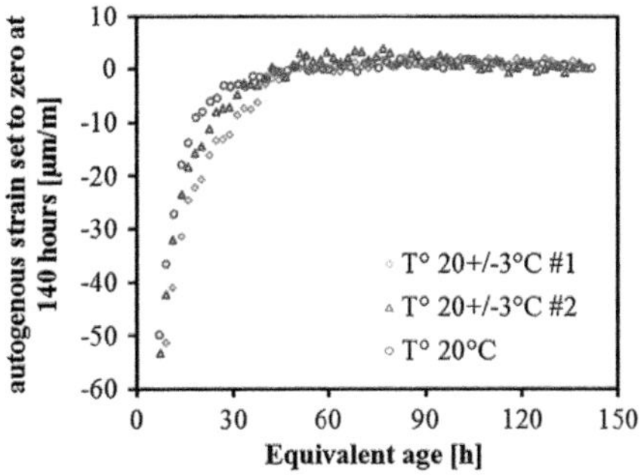

Fig. 5.23 Normalized autogenous strain at 140 h

It is then concluded that the new methodology developed here is able to monitor the evolution of the CTE and the autogenous strain since setting in one single test. The general scheme of the test protocol and the treatment of the results is presented in Fig. 5.24.

5.5 Investigations and Results

5.5.1 Sensitivity Analysis on the Determination of the CTE Induced by the Data Processing

As presented above, different methodologies have been developed and used in the past for the determination of the CTE since casting. At very early age, the CTE is very sensitive to how the data are processed. To illustrate and quantify the importance of the treatment of the data, the coefficient of thermal expansion of a concrete was computed by using six different methods.

The first method does not consider the evolution of the autogenous strain taking place during the thermal variation. The second method is inspired from the work

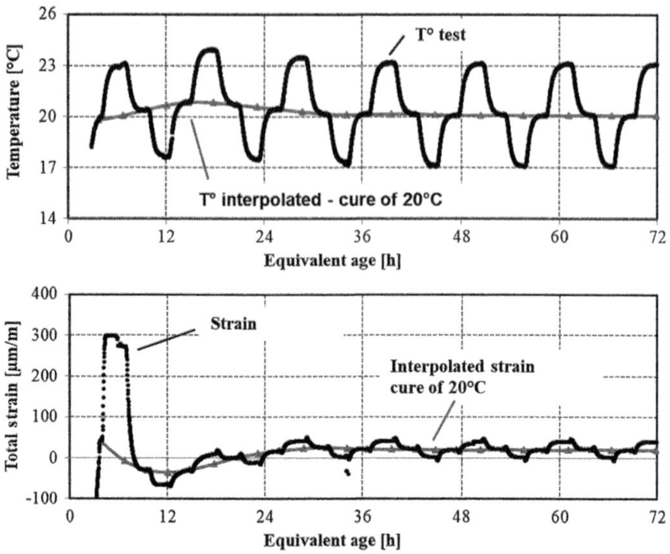

Fig. 5.24 Evolution of the temperature and the concrete strain during the test (Delsaute and Staquet 2017a)

of Maruyama and Teramoto (2011) for which the slope between the strain and the temperature of the concrete is computed between 10 and 90% of the thermal variation in order to define the CTE. Third, the method developed by Loser et al. (2010) has been adapted to the testing methodology developed at ULB. Three envelope curves considering data at 17–20 and 23 °C were computed as by Loser et al. (2010). The computation of the CTE has been carried out as the average of the CTE obtained between each envelop curve and the maturity of the sample was not considered in the computation. Ultimately, the novel method presented in this chapter was used with three sorts of interpolation: linear interpolation (method 4), spline interpolation (method 5) and Piecewise Cubic Hermite Interpolating Polynomial (PCHIP (Fritsch and Carlson 1980)—method 6). The studied concrete is a Very High Performance Self-Compacting Concrete (VHPSCC) which was designed in 2005 (Staquet et al. 2005) for pre-cambered beams. The decoupling between the autogenous and thermal strain is very complex for such composition due to the fast and intense development of the autogenous strain during the first 24 h after setting (Roziere et al. 2015).

In Fig. 5.25, the development of the CTE computed with each method is presented according to the equivalent age. These results correspond to the average obtained from measurement performed on 2 samples. The general trends observed are the same for each method. A strong decrease of the CTE is observed during a short period which is related to the setting of the material. At an age of around 15 h, the CTE reached a minimum and then increases slightly. Till an age of 20 h, the development of the CTE is significantly different according to the method used for the data processing. This is highlighted in Fig. 5.26 in which only results obtained before an equivalent age of 30 h

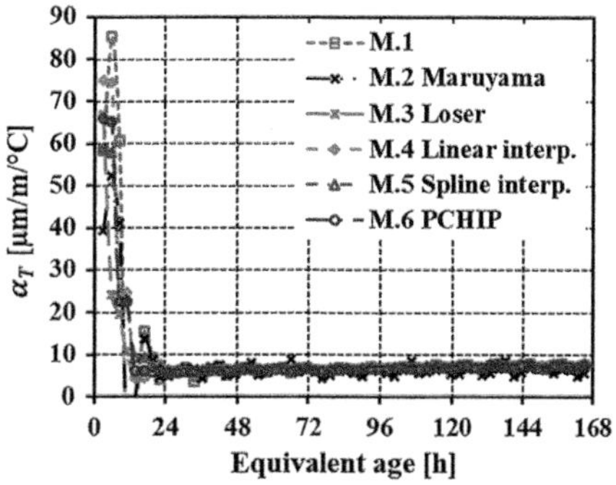

Fig. 5.25 Comparison of 6 methods for the determination of the CTE

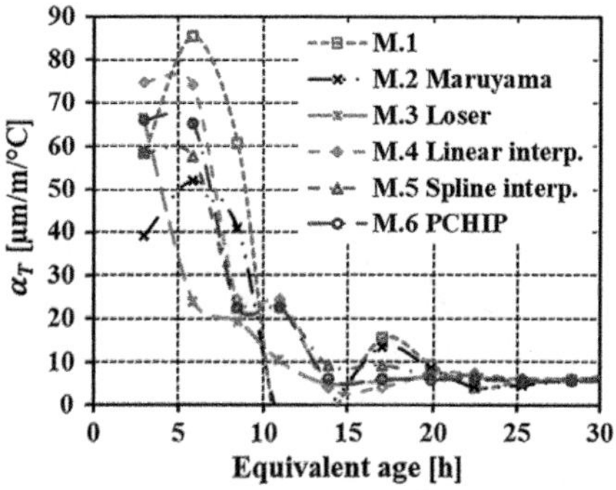

Fig. 5.26 Zoom during the first 30 h of the Fig. 5.25

are shown. In addition, in Fig. 5.27, the development of the autogenous strain and the autogenous strain rate is presented. CTE results got with the first and second methods show quite different trends in comparison to other methods of data processing. An increase and then a significant decrease of the CTE are calculated during the first 10 h. Negative CTE values are obtained between an equivalent age of 10 and 15 h. These non-physical variations are related to the non-consideration of autogenous deformations in the treatment of the results. Indeed, during this time interval, the autogenous strain rate is very intense (Fig. 5.27). When calculating the CTE, the

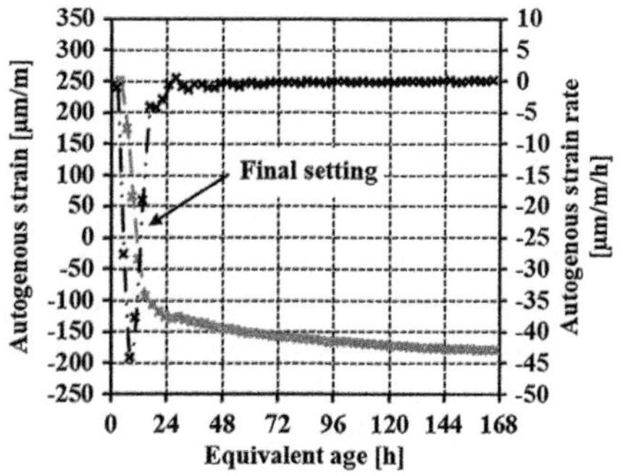

Fig. 5.27 Development of the autogenous strain and the autogenous strain rate

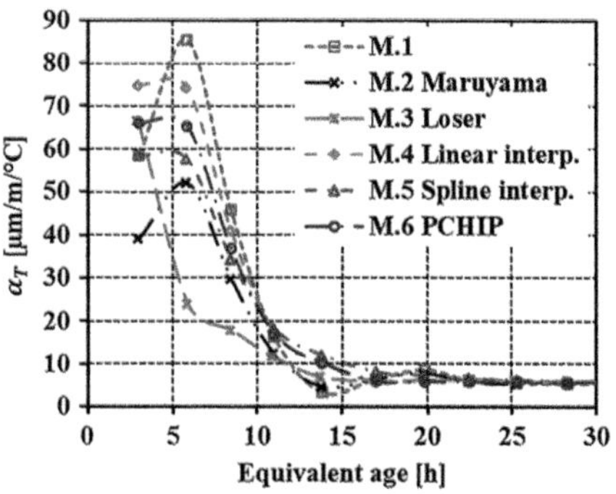

Fig. 5.28 Evolution of the CTE with the use of moving average on 4 points

concrete deformations due to temperature variations are underestimated during an increase in temperature and overestimated during a decrease in temperature. This underestimates and overestimates respectively the value of the CTE. The influence of the maturity of the concrete sample on the determination of the CTE is then studied with the third method. A faster decrease of the CTE is observed during the first 15 h. Then the value of the CTE is similar to the value obtained with the other methods. The influence of the interpolation method is studied with the methods 4–6. For each method, very similar value of the CTE is obtained after an age of 6 h and the decrease of the CTE takes place between an age of 6 and 15 h. Significant difference in the amplitude of the CTE between the three methods is only observed during the 20 first hours after casting.

The error in the computation of the CTE induced by the data processing used for the removal of the autogenous strain is reduced by applying 4 points moving average method on data obtained after the induction period (Fig. 5.28). It is generally observed that the spread in the results obtained is minimized and especially between an equivalent age of 10 and 20 h. No negative value of the CTE is then computed with the method 1 and 2. The early decrease of the CTE is still observed with the method 3.

The importance of the data processing in the determination of the coefficient of thermal expansion is quantified with the computation of the relative difference in the evolution of the CTE between the 6 methods. For each thermal variation, the mean value of the CTE obtained with the 6 methods of data processing $\alpha_{T,m}$ is computed. The relative change in the evolution of the CTE is given in Fig. 5.29. A maximal relative change of around 60% is obtained with the first and the second method at an equivalent age of 14 and 6 h respectively. At very early age, the relative difference is generally lower for the method 6. For later ages, a very limited relative difference is obtained with each method except for the method 2. The origin of this divergence comes from the difference of thermal inertia between the test rig, the tank and the concrete sample. Concrete strains are computed according to Eq. 5.6 for which it is assumed that the thermal dilation of the test apparatus occurs directly when thermal variation are applied. With such computation, it is not possible to correctly define the concrete strain when the temperature of the water in the tank changes. Therefore, the

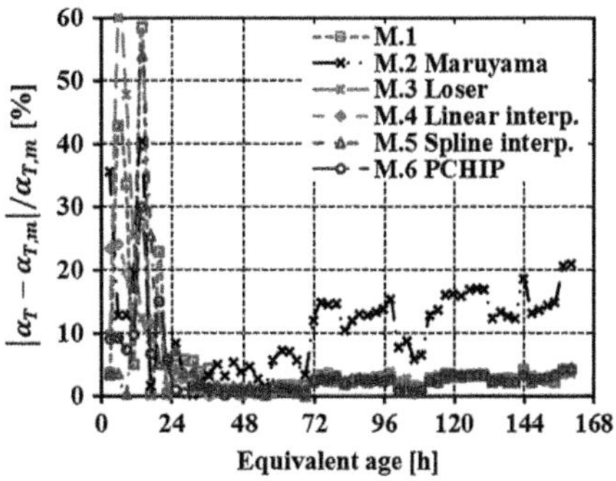

Fig. 5.29 Relative difference in the evolution of the CTE with the different method of data processing

second method of data processing cannot be used for the computation of the CTE at early age and at later ages. Such method can only be used if the concrete strains are well defined when the temperature varies. The use of this method leads here to an underestimation of the CTE during the hardened stage. In addition, for each method, the standard deviation s is computed according to Eq. 5.6 where N is the number of thermal variation applied during the test.

$$s = \sqrt{\sum_{i=1}^{N} \frac{\left(\alpha_T(t_i) - \alpha_{T,m}(t_i)\right)^2}{N}} \tag{5.9}$$

The standard deviation is given in Fig. 5.30 for data got till an equivalent age of 30 or 160 h. The methods developed for cement paste and mortar (methods 2 and 3) have higher standard deviation. This highlights the higher scattering in the results obtained when the autogenous strain development and the maturity of the concrete are not considered accurately. In addition, as presented by Loser et al. (2010), how concrete strains are interpolated when thermal variations are applied has a significant impact in the determination of the CTE. It is observed that the use of PCHIP decreases the standard deviation in comparison to linear interpolation and spline interpolation especially at early age. In conclusion, the method 6 of data

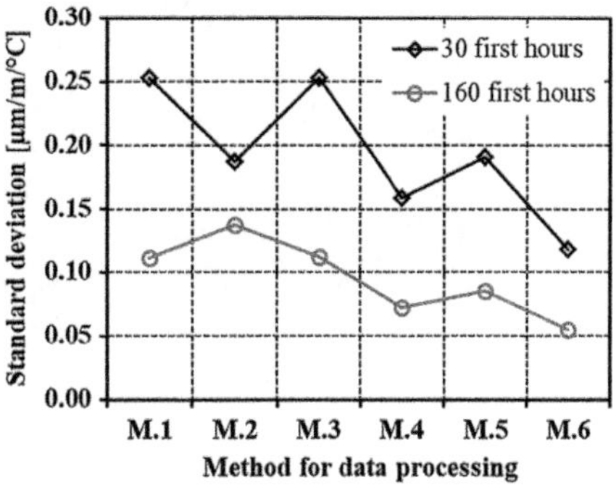

Fig. 5.30 Standard deviation of the CTE obtained with each method for test duration of 30 h and 168 h

processing is recommended for the determination of the CTE and the autogenous strain when thermal variations are applied on a sample.

5.5.2 *Extension to Cement Paste and Mortar Scale*

For the study of the multiscale aspect of the free deformation, a similar strategy has been developed for the monitoring of the CTE and the autogenous strain of cement paste and mortar. As presented in the Sect. 5.3.1, several devices were developed for the monitoring of the autogenous strain. One of the most used at the cement paste and mortar scale is the Autoshrink device developed by Mejlhede Jensen and Freiesleben Hansen (1995). A thermal regulation has been designed in order to apply thermal variations on the test rig (Milenkovic et al., no date; Delsaute 2016; Königsberger et al. 2018). The temperature of the sample is controlled by a flow of a specific liquid for thermal regulation circulating in a convolution system which surrounds the Autoshrink device. In order to improve the efficiency of the thermal regulation, the whole system is surrounded by a box with thermal insulation which limits exchanges with the ambient environment. The equipment is located in an air-conditioned room with a control system of the temperature and the humidity. The thermal regulation and the Autoshrink device are presented in Fig. 5.31. A thermal calibration of the whole test rig has been carried out by using a stainless steel rod for which the CTE is known. For the measurement of the displacement and the temperature, a Solartron network is used. It is composed by three displacement sensors Solartron LE2 and two temperature sensors. A thermocouple is used to monitor the temperature of one sample (embedded in the centre at mid-length) and one PT100 is used to monitor the temperature of the air inside the convolution system.

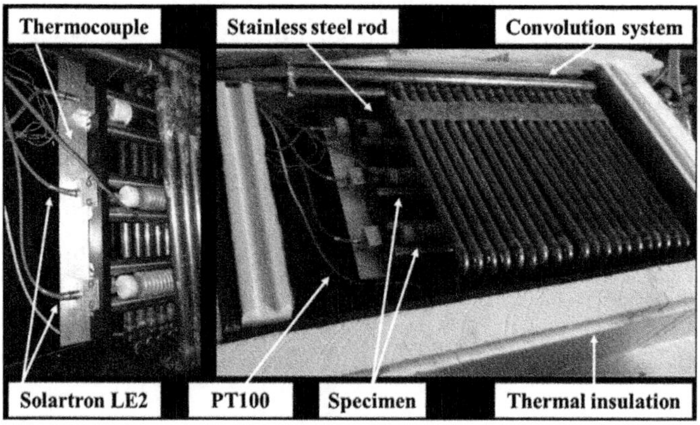

Fig. 5.31 Autoshrink device with thermal regulation developed at ULB (Delsaute 2016)

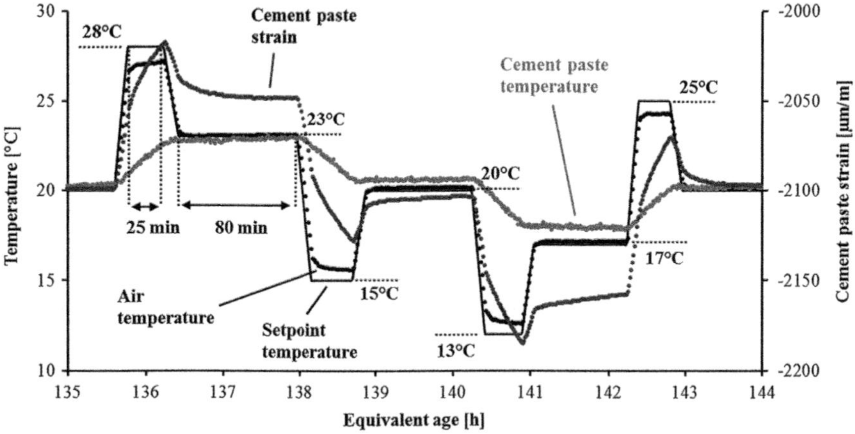

Fig. 5.32 Test protocol for cement paste and mortar with Autoshrink device

The thermal regulation is not performing as the one in the BTJADE system. For that reason, some adaptations have been made to the test protocol (Fig. 5.32). A first thermal variation of ±8 °C is applied in 10 min. The temperature is kept constant during 25 min. Then a thermal variation of ∓5 °C is applied in 10 min. Finally the temperature is kept constant during 80 min. The final magnitude of the temperature variation is ±3 °C and the duration of the isothermal phase is 80 min. Temperature stabilization in the cement paste or mortar is well reached after 80 min as shown in Fig. 5.32.

5.5.3 Correlation Between the Early Development of the CTE and the Autogenous Strain and the Setting

The setting period is commonly defined according to the results of penetration tests on cement pastes or mortars and corresponds to a progressive coalescence of a continuous path of hydrates. At the beginning of this period (initial setting), the concrete stiffness is almost inexistent while, at the end of this period (final setting), the concrete starts to stiffen (Hansen 2011). Correlations between the early development of the CTE, the autogenous strain and the setting had been made by several authors. Remarkable points were observed by different authors. Sant et al. (2006) have shown that final setting time can be defined using the rate of autogenous strain development in a cement paste. The final setting time was identified as the minimum of the derivative of the autogenous strain. Several authors (Loser et al. 2010; Wyrzykowski and Lura 2013a, b) have also associated the final setting time with the start of the decrease of the CTE from the liquid to the solid phase in cement paste. Delsaute and Staquet (2017a) have compared the evolution of the CTE and the autogenous strain to the initial and final setting obtained with penetration resistance [according to ASTM C304 standard (ASTM C403 2008)] and with ultrasonic measurement using P-wave and S-wave (Carette and Staquet 2015; Mohamed et al. 2017) for concrete composition with different W/C ratio and different type of binder. For each composition, the initial setting time has been associated to the moment when the value of the CTE does not correspond anymore to a liquid. The final setting time is associated to a minimum of the autogenous strain obtained before the swelling of the concrete. This value corresponds to a zero value of the derivative of the autogenous strain. Therefore, results of the autogenous strain and the CTE obtained by means of repeated thermal variation could be used for the determination of the initial and final setting time. However, the duration between two thermal variations must be reduced strongly at very early age in order to define both parameters accurately.

5.5.4 Further Recommendation for the Monitoring of the CTE and Autogenous Strain at Very Early Age

When the rate of the free strain is very high, as it is the case before (chemical shrinkage) and after (autogenous strain) setting, the computation of the CTE and the autogenous strain is very sensitive to how the data are processed (Sect. 5.5.1). To override this problem it is recommended to apply thermal variation faster at very early age even if their amplitude is lower in order to get a faster stabilization of the temperature in the sample. This will reduce the influence of the method used to treat the data. In addition, this will improve the assessment of the initial and final setting time. After this period, thermal variations of ± 3 °C each 160 min allow defining accurately the development of the autogenous strain and the CTE.

5.6 Conclusion and Outlook

The volume changes of cementitious materials in sealed conditions are governed at early age by the evolution of the coefficient of thermal expansion (CTE) and the autogenous strain. The most recent advances on the physical mechanisms associated to their development are reported. The recent devices and test protocols developed for the simultaneous monitoring of the coefficient of thermal expansion and the autogenous strain at three different scales (cement paste, mortar and concrete) are also presented. The general testing methodologies bear some similarities but major differences remain in the test set up designs, in the testing processes and also in the data processing. Based on the physical mechanisms, existing test facilities and test protocols, a new test protocol and its associated data processing has been developed for the monitoring of the autogenous strain and the CTE. This new methodology consists to apply repeated thermal variation of ± 3 °C each 160 min on a concrete sample with the device so-called BTJADE. Thermal and autogenous strains are distinguished by creating a fictive thermal cure at 20 °C from the experimental results by using interpolation method and by considering the maturity of the sample. A sensitivity analysis has been performed in order to highlight the importance of the consideration of the autogenous strain, the maturity of the sample and how data are interpolated between thermal variations in the determination of the CTE. The use of Piecewise Cubic Hermite Interpolating Polynomial is recommended for the interpolation of the concrete strain between thermal variations. An adaptation of this new testing methodology has been presented with the Autoshrink device for cement paste and mortar. Finally, recommendation on the test protocol and the data processing are proposed for the determination at very early age of the autogenous strain, the CTE and the initial and final setting time.

The new testing method based on repeated thermal variation is a field that can offer much to the characterization of cement-based materials especially for the characterization of cement based materials composed of new raw materials (e.g. recycled aggregate, fibres and geopolymer). A lot of meaningful research efforts have emerged recently showing the capability of this new testing methodology to define with one test the initial and final setting time, the autogenous strain and the CTE. However these works were limited to laboratory temperature condition (20 °C), binder composed of mainly ordinary Portland cement and sealed condition. Therefore suggestions for future research include:

- the measurement of the internal relative humidity of the sample,
- the optimization of the time duration and amplitude of the thermal variation according to the maturity of the material,
- the decoupling between the instantaneous and delayed terms of the CTE,
- more sophisticated tests with complex histories of temperature,
- the influence of mineral additions (e.g. slag and fly ash), admixtures (e.g. superabsorbent polymer and superplasticizer) and the porosity of the aggregate at very early age on the development of the autogenous strain and the CTE,
- the influence of external drying during the hardening process on the development of the CTE and the autogenous strain.

References

Alungbe, G. D., Tia, M., & Bloomquist, D. G. (1992). Effects of aggregate, water/cement ratio, and curing on the coefficient of linear thermal expansion of concrete. *Transportation Research Record (TRR), 1335*, 44–51.

ASTM C403. (2008). Standard test method for time of setting of concrete mixtures by penetration resistance. *American Society of Testing and Materials*. https://doi.org/10.1520/c0403_c0403m-08.

ASTM C531-00. (2012). Standard test method for linear shrinkage and coefficient of thermal expansion of chemical-resistant mortars, monolithic surfacings and polymer concretes. *ASTM International*. 10.1520/C0531-00R12.Copyright.

ASTM Standard C1698. (2014). Test method for autogenous strain of cement paste and mortar. i(Reapproved 2014) (pp. 1–8). https://doi.org/10.1520/c1698-09r14.

Bangham, D. H., & Fakhoury, N. (1931). CLXXV.—The translation motion of molecules in the adsorbed phase on solids. *Journal of the Chemical Society*, 1324–1333. https://doi.org/10.1039/jr9310001324.

Barcelo, L., et al. (1999). Linear vs. volumetric autogenous shrinkage measurement: material behaviour or experimental artefact. In *Self-desiccation and its importance in concrete technology* (pp. 109–125). Available at: http://scholar.google.com/scholar?hl=en&q=Linear+vs.+volumetric+autogenous+shrinkage+measurements:+material+behaviour+or+experimental+artefact&btnG=&as_sdt=1,22&as_sdtp=#0.

Baroghel-Bouny, V. (1994). *Caractérisation des pâtes de ciment et des bétons, Méthodes, Analyse, Interprétation*. Ph.D. thesis, Ecole Nationale des Ponts et Chaussées.

Baroghel-Bouny, V., et al. (2006). Autogenous deformations of cement pastes. *Cement and Concrete Research, 36*(1), 123–136. https://doi.org/10.1016/j.cemconres.2004.10.020.

Bažant, Z. P. (1970). Delayed thermal dilatations of cement paste and concrete due to mass transport. *Nuclear Engineering and Design, 14*(2), 308–318. https://doi.org/10.1016/0029-5493(70)90108-1.

Bažant, Z. P. (1972). Thermodynamics of hindered adsorption and its implications for hardened cement paste and concrete. *Cement and Concrete Research, 2*(1), 1–16. https://doi.org/10.1016/0008-8846(72)90019-1.

Beltzung, F., & Wittmann, F. H. (2005). Role of disjoining pressure in cement based materials. *Cement and Concrete Research, 35*(12), 2364–2370. https://doi.org/10.1016/j.cemconres.2005.04.004.

Benboudjema, F., et al. (2019). Mechanical properties. In E. M. R. Fairbairn & M. Azenha (Eds.), *Thermal cracking of massive concrete structures—State of the art report of the RILEM Technical Committee 254-CMS* (pp. 69–114). https://doi.org/10.1007/978-3-319-76617-1_4.

Bentz, D. P. (2007). Transient plane source measurements of the thermal properties of hydrating cement pastes. *Materials and Structures, 40*(10), 1073–1080. https://doi.org/10.1617/s11527-006-9206-9.

Bentz, D. P., Garboczi, E. J., & Quenard, D. (1998). Modelling drying shrinkage in reconstructed porous materials: application to porous Vycor glass. *Modelling and Simulation in Materials Science and Engineering, 6*(1998), 211–236. https://doi.org/10.1088/0965-0393/6/3/002.

Bjontegaard, O. (1999) *Thermal dilation and autogenous deformation as driving forces to self-induced stresses in high-performance concrete*.

Bjøntegaard, Ø. (2011) *Basis for and practical approaches to stress calculations and crack risk estimation in hardening concrete structures—State of the art*.

Bjøntegaard, Ø., et al. (2015). *RILEM Technical Committee 195-DTD Recommendation for Test Methods for AD and TD of Early Age Concrete*. Dordrecht: Springer Netherlands (RILEM State-of-the-Art Reports). https://doi.org/10.1007/978-94-017-9266-0.

Bjøntegaard, Ø., Hammer, T., & Sellevold, E. J. (2004). On the measurement of free deformation of early age cement paste and concrete. *Cement & Concrete Composites, 26*(5), 427–435. https://doi.org/10.1016/S0958-9465(03)00065-9.

Bjøntegård, Ø., & Sellevold, E. J. (2001). Interaction between thermal dilation and autogenous deformation in high performance concrete. *Materials and Structures, 34*(239), 266–272. https://doi.org/10.1617/13731.

Bouasker, M. (2007). *Etude Numerique Et Experimentale Du Retrait Endogene Au Tres Jeune Age Des Pates De Ciment Avec Et Sans Inclusions*. Ph.D. thesis, Université de Nantes.

Bouasker, M., et al. (2008). Chemical shrinkage of cement pastes and mortars at very early age: Effect of limestone filler and granular inclusions. *Cement & Concrete Composites, 30*(1), 13–22. https://doi.org/10.1016/j.cemconcomp.2007.06.004.

Boulay, C. (2003) Determination of the coefficient of thermal expansion. In A. Bentur (Ed.), *Early age cracking in cementitious systems—Report of RILEM Technical Committee 181-EAS—Early age shrinkage induced stresses and cracking in cementitious systems* (pp. 217–224). RILEM Publications SARL.

Boulay, C., et al. (2010). Quasi-adiabatic calorimetry for concretes : Influential factors. *Bulletin des Laboratoires des Ponts et Chauss ees*, 19–36. Available at: https://hal.archives-ouvertes.fr/hal-00562100%5Cn.

Boulay, C. (2012). Test rig for early age measurements of the autogenous shrinkage of a concrete. In *Proceedings of the RILEM-JCJ International Workshop ConCrack 3* (pp. 111–122).

Boulay, C., et al. (2014). How to monitor the modulus of elasticity of concrete, automatically since the earliest age? *Materials and Structures, 47*(1–2), 141–155. https://doi.org/10.1617/s11527-013-0051-3.

Briffaut, M., et al. (2012). Effects of early-age thermal behaviour on damage risks in massive concrete structures. *European Journal of Environmental and Civil Engineering, 16*(5), 589–605. https://doi.org/10.1080/19648189.2012.668016.

Brooks, J., & Megat Johari, M. (2001). Effect of metakaolin on creep and shrinkage of concrete. *Cement & Concrete Composites, 23*(6), 495–502. https://doi.org/10.1016/S0958-9465(00)00095-0.

Cagnon, H., et al. (2016). Effects of water and temperature variations on deformation of limestone aggregates, cement paste, mortar and High Performance Concrete (HPC). *Cement & Concrete Composites, 71,* 131–143. https://doi.org/10.1016/j.cemconcomp.2016.05.013.

Campbell-Allen, D., & Thorne, C. P. (1963). The thermal conductivity of concrete. *Magazine of Concrete Research, 15*(43), 39–48. https://doi.org/10.1680/macr.1963.15.43.39.

Carette, J., & Staquet, S. (2015). Monitoring the setting process of mortars by ultrasonic P and S-wave transmission velocity measurement. *Construction and Building Materials, 94,* 196–208. https://doi.org/10.1016/j.conbuildmat.2015.06.054.

Charron, J. P. (2003). *Contribution à l'étude du comportement au jeune âge des matériaux cimentaires en conditions des déformations libre et restreinte*. Ph.D. thesis, Université Laval.

Chaunsali, P., & Mondal, P. (2016). Physico-chemical interaction between mineral admixtures and OPC–calcium sulfoaluminate (CSA) cements and its influence on early-age expansion. *Cement and Concrete Research, 80,* 10–20. https://doi.org/10.1016/j.cemconres.2015.11.003.

Coussy, O. (2003). The equivalent pore pressure and the swelling and shrinkage of cement-based materials. *Materials and Structures, 37*(265), 15–20. https://doi.org/10.1617/14080.

Craeye, B., et al. (2010). Effect of mineral filler type on autogenous shrinkage of self-compacting concrete. *Cement and Concrete Research, 40*(6), 908–913. https://doi.org/10.1016/j.cemconres.2010.01.014.

Cusson, D., & Hoogeveen, T. (2006). Measuring early-age coefficient of thermal expansion in high-performance concrete. In *International Rilem Conference* (pp. 321–330). https://doi.org/10.1617/2351580052.034.

Cusson, D., & Hoogeveen, T. (2007). An experimental approach for the analysis of early-age behaviour of high-performance concrete structures under restrained shrinkage. *Cement and Concrete Research, 37*(2), 200–209. https://doi.org/10.1016/j.cemconres.2006.11.005.

Darquennes, A. (2009). *Comportement au jeune âge de bétons formulés à base de ciment au laitier de haut fourneau en condition de déformations libre et restreinte*. Ph.D. thesis, Université Libre de Bruxelles.

Darquennes, A., Staquet, S., & Espion, B. (2011). Determination of time-zero and its effect on autogenous deformation evolution. *European Journal of Environmental and Civil Engineering, 15*(7), 1017–1029. https://doi.org/10.1080/19648189.2011.9695290.

Day, R. L. (1974). *Thermal deformation of cement paste*. Canada: The university of Calgari.

De Schutter, G., & Taerwe, L. (1995). Specific heat and thermal diffusivity of hardening concrete. *Magazine of Concrete Research, 47*(172), 203–208. https://doi.org/10.1680/macr.1995.47.172.203.

Delsaute, B. (2016). *New approach for monitoring and modelling of the creep and shrinkage behaviour of cement pastes, mortars and concretes since setting time*. Université Libre de Bruxelles (BATir) and Université Paris Est (IFSTTAR).

Delsaute, B., Boulay, C., & Staquet, S. (2016). Creep testing of concrete since setting time by means of permanent and repeated minute-long loadings. *Cement & Concrete Composites, 73,* 75–88. https://doi.org/10.1016/j.cemconcomp.2016.07.005.

Delsaute, Brice, & Staquet, S. (2017a). Decoupling thermal and autogenous strain of concretes with different water/cement ratios during the hardening process. *Advances in Civil Engineering Materials, 6*(2), 22. https://doi.org/10.1520/ACEM20160063.

Delsaute, B., & Staquet, S. (2017b). Impact of recycled aggregate in concrete on the evolution of the free deformation. In S. Staquet & D. G. Aggelis (Eds.), *Proceedings of the Second International RILEM/COST Conference on Early Age Cracking and Serviceability in Cement-based Materials and Structures*, Brussel (pp. 197–202).

Delsaute, B., & Staquet, S. (2018a). Impact de la présence de granulats et de sables recyclés sur le développement des propriétés du béton depuis la prise. In *NoMaD 2018* (p. 8).

Delsaute, B., & Staquet, S. (2018b). Influence of recycled aggregate and recycled sand on the development of the early age properties of concrete since setting. In *SynerCrete'18 International Conference on Interdisciplinary Approaches for Cement-based Materials and Structural Concrete* (pp. 231–236).

Delsaute, B., & Staquet, S. (2019). Development of strain-induced stresses in early age concrete composed of recycled gravel or sand. *Journal of Advanced Concrete Technology, 17*(6), 319–334. https://doi.org/10.3151/jact.17.319.

Delsaute, B., & Staquet, S. (2020). Impact of recycled sand and gravels in concrete on volume change. *Construction and Building Materials, 232* (under rev).

Emanuel, J. H., & Hulsey, J. L. (1977). Prediction of the thermal coefficient of expansion of concrete. *ACI Journal Proceedings, 74*(4), 149–155.

Eppers, S., & Mueller, C. (2008). Autogenous shrinkage and time-zero of UHPC determined with the shrinkage cone. In *8th International Conference on Creep, Shrinkage and Durability Mechanics of Concrete and Concrete Structures* (pp. 709–714). https://doi.org/10.1201/9780203882955.ch94.

Esping, O. (2008). Effect of limestone filler BET(H2O)-area on the fresh and hardened properties of self-compacting concrete. *Cement and Concrete Research, 38*(7), 938–944. https://doi.org/10.1016/j.cemconres.2008.03.010.

Feldman, R. F., & Sereda, P. J. (1968). A model for hydrated Portland cement paste as deduced from sorption-length change and mechanical properties. *Matériaux et Constructions, 1*(6), 509–520. https://doi.org/10.1007/BF02473639.

Ferraris, C. F., & Wittmann, F. H. (1987). Shrinkage mechanisms of hardened cement paste. *Cement and Concrete Research, 17*(3), 453–464. https://doi.org/10.1016/0008-8846(87)90009-3.

fib. (2013). *Code-type models for structural behaviour of concrete: Background of the constitutive relations and material models in the fib Model Code for Concrete Structures 2010, fib Bulletin 70.*

Fritsch, F., & Carlson, R. (1980). Monotone piecewise cubic interpolation. *SIAM J. Numer. Anal., 17*(2), 238–246. https://doi.org/10.1137/0717021.

Garcia Boivin, S. (1999). *Retrait au jeune âge du béton. Développement d'une méthode expérimentale et contribution à l'analyse physique du retrait endogène*. Ph.d. Thesis, ENPC/LCPC.

Grasley, Z. C., & Lange, D. (2006). Thermal dilation and internal relative humidity of hardened cement paste. *Materials and Structures, 40*(3), 311–317. https://doi.org/10.1617/s11527-006-9108-x.

Gutsch, A.-W. (2000). *Stoffeigenschaften jungen Betons – Versuche und Modelle.*

Hammer, T. A. (2003). Measurement methods for testing of early age autogenous strain. In *Early age cracking in cementitious systems—Report of RILEM Technical Committee 181-EAS—Early age shrinkage induced stresses and cracking in cementitious systems* (pp. 207–215).

Hammer, T. A., Bjøntegaard, Ø. (2006). Testing of autogenous deformation (AD) and thermal dilation (TD) of early age mortar and concrete—recommended test procedure. In *International RILEM Conference on Volume Changes of Hardening Concrete: Testing and Mitigation* (pp. 341–346). RILEM Publications. https://doi.org/10.1617/2351580052.036.

Hammer, T. A., Bjøntegaard, Ø., & Sellevold, E. J. (2003). Measurement methods for testing of early age autogenous strain. In: *Early age cracking in cementitious systems—Report of RILEM Technical Committee 181-EAS—Early age shrinkage induced stresses and cracking in cementitious systems* (pp. 217–228).

Hansen, W. (2011). Report on early-age cracking. *Concrete International, 33*(3), 2–5.

Instruments, T. (2017). *Tam air isothermal calorimetry*. Available at: http://www.tainstruments.com/pdf/brochure/TAM-AIR-brochure.pdf. Accessed February 14, 2017.

Jensen, O. M., & Hansen, P. F. (1999). Influence of temperature on autogenous deformation and relative humidity change in hardening cement paste. *Cement and Concrete Research, 29*(4), 567–575. https://doi.org/10.1016/S0008-8846(99)00021-6.

Jensen, O. M., & Hansen, P. F. (2001). Autogenous deformation and RH-change in perspective. *Cement and Concrete Research, 31*(12), 1859–1865. https://doi.org/10.1016/S0008-8846(01)00501-4.

Johnson, W. H., & Parsons, W. H. (1944). Thermal expansion of concrete aggregate materials. *Journal of Research of the National Bureau of Standards, 32*(3), 101–126. https://doi.org/10.6028/jres.032.002.

Kada, H., et al. (2002). Determination of the coefficient of thermal expansion of high performance concrete from initial setting. *Materials and Structures, 35*(245), 35–41. https://doi.org/10.1617/13684.

Koenders, E. A. B. (1997). *Simulation of volume changes in hardening cement-based materials*. Delft. Available at: http://repository.tudelft.nl/view/ir/uuid:1dbcb7fb-3f8f-466b-8517-b2235ad4912f/.

Kohno, K., et al. (1999). Effects of artificial lightweight aggregate on autogenous shrinkage of concrete. *Cement and Concrete Research, 29*(4), 611–614. https://doi.org/10.1016/S0008-8846(98)00202-6.

Königsberger, M., Delsaute, B., & Staquet, S. (2018). Evolution of thermal expansion coefficients during hydration of cement paste: Insights from experimental monitoring and thermo-poro-micromechanical modeling. In *ICEM18*, Brussels.

Kovler, K. (1994). Testing system for determining the mechanical behaviour of early age concrete under restrained and free uniaxial shrinkage. *Materials and Structures, 27*(6), 324–330. https://doi.org/10.1007/BF02473424.

Kronlöf, A., Leivo, M., & Sipari, P. (1995). Experimental study on the basic phenomena of shrinkage and cracking of fresh mortar. *Cement and Concrete Research, 25*(8), 1747–1754. https://doi.org/10.1016/0008-8846(95)00170-0.

Kumar Mehta, P., & Monteiro, P. J. M. (2006). *Concrete: Microstructure, properties, and materials* (4th ed.). Third. The McGraw-Hill Companies.

Laplante, P., & Boulay, C. (1994). Evolution du coefficient de dilatation thermique du béton en fonction de sa maturité aux tout premiers âges. *Materials and Structures, 27*(10), 596–605. https://doi.org/10.1007/BF02473129.

Le Chatelier, H. (1900). Sur le changement de volume qui accompagne le durcissement des ciments. *Bulletin de la Société de l'encouragement pour l'industrie nationale, 5*(5), 54–57.

Le Roy, R. (1995). *Déformations instantanées et différées des bétons à hautes performances*. Ph.D. thesis, Ecole Nationale des Ponts et Chaussées, Paris, France.

Liu, Z., & Hansen, W. (2016). Aggregate and slag cement effects on autogenous shrinkage in cementitious materials. *Construction and Building Materials, 121*, 429–436. https://doi.org/10.1016/j.conbuildmat.2016.06.012.

Lokhorst, S. J. (1998). *Deformational behaviour of concrete influenced by hydration related changes of the microstructure*.

Loser, R., Münch, B., & Lura, P. (2010). A volumetric technique for measuring the coefficient of thermal expansion of hardening cement paste and mortar. *Cement and Concrete Research, 40*(7), 1138–1147. https://doi.org/10.1016/j.cemconres.2010.03.021.

Loukili, A., et al. (2000). A new approach to determine autogenous shrinkage of mortar at an early age considering temperature history. *Cement and Concrete Research, 30*(6), 915–922. https:// doi.org/10.1016/S0008-8846(00)00241-6.

Lura, P., & Durand, F. (2006). Volume changes of hardening concrete : testing and mitigation. In O. M. Jensen, P. Lura, & K. Kovler (Eds.), *Concrete* (pp. 57–65). Lyngby.

Maruyama, I., & Teramoto, A. (2011). Impact of time-dependant thermal expansion coefficient on the early-age volume changes in cement pastes. *Cement and Concrete Research, 41*(4), 380–391. https://doi.org/10.1016/j.cemconres.2011.01.003.

Maruyama, I., & Teramoto, A. (2012). Effect of water-retaining lightweight aggregate on the reduction of thermal expansion coefficient in mortar subject to temperature histories. *Cement and Concrete Composites, 34*(10), 1124–1129. https://doi.org/10.1016/j.cemconcomp.2012.08.003.

Maruyama, I., Teramoto, A., & Igarashi, G. (2014). Strain and thermal expansion coefficients of various cement pastes during hydration at early ages. *Materials and Structures, 47*(1–2), 27–37. https://doi.org/10.1617/s11527-013-0042-4.

Mehta, P. K. (1973). Mechanism of expansion associated with ettringite formation. *Cement and Concrete Research, 3*(1), 1–6. https://doi.org/10.1016/0008-8846(73)90056-2.

Mejlhede Jensen, O., & Freiesleben Hansen, P. (1995). A dilatometer for measuring autogenous deformation in hardening portland cement paste. *Materials and Structures, 28*(7), 406–409. https://doi.org/10.1007/BF02473076.

Min, D., & Mingshu, T. (1994). Formation and expansion of ettringite crystals. *Cement and Concrete Research, 24*(1), 119–126. https://doi.org/10.1016/0008-8846(94)90092-2.

Mindess, S., Young, J. F., & Darwin, D. (2003). *Concrete*. Upper Saddle River, NJ, USA: Prentice Hall, Pearson Education, Inc.

Milenkovic, N., et al. (no date). Advanced characterisation of the early age behaviour of bulk hydrophobic mortars. *Construction and Building Materials*.

Mitani, H. (2003). *Variations volumiques des matrices cimentaires aux très jeunes âges : approche expérimentale des aspects physiques et microstructuraux*. Ph.D. thesis, Ecole Nationale des Ponts et Chaussées.

Mitchell, L. (1953). Thermal expansion tests on aggregates, neat cements, and concretes. *Proceedings—American Society for Testing and Materials*.

Mohamed, M. S., et al. (2017). Applicability of ultrasonic measurement on the monitoring of the setting of cement pastes: effect of water content and mineral additions. *Advances in Civil Engineering Materials, 6*(2). https://doi.org/10.1520/acem20160062.

Mounanga, P., Khelidj, A., & Bastian, G. (2004). Experimental study and modelling approaches for the thermal conductivity evolution of hydrating cement paste. *Advances in Cement Research, 16*(3), 95–103. https://doi.org/10.1680/adcr.2004.16.3.95.

Nilsson, L. (1987). Temperature effects in RH measurements on concrete—some preliminary studies. In *Symposium and day of building physics* (pp. 456–462). Lund, Sweden: Swedish Council for Building Research.

Odler, I., & Colán-Subauste, J. (1999). Investigations on cement expansion associated with ettringite formation. *Cement and Concrete Research, 29*(5), 731–735. https://doi.org/10.1016/S0008-8846(99)00048-4.

Ozawa, M., & Morimoto, H. (2006). Estimation method for thermal expansion coefficient of concrete at early ages. In *International RILEM Conference on Volume Changes of Hardening Concrete: Testing and Mitigation* (pp. 331–3390. https://doi.org/10.1617/2351580052.035.

Paillere, A. M., & Serrano, J. J. (1976). Appareil d'étude de la fissuration du béton. *Bulletin de liaison du Laboratoire Central des Ponts et Chaussées, 83*, 29–38.

Persson, B. (2002). Self-desiccation and chloride migration. In *The Third International Research Seminar in Lund* (pp. 175–194).

Powers, T. C. (1965). Mechanisms of shrinkage and reversible creep of hardened cement paste. In *The structure of concrete and its behaviour under load*, London (pp. 319–344).

Powers, T. C. (1968). The thermodynamics of volume change and creep. *Materiaux et constructions, 1*(6), 487–507. https://doi.org/10.1007/BF02473638.

Radjy, F., Sellevold, E. J., & Hansen, K. K. (2003). Temperature data for water sorption in hardened cement paste: Enthalpy, entropy and sorption isotherms at different temperatures. In *Report R057* (p. 58). Available at: papers2://publication/uuid/345F4693-1D32-4D3D-A911-CA13EA5D47AA.

Roziere, E., et al. (2015). Experimental assessment of autogenous shrinkage. In *CONCREEP 2015: Mechanics and Physics of Creep, Shrinkage, and Durability of Concrete and Concrete Structures—Proceedings of the 10th International Conference on Mechanics and Physics of Creep, Shrinkage, and Durability of Concrete and Concrete Structure*. https://doi.org/10.1061/9780784479346.117.

Sant, G., Lura, P., & Weiss, J. (2006). A discussion of analysis approaches for determining "time-zero" from chemical shrinkage and autogenous strain measurements in cement paste. In *International RILEM Conference on Volume Changes of Hardening Concrete: Testing and Mitigation*, Lyngby (pp. 375–383). Available at: http://www.rilem.net/gene/main.php?base=500218&id_publication=57&id_papier=2271.

Sant, G., et al. (2011). The origin of early age expansions induced in cementitious materials containing shrinkage reducing admixtures. *Cement and Concrete Research, 41*(3), 218–229. https://doi.org/10.1016/j.cemconres.2010.12.004.

Scherer, G. W. (2000). Measuring permeability of rigid materials by a beam-bending method: I, Theory. *Journal of the American Ceramic Society, 83*(9), 2231–2239.

Scherer, G. W. (2003). Characterization of saturated porous bodies. *Materials and Structures, 37*(265), 21–30. https://doi.org/10.1617/14079.

Sellevold, E. J., & Bjøntegaard, Ø. (2006). Coefficient of thermal expansion of cement paste and concrete: Mechanisms of moisture interaction. *Materials and Structures, 39*, 809–815. https://doi.org/10.1617/s11527-006-9086-z.

Siddiqui, M. S., & Fowler, D. W. (2014). Optimizing coefficient of thermal expansion of concrete and its importance on concrete structures. *Construction Materials and Structures*, 47–56. https://doi.org/10.3233/978-1-61499-466-4-47.

Siddiqui, M. S., & Fowler, D. W. (2015). Effect of internal water pressure on the measured coefficient of thermal expansion of concrete. *Journal of Materials in Civil Engineering, 27*(4), 04014151. https://doi.org/10.1061/(ASCE)MT.1943-5533.0001095.

Staquet, S., et al. (2005). *Expérimentation sur poutres préfléchies à talon BTHP*, Paris.

Staquet, S., et al. (2019). Mixture proportioning for crack avoidance. In *Thermal cracking of massive concrete structures—State of the Art Report of the RILEM Technical Committee 254-CMS* (pp. 115–151). https://doi.org/10.1007/978-3-319-76617-1_5.

Stefan, L. (2009). *Étude Expérimentale Et Modélisation De L'Évolution Des Propriétés Mécaniques Au Jeune Âge Dans Les Matériaux Cimentaires*. Ecole normale supérieure de Cachan. Available at: http://tel.archives-ouvertes.fr/tel-00624989/.

Stefan, L., et al. (2018). Influential factors in volume change measurements for cementitious materials at early ages and in isothermal conditions. *Cement and Concrete Composites, 85*, 105–121. https://doi.org/10.1016/j.cemconcomp.2017.10.007.

Viviani, M., Glisic, B., & Smith, I. F. C. (2006). System for monitoring the evolution of the thermal expansion coefficient and autogenous deformation of hardening materials. *Smart Materials and Structures, 15*(6). https://doi.org/10.1088/0964-1726/15/6/n01.

Viviani, M., Glisic, B., & Smith, I. F. C. (2007). Separation of thermal and autogenous deformation at varying temperatures using optical fiber sensors. *Cement and Concrete Composites, 29*(6), 435–447. https://doi.org/10.1016/j.cemconcomp.2007.01.005.

Wang, H., et al. (2018). May reversible water uptake/release by hydrates explain the thermal expansion of cement paste?—Arguments from an inverse multiscale analysis. *Cement and Concrete Research*. https://doi.org/10.1016/j.cemconres.2018.05.008.

Wittmann, F. (1968). Surface tension, shrinkage, and strength of hardened cement paste. *Materiaux et constructions, 1*(6), 547–552. https://doi.org/10.1007/BF02473643.
Wittmann, F., & Lukas, J. (1974). Experimental study of thermal expansion of hardened cement paste. *Matériaux et Constructions, 7*(4), 247–252. https://doi.org/10.1007/BF02473853.
Wyrzykowski, M., & Lura, P. (2013a). Controlling the coefficient of thermal expansion of cementitious materials—A new application for superabsorbent polymers. *Cement and Concrete Composites, 35*(1), 49–58. https://doi.org/10.1016/j.cemconcomp.2012.08.010.
Wyrzykowski, M., & Lura, P. (2013b). Moisture dependence of thermal expansion in cement-based materials at early ages. *Cement and Concrete Research, 53,* 25–35. https://doi.org/10.1016/j.cemconres.2013.05.016.
Xi, Y., Bažant, Z. P., & Jennings, H. M. (1994). Moisture diffusion in cementitious materials Adsorption isotherms. *Advanced Cement Based Materials, 1*(6), 248–257. https://doi.org/10.1016/1065-7355(94)90033-7.
Yeon, J. H, Choi, S., & Won, M.C. (2009). Effect of relative humidity on coefficient of thermal expansion of hardened cement paste and concrete. *Transportation Research Record: Journal of the Transportation Research Board, 2113*(09), 83–91. https://doi.org/10.3141/2113-10.
Yeon, J. H., Choi, S., & Won, M. C. (2013). 'In situ measurement of coefficient of thermal expansion in hardening concrete and its effect on thermal stress development. *Construction and Building Materials, 38,* 306–315. https://doi.org/10.1016/j.conbuildmat.2012.07.111.
Zhutovsky, S., & Kovler, K. (2017). Application of ultrasonic pulse velocity for assessment of thermal expansion coefficient of concrete at early age. *Materials and Structures, 50*(1), 5. https://doi.org/10.1617/s11527-016-0866-9.
Ziegeldorf, S. W., Kleiser, K., & Hilsdorf, H. K. (1978). Effect of thermal expansion of aggregate on thermal expansion of concrete. Budapest. In *Colloque Internationale sur les materiaux granulaire/International symposium on aggregates and filler* (pp. 452–464).

Chapter 6
Testing Concrete Since Setting Time Under Free and Restrained Conditions

Brice Delsaute and Stéphanie Staquet

Abstract Restrained deformations in concrete structures induce the development of stresses and sometimes cracking. Therefore the whole service life of concrete structures is influenced by the appropriate consideration of the early age behaviour of concrete under restrained conditions. For that reason, several devices were developed in the past to characterize the risk of cracking of cement-based materials. In the 1990s, a new experimental concrete testing concept has been designed: the Temperature Stress Testing Machine (TSTM). The device is able to monitor several parameters such as the age of cracking, the stiffness development and the stress relaxation. Between 1990 and 2010, less than fifteen laboratories worldwide have developed or acquired this kind of experimental equipment. All devices bear some similarities but major differences remain in the test set up designs, in the testing processes and also in the scale of the material used. This chapter summarizes all existing technics used to assess the behavior of concrete under restrained deformation, a comparison of the different existing TSTM is presented. New advanced techniques aim at monitor the cracking risk of cement-based materials such as the active ring test or the elliptical ring test are also presented.

Keywords Restrained deformation · Cracking · Shrinkage · Early age · Creep/relaxation

6.1 Introduction

As explained in detail in the Chaps. 4 and 5, concrete has the particularity to be a complex material for which its properties continuously change. It evolves from a

---

B. Delsaute (✉) · S. Staquet
Service BATir, LGC, Université Libre de Bruxelles (ULB), Brussels, Belgium
e-mail: bdelsaut@ulb.ac.be

M. Serdar et al. (eds.), *Advanced Techniques for Testing of Cement-Based Materials*, Springer Tracts in Civil Engineering, https://doi.org/10.1007/978-3-030-39738-8_6

nearly liquid state to a viscoplastic material within a few hours, followed by the setting of the concrete. Then the mechanical properties start to develop and the material exhibits viscoelastic behaviour. During the first days after mixing, the evolution of the concrete properties is very intense. This period is called the early age. Finally, the concrete properties continue to evolve on a period counted in years. Concrete has also another important specificity coming from its time-dependent properties. At early age, the development of the free deformation induced by desiccation (drying shrinkage) and hydration (thermal and autogenous deformation) are generally partly or fully restrained which leads to the development of internal stresses. For the determination of the stress under restrained condition, the development of the elastic and creep properties are important parameters. To model the behaviour of the concrete, the total strain of the concrete ε_{tot} are decomposed as the sum of 5 terms: the autogenous strain ε_{au}, the thermal strain ε_{th}, the drying strain ε_{ds}, the elastic strain ε_{el} and the creep strain ε_{cr} (Eq. 6.1). Creep strains are also divided in two terms. The first term is related to the behaviour of concrete under sealed condition and the second term is caused by the drying phenomenon. Both are called basic creep and drying creep respectively. The constitutive law related to the viscoelastic behaviour of cement based materials is given in Eq. 6.2 for a constant uniaxial stress σ where t is the age of the concrete, t′ the age of the concrete at loading, $J(t, t')$ is the compliance function, $E(t')$ is the elastic modulus and $C(t, t')$ is the specific creep.

$$\varepsilon_{tot} = \varepsilon_{au} + \varepsilon_{th} + \varepsilon_{ds} + \varepsilon_{el} + \varepsilon_{cr} \tag{6.1}$$

$$\varepsilon_{el}(t') + \varepsilon_{cr}(t, t') = \sigma \cdot J(t, t') = \sigma \cdot \left(\frac{1}{E(t')} + C(t, t')\right) \tag{6.2}$$

For concrete structures, the evolution of the restrained strains and the associated stress development depend on the type of structure (thin or massive), the exposition of the structure to the environment (e.g. sealed by the formwork or exposed to drying) and the composition of the concrete (e.g. type of cement, W/C …). For illustration, two opposite situations are presented in the Fig. 6.1. On the left, the first case corresponds to a thin concrete element exposed to drying since the casting. This is for example the case of concrete slab for which normal strength concrete is generally used. In such case, low thermal changes are caused by hydration. Thus the development of the free deformation is mainly driven by the autogenous strain and the drying shrinkage (Bendimerad et al. 2020; Delsaute et al. n.d.). This corresponds to a global shrinkage of the concrete element as shown in Fig. 6.1, left. As a result, tensile stress is induced in the concrete element when the free deformation is restrained. The second case refers to a massive structure for which the free deformations are mainly composed of the autogenous and thermal strain. Two periods are observed in the development of the free deformation: a heating period followed by a cooling one (Fig. 6.1, right) as explained in (Delsaute and Staquet 2019; Delsaute et al. 2017). The heating period begins just after the final setting when the mechanical properties of concrete start to develop. During and after the setting, the heat flow of

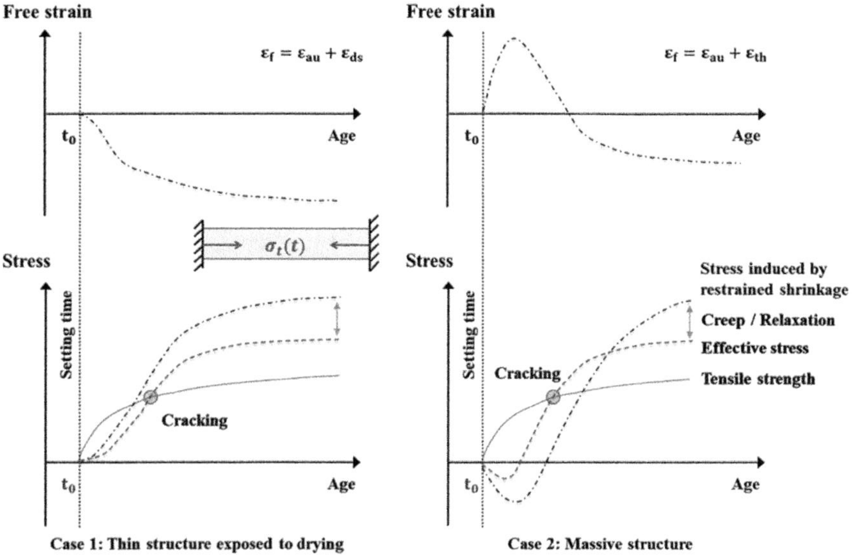

Fig. 6.1 Evolution of the stress induced by the restriction of the concrete free strain for thin structure exposed to drying (left) and massive structure [right (Delsaute et al. 2017)]

the cement is very intense which leads to an increase of the temperature inside the concrete element and as a result to an increase of the thermal strain. In the meantime, autogenous strain starts to develop. No systematic tendency can be given for the autogenous strain, because during the heating period, autogenous deformation results in swelling or shrinkage according to the mixtures proportions and content (Benboudjema et al. 2019; Bentur 2003b; Staquet et al. 2019). However thermal strains are generally higher than autogenous strain (especially for massive structures) and thus a general swelling of the concrete occurs during the heating period. The cooling period starts when the heat flow of the cement decreases strongly or when the formwork is removed (according to the massivity). During this period, both autogenous and thermal strains decrease. As a result of the restriction of the strains of concrete, stresses are induced (Fig. 6.1, right). During the heating period, concrete element is in compression and inversely during the cooling period the concrete is submitted to tension.

The impact of viscous properties at very early age, in case of restrained deformation, is highlighted on the development of the stress as shown in Fig. 6.1 (red curve, viscoelastic stress). In the case of a thin concrete element exposed to drying, creep and relaxation have the positive effect to reduce the development of the tensile stress, especially at very early age when the creep/relaxation phenomena are very important. In the case of massive concrete element, in a general view, no consideration of creep/relaxation leads to a global overestimate of the stress. Thus creep/relaxation seems to play a general positive role for the design of massive concrete structures at early age. However, at very early age, the creep/relaxation amplitude is very significant and reduces strongly the compressive stresses. Then, during the cooling period, stresses switch rapidly in tension. During this period, an underestimation of the creep/relaxation phenomenon leads to an underestimation of the tensile stresses which can cause cracking in the concrete structure. Hence it is important to consider correctly the creep and relaxation phenomena since final setting time for the study of the behavior of cement based materials under restrained conditions. All the concrete properties depend on the composition of the concrete. For high strength concrete, the development of the autogenous and thermal strain is higher and thus those compositions are more sensitive to a cracking risk.

The cracking sensitivity of cement based materials is therefore defined by the type of structure, the environmental conditions and the concrete compositions. Moreover the early age and long term concrete properties are dependent of the massivity of the structure and the environmental exposition. This makes the study of the cracking sensitivity of concrete structure a challenging task when considering the evolution of the whole concrete properties for the determination of the stress inside a concrete element when its deformation is restrained. For that reason, several tests have been developed to estimate the cracking risk of cementitious materials, such as the ring tests, the plate tests and the passive longitudinal tests (Bentur 2003a). Generally, these experimental tests allow determining the number of cracks, their width and the stresses evolution inside the concrete sample. However, this evolution strongly depends on the geometry and the stiffness of the material used to obtain the restrained conditions. Thus, these devices can be used for qualitative means which are useful for comparison of the cracking sensitivity of different concrete compositions but are not sufficient alone for the study of the mechanisms involved in the cracking risk. To eliminate these drawbacks, an active longitudinal test named Temperature Stress Testing Machine (TSTM) was developed from the beginning of the 1980s. With this device, a full or a partial degree of restriction of the concrete sample can be obtained by means of the application of a compressive/tensile stress on a dog-bone shaped concrete specimen to keep its length constant with a manual or an automatic system.

The present chapter summarizes the analysis of the development of the TSTM's at the Université Libre de Bruxelles. Based on this analysis, a revisited TSTM and experimental procedure have been developed to study the sensitivity to cracking of concretes. The present chapter is structured in five main sections (apart from the present introduction). The second section of this chapter is related to the presentation of the tests rig developed for the determination of the stress of concrete under restrained condition. The third section explains the principle of a TSTM, the history of the TSTM development in a dozen of other laboratories as well as a synthesis of the key elements in the design of a TSTM. In the fourth section, the development and the new technical advances performed on the TSTM during this last decade at ULB are presented. Several new applications of the TSTM are introduced in the fifth section. Finally, conclusions are given in the sixth and ultimate section of the present chapter.

6.2 Test Rig Designed for the Study of the Risk of Cracking of Cement Based Materials

Restrained shrinkage tests can be divided mainly into two categories: passive and active restrained shrinkage tests. In case of passive restrained shrinkage test, there is no control of the degree of restraint of the concrete strain in the test setup. The degree of restraint is driven by the difference of stiffness of the restraining frame and the concrete one. For that reason, active devices, for which the restraining part of the device is controlled (in terms of displacement, force or temperature controlled), have been developed.

6.2.1 Passive Restrained Shrinkage Test

As presented by Bentur and Kovler in (Bentur 2003a), mainly four passive restrained shrinkage tests have been developed, namely the plates test, the substrate restrained test, the longitudinal test and the ring test.

6.2.1.1 Plate Test

The plate test is mainly used in the study of plastic shrinkage and it quantifies the performance of concrete by estimating the extent of its cracking by means of the average and maximum width of the crack, of their total length, their total area and number per unit area (Bentur 2003a). For this test, the concrete is poured into a slab mold (Fig. 6.2). In addition to the relative friction in the contact zone between the mold and the slab, the restrained is mainly induced by the presence of steel reinforcement embedded in the slab and fixed to the edges of the plate.

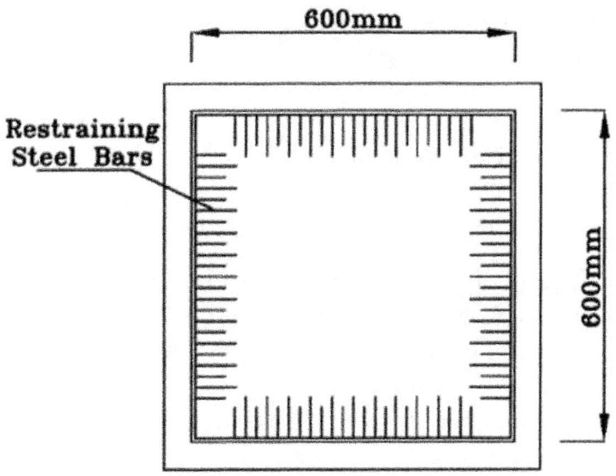

Fig. 6.2 Scheme of the plate test (Yokoyama et al. 1994)

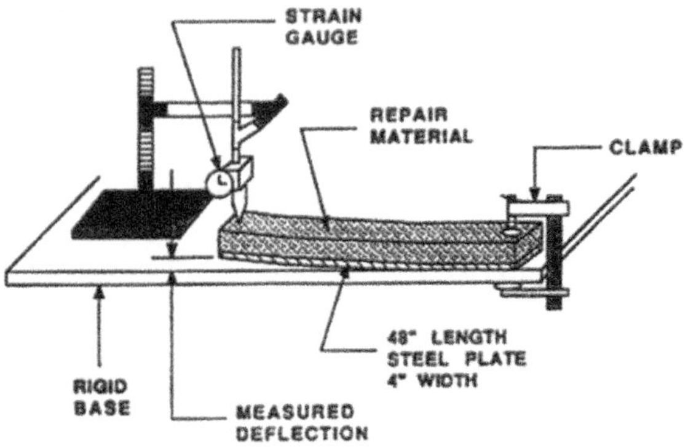

Fig. 6.3 Substrate restrained tests developed for the study of the curling (Bentur 2003a)

6.2.1.2 Substrate Restrained Test

Substrate restrained tests were developed to simulate the restraint of the deformation of fresh mortar or concrete used for the reparation or the resurfacing of concrete structure. Two types of test were developed in order to study the risk of cracking or curling. To study the risk of cracking, a fresh layer of mortar or concrete is casted on a hardened concrete layer (Banthia et al. 1993, 1995, 1996) or a steel layer (Vaysburd et al. 2001) and is exposed to ambient air (Vaysburd et al. 2001) or to a controlled temperature and relative humidity (Banthia et al. 1993, 1995, 1996). This quantifies the performance of mortar/concrete by estimating the extent of its cracking by means of the number of cracks, the average and maximum width of the crack, time of cracking and detachment from the substrate layer. To evaluate the risk of curling, a rectangular layer of the material is casted alone or on a steel substrate and the top part of the specimen is exposed to ambient environment. One end of the specimen is embedded while the other one is free. The vertical displacement of the free extremity or at mid-span is measured to evaluate the performance of the material as illustrated in Fig. 6.3 (Vaysburd et al. 2001). At the structural scale, similar test setups were designed for the study of the curling phenomenon induced by the differential shrinkage between the bottom and the top surface of slab [see e.g. (Jaafri et al. 2019)].

6.2.1.3 Longitudinal Test

Two passive longitudinal tests have been designed:

- Longitudinal-qualitative: In this test, the longitudinal geometry of the apparatus is used as a restriction to the free deformations of the concrete element and it only allows quantifying the cracking by measuring the width of the cracks and the total length of cracking.
- Longitudinal-quantitative passive: In addition to cracking measurement, this device is able to determine the restriction forces and stresses developed within

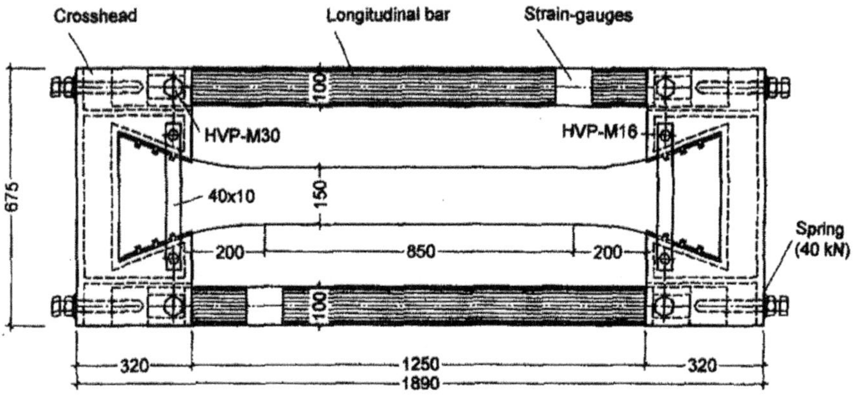

Fig. 6.4 Longitudinal passive restrained shrinkage test (RILEM TC 119-TCE 1997)

the material. This test is considered as passive because the restriction is obtained by means of external bars of constant length (Fig. 6.4). In this device, concrete deformations are not completely restrained and thus the length of the specimen is still slightly variable. Deformation gauges are placed on the metal columns and quantify the existing displacement. The restraining stresses are therefore calculated from the strain gauge data with consideration of the steel properties of the outer columns of the device and the dimensions of the section of the concrete specimen. In addition, for the study of mass concrete element, insulation can be placed around the sample and the mold can be regulated in temperature (Bentur 2003a).

Longitudinal-quantitative passive tests were also developed at the structural scale. In the frame of the French national project CEOS (CEOS.fr n.d.; Delsaute et al. 2013), 3 blocks subject to restrained shrinkage were tested. The dimension of each body is 6.1 m × 1.6 m × 0.8 m (Fig. 6.5). These tests were performed in order to improve the design of massive concrete structure, specifically the crack pattern of special

Fig. 6.5 Restrained shrinkage test at structural scale (CEOS.fr n.d.)

concrete structures. For that reason, reinforcement ratio (0.6 and 2%) and concrete cover (50 and 70 mm) vary from one block to another.

6.2.1.4 Passive Ring Test

The ring test is the most common device used for the study of the restrained shrinkage. Plastic shrinkage cracking and shrinkage cracking can be assessed with the ring test. The device is composed of a ring (generally in steel) around which a concrete ring is casted. During the test, the free shrinkage of the concrete ring is restrained by the inner ring. It quantifies the performance of the material by estimating the age of cracking, number of crack and the average and maximum width of the cracks. When lateral faces of the concrete ring are exposed to drying, the test simulate the behavior of an infinitely long structure subjected to restriction (Weiss and Shah 2002). As highlighted by Kanavaris et al. (2019), the main parameters varying in the design of a ring test are related to the test apparatus, the environment and the test protocol. For example, the dimensions of both rings depend on the standards and the type of material studied (i.e. cement paste, mortar or concrete). All variable parameters are synthesized in the Table 6.1.

For illustration, the dimensions of the ring test according to the AASHTO standard (Darquennes et al. 2006) are shown in Fig. 6.6.

Based on the geometry of the test and the measurement of the deformation in the internal part of the inner ring, analytical solutions were developed for the calculation of the elastic hoop stress of concrete [see e.g. (Hossain and Weiss 2004, 2006; Kovler and Bentur 2009; Weiss et al. 2000)]. A comparison of the different analytical approach fund in the literature is performed in (Kanavaris et al. 2019). Recently, a critical review of the development of the ring test was carried out by Kanavaris et al.

Table 6.1 Main parameters in the design of a ring test

Test apparatus	Environment	Test protocol
• Dimension of the restraining core (inner and outer radius, hollow/solid and height) • Dimension of the restrained ring (inner and outer radius and height) • Nature of the restraining ring (e.g. stainless steel, Invar) • Nature of the restrained ring (cement paste, mortar or concrete) • Degree of restraint	• Curing temperature (ambient or controlled) • Relative humidity (ambient or controlled)	• Formwork removal age • Drying direction (circumferential, top and bottom, top, all 3 sides exposed or all 3 sides sealed)

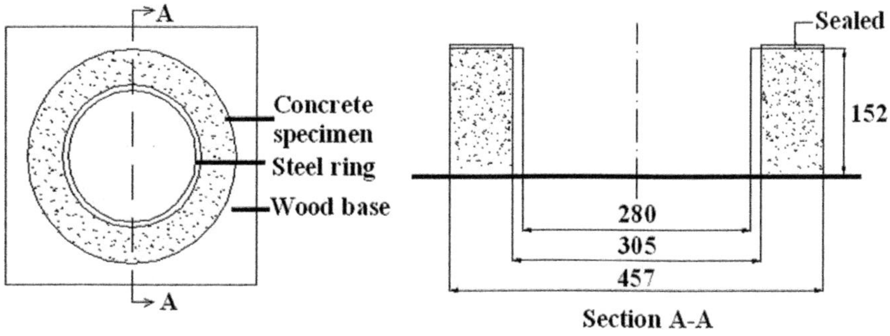

Fig. 6.6 Ring test according to AASHTO standard (Darquennes et al. 2006)

(2019). From this review, new recommendations in the testing method are addressed according to the type of application.

Other alternatives to the conventional ring test were developed to accelerate the development of cracking. This improvement is interesting since cracking can occur after several months for Portland cement-based concretes [see e.g. (Darquennes et al. 2006)]. For example, Dahl (Bentur 2003a) encloses the concrete ring between two steel rings. The outer ring is provided with ribs causing an increase in the stresses within the material. Other authors (Turcry 2004) have notched their concrete ring, which allows obtaining relatively short material cracking times. Recently, several authors have also proposed to replace the steel ring by an elliptical ring for which stress development is higher in the minor radius of the ellipse (Dong et al. 2018; He et al. 2004; Zhou et al. 2014). For the consideration of the restrained of the swelling deformation, Weiss et al. (2008) have added an outer restraining steel ring. This device is called dual ring test.

6.2.2 Active Restrained Shrinkage Test

In order to control the degree of restraint of the concrete strain during the test, active restrained shrinkage tests have been developed. Three methods were adopted: the control in displacement, in force and/or in temperature of the concrete specimen. Mainly, two types of device were developed: active ring tests and active longitudinal tests.

6.2.2.1 Active Ring Test

The active feature of the ring test is related to controlled temperature changes imposed during the test. As the coefficients of thermal expansion (CTE) of the restraining ring(s) and the concrete ring are different, a change in temperature restrains the deformation or imposes an expansion of the concrete ring. First, Kovler et al. (1993) have replaced the hollow steel ring by an active-expansive core with a CTE higher than concrete. When increasing the temperature, additional tensile stresses were therefore induced in the concrete ring which shortened the age of cracking. Then Schlitter et al. (2010a, b) have developed a dual ring test for which both restraining ring are made of Invar. The whole device is placed in a highly insulated chamber which is controlled in temperature. As the CTE of Invar is very low, the restraining boundaries of the concrete ring are nearly constant. For the study of massive concrete element, Briffaut et al. (2011a, b) used a thermally-active ring. The restraining ring is made of brass (the CTE is about 3 times higher than concrete) and is regulated in temperature thanks to water circulation into it. Contrary to Schlitter's system, the restraining boundaries are moving during the test. Recently, Bourchy (2018) has developed a dual ring test for which both restraining rings are made of Invar and are thermally-controlled (the temperature evolution correspond to the temperature recorded during a quasi-adiabatic calorimetry test).

6.2.2.2 Active Longitudinal Test

The active part of the longitudinal test is related to the control in force or in displacement of one of the extremity of the concrete specimen. Two configurations of the specimen were developed: vertical and horizontal test rig. These devices, named TSTM (Temperature Stress Testing Machine), are described in detail in the next paragraph.

6.3 The TSTM (Temperature Stress Testing Machine)

6.3.1 Principle of TSTM's

The determination of the different concrete parameters is based on the hypothesis of independence of the deformation components. The total deformation is assumed to be equal to the sum of the elastic ε_e, creep ε_{cr}, thermal ε_{th} and shrinkage ε_{sh} deformations, as expressed in Eq. 6.1. The degree of restriction of deformations is equal to zero if this sum is null. For the study of the evolution of the different parameters, compensation cycles are used with a threshold value of deformation ε_l equal to a few microns (or microstrains) for applying the restriction conditions during the experimental test (Fig. 6.7). As soon as this value is exceeded, a load is applied on the concrete sample to cancel the deformation. Then, the concrete sample can shrink again and the load is kept constant but the stress/strength ratio increase with the time. Then another compensation cycle begins (Fig. 6.8). So the cycles succeed to themselves during the entire test until cracking of the sample occurs.

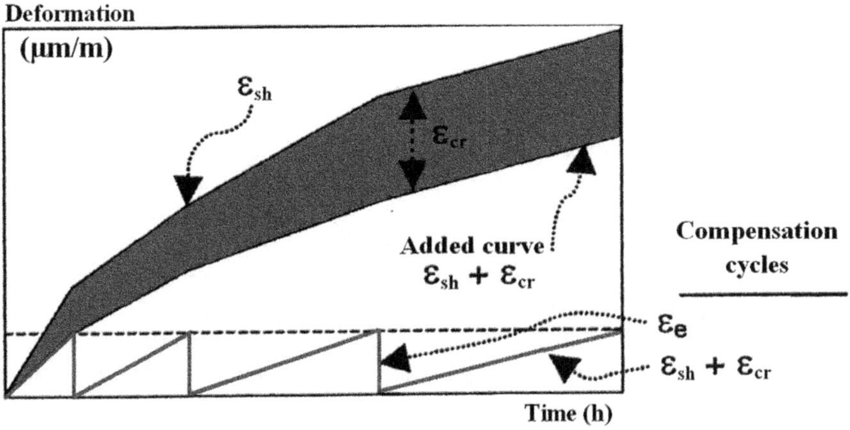

Fig. 6.7 Schematic description of the compensating cycles to obtain the restriction conditions with the TSTM (Kovler 1994)

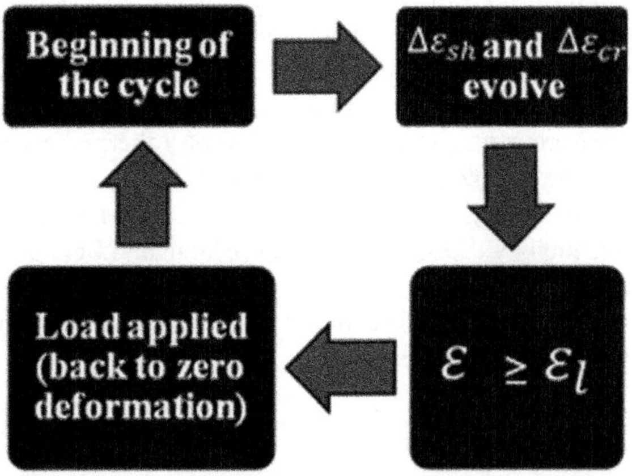

Fig. 6.8 Steps inside each compensating cycle

Following these compensation cycles, the sum of all elastic deformations is equal to the sum of the restrained deformations, i.e. creep and free strain (Fig. 6.7). From these experimental results, the creep deformation can be calculated as the difference between the curves of free strain and restrained deformations. Another parameter can also be estimated from this test: the relaxation stresses.

6.3.2 History of the Development of TSTM's

One of the oldest TSTM equipment was developed in the beginning of the 1980s by Springenschmid et al. (1994) at the Technical University of Munich in Germany. In this equipment, the dog-bone sample is characterized by a square cross section 150 × 150 mm^2 and a length equal to 1500 mm (Schöppel et al. 1994). The restriction conditions were obtained for a deformation threshold equal to 1 μm and were applied when the material is characterized by a significant stiffness. This time is the moment when the stresses begin to develop inside the concrete sample. For the experimental tests performed with this TSTM, the displacement was measured by means of steel bars placed in the concrete sample in its central part at an interval equal to 500 mm. The same device was used for measuring the free shrinkage. This boundary condition was obtained by cancelling the force applied on the concrete sample when the stress inside the concrete reaches a value equal to 0.01 MPa. It appears that this first device is already characterized by two main parameters limiting the artifacts of restrained shrinkage measurements: a thermal regulation of the concrete sample and a system of deformation measurement directly placed on the concrete sample. At the same period, Paillière et al. (1989) also developed a TSTM equipment in which a dog-bone sample was tested in vertical position. However, this device cannot be used at early age because its specimen must be firstly casted in a horizontal mold before its placement in the TSTM. Moreover, this device is not equipped with a thermal regulation and the vertical position of the sample can lead to an early cracking localized in its upper part as a consequence of the non-uniform stress distribution arising from the hanging self-weight. These first TSTM devices allow underlining several major parameters which must be taken into account for their design: the position of the deformation measurement system, the presence of a thermal regulation on the concrete sample and the sample position (Fig. 6.9).

At the beginning of the 1990s, a TSTM was set up at the Israel Institute of Technology (Technion) at Haifa (Bloom and Bentur 1995; Igarashi et al. 2000). The free and restrained shrinkages were simultaneously measured on samples that differ only by one of their extremities, which is free for the sample used for measuring the free shrinkage and which is piloted in the other case. With this equipment, the

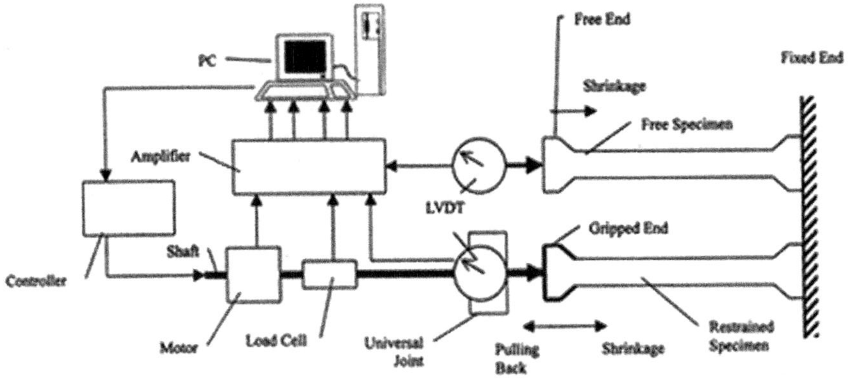

Fig. 6.9 System for monitoring free and restrained shrinkage at the Israel Institute of Technology (Igarashi et al. 2000)

deformation was measured by means of LVDT sensors placed on the moving head of the restrained sample. It has been demonstrated that this kind of method can generate some artifacts of displacement measurements (Altoubat and Lange 2002). Indeed, the interaction between the grip and the concrete sample can pollute the displacement measurements and induces a decrease in the restrained degree of sample. Moreover, a sliding between the sample and the grip is always possible. During these tests, the thermal regulation was provided by means of the room air conditioning only.

Several tests on TSTM's devices were carried out for several applications at the Norwegian University of Sciences and Technology (NTNU) at Trondheim in Norway (Bjontegaard 1999; Ji et al. 2018) and at the Technical University of Delft (DTU) in the Netherlands (Lokhorst 1998; Lura 2003). Tests on high performance concrete were realized at NTNU. Bjøntegaard investigated the effect of different isothermal curing temperatures and realistic histories of temperature on high performance concretes (Bjontegaard 1999). For this purpose, they equipped the TSTM mold with a thermal regulation system. This system consists of copper pipes disposed along the mold. It allows obtaining an isothermal temperature of 20 °C after about 1–2 h. Moreover, the temperature increase due to the hydration of a Portland cement concrete sample is limited to 1.5 °C. However, a thermal gradient in the longitudinal and transverse axis always exists because the temperature of the liquid coolant varies from its initial temperature during its travel following its calorific losses. They also demonstrated that the value of the deformation threshold directly affects the value of Young's modulus calculated from each compensative cycle. From these first experimental results, it seems better to evaluate the value of the elastic modulus from complementary tests.

More recently, several university laboratories, namely at the University of Illinois at Urbana Champaign in the USA (Altoubat and Lange 2001), the Swiss Federal Institute of Technology in Lausanne in Switzerland (Kamen et al. 2008), the Monash University in Australia (Aly and Sanjayan 2008), the University of Tokyo in Japan (Kishi and Lin 2008) and the Polytechnical School of Sao Paulo in Brazil (Melo Neto et al. 2007), equipped themselves with a TSTM equipment. Although these equipments bear some similarities (a dog-bone sample, a horizontal position for the specimen …), it appears that their design and their test method differ by several points. These differences are indicated in the Table 6.2 and summarized below:

- Position (top or sides of the straight part of dog-bone specimen) and accuracy of the LVDT sensors, as well as the gauge length generally included between 500 and 750 mm.
- Size of the sample: length of the straight part (about 1000 mm), cross section (from $40 \times 40\ mm^2$ to $150 \times 150\ mm^2$).
- Existence of a thermal regulation system around the specimen for stabilizing the temperature at the beginning of the test, keeping an isothermal temperature during the test and limiting the thermal gradient in the sample.
- System for controlling the displacement of concrete sample during the compensation cycles (manual or automatic).
- Kind of device for monitoring the free shrinkage (rectangular or dog-bone sample).

Table 6.2 Differences between TSTM systems

References	Springenschmid et al. (1994)	Igarashi et al. (2000), Bloom and Bentur (1995)	Bjontegaard (1999)	Grazia (1999), Charron (2003)	Lokhorst (1998), Lura (2003)
Test start	0.01 MPa	24 h	Setting time	24 h	8 h
Deformation threshold (μm/m)	2	5	0.86	6	6
Sample cross section	150 × 150 mm	40 × 40 mm	90 × 100 mm	50 × 50 mm	150 × 150 mm
Sample length (mm)	1500	1000	1000	1000	1000
Gauge length (mm)	500	–	700	750	750
Deformation measurement	Side	Mobile head	Side	Top	Side
Thermal regulation	Yes	Room conditioning	Yes	Yes	Yes

- Value of the deformation threshold used for obtaining the restriction conditions which varies from a value close to 0 (0.1; 0.6 μm/m) to a value equal to a few microstrains (4; 6 μm/m).
- Starting time of the restrained shrinkage test: setting of material, stress threshold, material age …

These TSTM devices allowing the study of the visco-elastic behaviour of concrete at early age respect the recommendations of the technical committee Rilem 42-CEA (Rilem 42-CEA 1981) for uniaxial tensile testing of concrete at early age:

1. The experimental tests are performed horizontally to avoid the dead load effect of specimen.
2. The friction between the sample and the mold is minimized by using specific material like the combination of Teflon and plastic film.
3. The sample is prismatic and its extremities are larger than its central part.
4. The length of the longitudinal central part of the sample is at least four times higher than that of its cross section.

So, these different parameters have been taken into account in the design with the integration of a thermal regulation inside the mold of the TSTM sample at the laboratory of the Université Libre de Bruxelles (ULB). Moreover, an appropriate test method for studying the restrained shrinkage has been also chosen to limit the early cracking of the sample. The details of the revisited TSTM system are presented in the next paragraph.

6.4 Design Testing System

6.4.1 Test Setup

A revisited TSTM system was developed (since 2006) in the laboratory of civil engineering at the ULB for testing concrete since setting time under free and restrained conditions. For this purpose, the testing machine is a Walter + Bay LFMZ 400 kN electromechanical testing setup. The machine is totally programmable and controlled (force and displacement of each sensor) by computer. The machine is composed by a fixed steel head, a central unidimensional part and a moving end. The moving end is controlled by a motor moving the steel head.

The transition area between the ends and the central part is characterized by a rounded shape in order to minimize a possible stress concentration and the risk of premature cracking in this zone. In the central part where the measurements of the displacements are taken, the stress field has been computed with finite element analysis and it was shown that the stress filed is homogenous (Baesens and Delsaute 2010). The shape of the mold is a dog-bone (Table 6.3). The dimensions of the cross section are $100 \times 100\ \text{mm}^2$ in the central part and $300 \times 100\ \text{mm}^2$ at the ends. The total length of the straight part equals to 1000 mm. With these dimensions, it is possible to perform tests for concrete with a maximal aggregate size of 20 mm. The shape of the mold induces stress concentration in the junction between the head and the straight part of the specimen. The failure of the sample takes often place in this part of the sample. For the study of the creep at high stress level, the stress concentration at the junction between the head and the straight part of the sample has to be decreased. For these reasons, the geometry of the head was replaced (Fig. 6.10).

The fresh concrete is cast inside the mold until a level of 100 mm. One T type thermocouple is placed in the middle of the sample during the casting, at each end and at mi-length of the central part of the mold. The temperature of the sample is controlled by a flow of a specific liquid for thermal regulation circulating on all the sides of the specimen. A plastic sheet is placed, before casting, in the mold to ensure sealed conditions. Moreover, the plastic sheet helps also to reduce, with the presence of Teflon, the friction between the sample and the mold. The walls of the mold are made of aluminum (Fig. 6.11) which was chosen for its high thermal conductivity (237 W/m K) and its low density (2700 kg/m^3). The deflection of the mold is very limited (±2 mm) and the isothermal conditions are ensured thanks to a fast heat

Table 6.3 Dimensions of the specimen in the TSTM device

Length of the span (m)	1
Width of the span (m)	0.1
Radius of curvature (m)	0.5
Width of head (m)	0.3
Length of head (m)	0.1

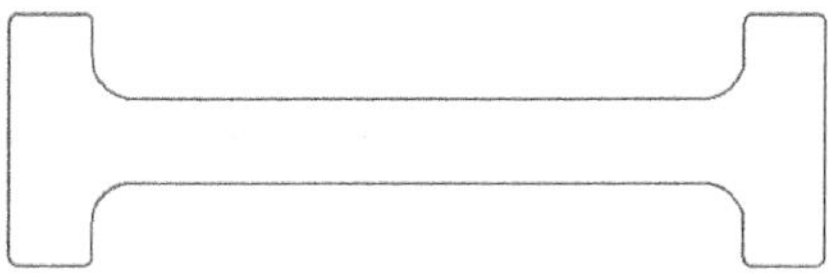

Fig. 6.10 New geometry of the head

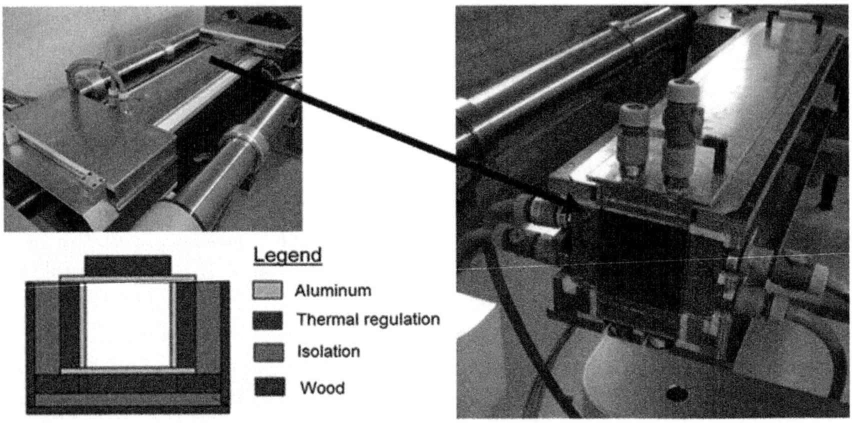

Fig. 6.11 Thermal insulation

transfer with the zinc boxes. A specific liquid for thermal regulation flows inside zinc boxes beams which are placed all around the central part and on the ends and under them. Each box is independent and has its own water input and output system. A thermal insulation limits exchanges with the ambient environment. The equipment is located in an air-conditioned room with a control system of the temperature and the relative humidity.

In addition to the creep and elastic deformations, thermal and shrinkage deformation must also be known. For this purpose, a dummy mold was built for the measurement of the thermal and free shrinkage deformations. This mold has exactly the same geometry as the first one. The only difference is the total free movement of one of the ends (Fig. 6.12).

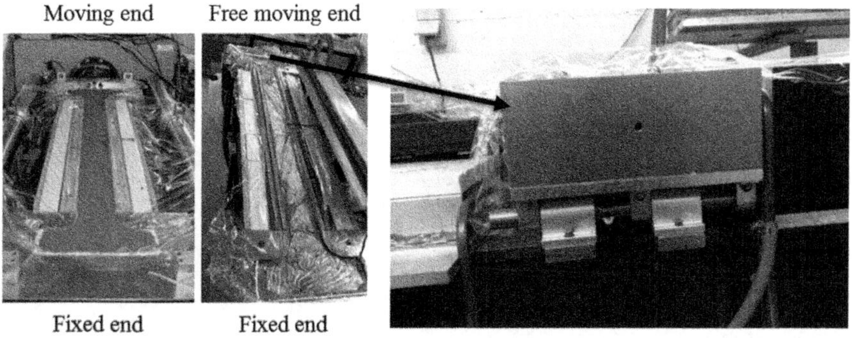

Fig. 6.12 Left to right: mold (boundary conditions)—dummy mold (free moving end thanks to ball bearings)

In this equipment, several instrumentations were developed at ULB for the monitoring of the concrete deformation. For the first instrumentation, the displacement is measured by means of two transducers (Solartron LE12). The displacement sensors are placed on invar supports which are fixed on a rigid frame made of aluminum tubes externally supported by the TSTM (Fig. 6.13). The distance between both sensors is 750 mm (where the stress field is homogenous in the sample). These sensors are characterized by a measuring range of 12 mm and accuracy of about 0.4 μm.

Invar rods are anchored in the concrete at a depth of 50 mm. The link between the concrete displacement and the sensor is then assured. These invar rods are characterized by a low dilatation coefficient limiting the effect of ambient temperature on the deformations measurement. They are kept in their initial position thanks to a brass ring before the beginning of the test. The horizontality of the TSTM device is an advantage for the efficiency of the casting. Moreover, sensors are well anchored at mid-height and not in the superficial layer of the sample, what could have induced an error by a possible different behavior between the core layer and the superficial layer (less aggregates presence) (Fig. 6.14).

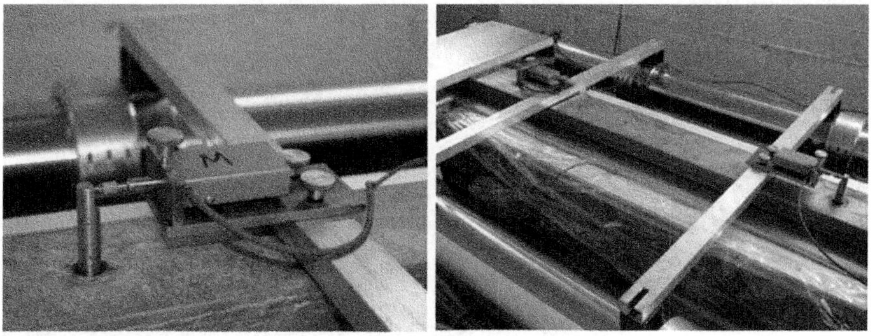

Fig. 6.13 Displacement sensors in the central part of the specimen

Fig. 6.14 Stem (Invar, threaded stem in steel) with plastic plug and brass ring (**a**)—incorporate stem in the concrete (**b**)—embedding system in steel of the Invar rod (**c**) (Darquennes 2009)

In order to improve the displacement measurement since setting time and to perform relaxation test, different changes were carried out on the instrumentation. The aluminum tubes connecting the columns of the test device and the displacement sensors have been replaced by solid steel bars for which the thermal inertia is higher and the CTE is lower. The influence of thermal variation in the room is therefore lower on the measurement of displacements. Displacement sensors in the central part of the specimen were replaced by Foucault current's sensors (without contact sensor) (Fig. 6.15). Sensors have a resolution of 0.014 μm and a measurement range of 2 mm.

These sensors have two advantages. Firstly, the absence of contact between the sensors and the mold avoids measurement artifacts induced by the spring of the LVDT sensor when the stiffness of the concrete is very low. Secondly, an instantaneous Volt conversion in micrometer of the sensors allows piloting the TSTM system with the displacement directly measured in concrete. A real time subtraction between the

Fig. 6.15 Displacement sensor without contact

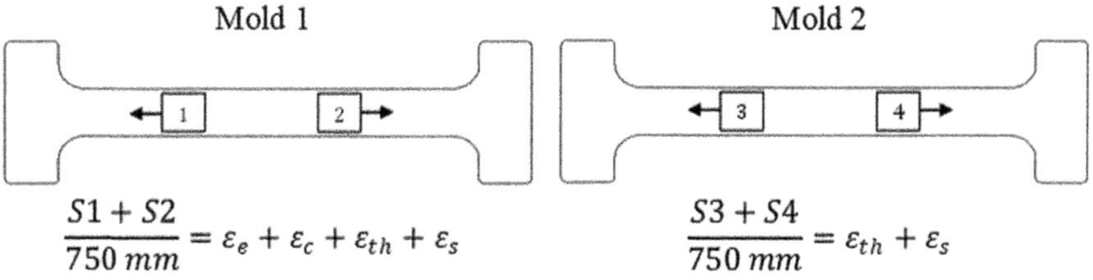

Fig. 6.16 For relaxation test, TSTM device is controlled by (S1 + S2)–(S3 + S4) which corresponds to creep and elastic deformations

deformation of the specimen in the TSTM and the specimen in the companion (passive) mold allows piloting the TSTM system on basis of the mechanical parameters excluding the free deformations due to thermal variations ε_{th} and due to shrinkage ε_{sh}. So, the direct measurement of the elastic ε_e and creep ε_c deformations or the relaxation stresses is possible with the simultaneous use of the first and the second mold as illustrated in Fig. 6.16.

Without contact sensors are very sensitive to the air conditioning. A good protection of the sensor is needed. Figure 6.17 shows the perturbation caused by the air conditioning on the sensor when repeated loadings in tension are applied on a concrete sample (Delsaute 2016). The different "waves" are in direct relation with the air conditioning. The use of displacement sensor with contact as LVDT decreases the impact of the air conditioning. However this problem was solved by protecting the sensor from the air conditioning. A wood box was designed in order to protect completely the sensor from the ambient air.

For some tests, the failure occurs in the straight part of the specimen where the anchorage is placed. To decrease the risk of failure and to avoid a preliminary damage in this zone, the plastic plug was removed to limit the reduction of the concrete cross section. However different problems were observed for the measurement of the

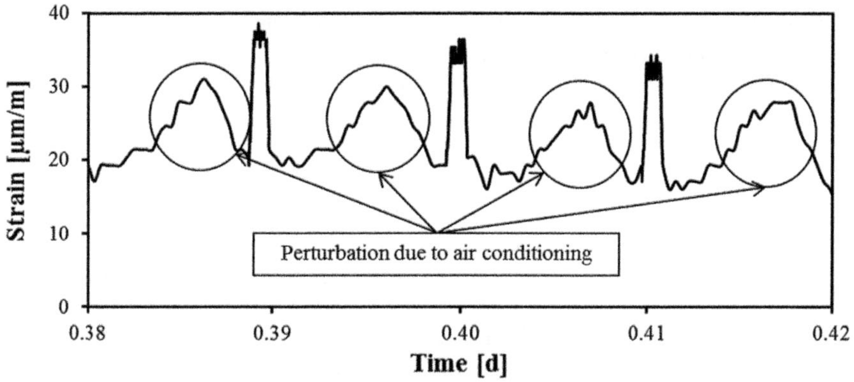

Fig. 6.17 Perturbation on the sensor of displacement caused by the air-conditioning

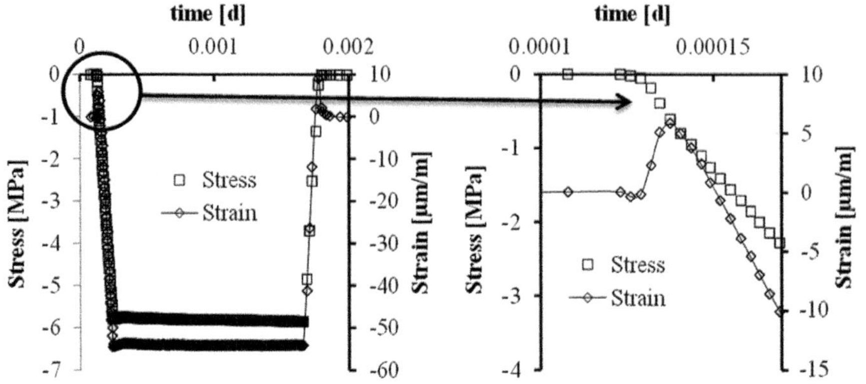

Fig. 6.18 Illustration of adherence/deterioration in the concrete around the stem anchorage in the sample

Young's modulus in tension and in compression. Damage by fatigue phenomenon is observed. A problem of adherence/deterioration was also observed for the anchorage of the stem as illustrated in Fig. 6.18 for one loading. During the beginning of loading, when a compressive load is applied to the sample, the stress increases as expected but the strain increases as if a tensile load is applied. Then the strain develops in the good direction. This problem is assumed to be a wobble in the sample due to damage around the anchorage. This phenomenon was observed in different concrete when repeated loadings are applied. The higher the W/C ratio and sooner this phenomenon appeared. For repeated loading, internal anchorage has to be disallowed.

For these reasons, a new experimental measurement tool was designed for the measurement of the longitudinal strain. This tool is an extensometer which is adapted to the TSTM device. The transversal measurement is added for the study of the Poisson's ratio and the creep dilatancy. Here are the different criteria which are on the basis of the design of the extensometer:

- For the limitation of the thermal dilation of the extensometer, all structural elements of the extensometer are in INVAR©.
- All elements for the measurement of longitudinal and transversal displacements have to be protected from the air conditioning.
- Sensors with a high accuracy are needed for the measurement of the transversal displacement. For this purpose, Solartron LE12 sensors (digital sensor) are used.
- External anchorages are used to avoid internal damage in the section of the concrete specimen. Three elastic anchorages from the J2P company (Boulay and Colson 1981) have been used to assure a good contact between the extensometer and the sample.
- Measurement of the transversal displacement has to be done in the same section as the anchorage to follow the longitudinal displacement and avoid artifacts of measurements.

Fig. 6.19 New extensometer designed for the TSTM device

- A free lateral displacement of the specimen has to be allowed during the test to avoid confinement effect when a compressive load is applied.

From these criteria, an extensometer composed of two parts with a U shape has been designed. Figure 6.19 shows the new extensometer.

For the study of the restrained shrinkage since setting time, such instrumentation cannot be used. Indeed, in that case the test must start just after the setting. The implementation of the extensometer takes time and needs to remove the formwork. For this reason, all tests can be performed with the new extensometer except test of restrained shrinkage. For restrained shrinkage, anchorage rod system is used.

Finally, for the study of the restrained shrinkage under drying condition, the upper part and/or the lateral part of the mold and the thermal regulation is removed in order to expose the sample to the controlled environment (temperature and relative humidity). To simulate the case of a slab, only the upper part of the thermal regulation is removed (Bendimerad et al. 2020; Delsaute et al. n.d.) while the case of the formwork removal of a wall is simulated by removing the lateral faces of the mold and the thermal regulation.

6.4.2 Test Protocol for the Restrained Shrinkage

Historically, the TSTM was developed for the study of the restrained shrinkage. A test method has been defined to measure the restrained shrinkage (Fig. 6.20). At the beginning of the test, the sample placed in the TSTM is initially restraint by the stiffness of the frame with the motor being turned off until the stress inside the concrete reaches a threshold value of 0.01 MPa. At that moment, the displacement transducer readings are set to zero and the concrete sample can deform freely until the recorded deformation in the central part of the specimen reaches a deformation

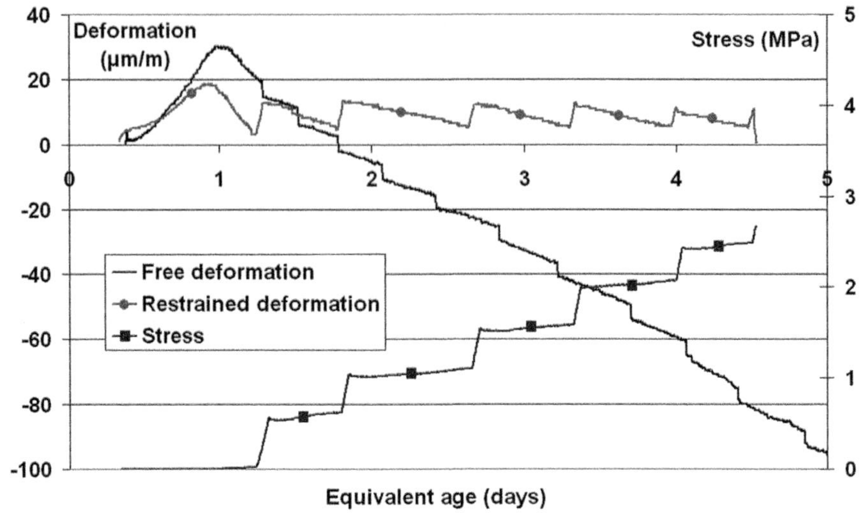

Fig. 6.20 Free and restrained shrinkage evolution of a Portland cement concrete and stress build-up in the concrete sample (Darquennes 2009)

threshold corresponding to 6.7 µm/m. This moment coincides with the end of the first cycle. From this time on, the load is adjusted to pull the specimen back to its initial length. Then, the applied load at the end of the adjustment process is kept constant throughout the following cycles until the sample deformation reaches again the deformation threshold value. The test goes on until the specimen cracks. This kind of experimental process is quite similar to the one used by Charron (2003) and allows avoiding an early cracking of the specimen. During this test, four parameters were continually monitored: the load applied on the sample, its deformation, its temperature and the displacement of the moving head. For concrete composition with high W/C ratio (Delsaute and Staquet 2017; Delsaute et al. 2016c) or high content of slag (Carette et al. 2018b) or in case of high increase of temperature due to hydration, a swelling is observed just after the final setting time. To consider the restrained of the swelling strain, the test protocol has been adapted (Carette et al. 2018a). The test starts at the final setting time. At that time the strains are set to zero. When the strain in the central part of the specimen reaches a strain threshold corresponding to −6.7 µm/m or +6.7 µm/m, a tensile or a compressive load is applied to put the specimen back to its initial length.

6.5 Applications

The use of TSTM devices was mainly related to the study of the restrained shrinkage at the laboratory scale on concrete as presented in the previous sections. However, during this last decade, other kinds of TSTM devices were developed whether for industrial applications or for a better physical understanding of the physical mechanisms related to early age cracking of cementitious materials. Several new kinds of device and associated applications are presented succinctly below.

6.5.1 Monitoring of the Viscoelastic Properties Since Setting

As presented in Chap. 4, repeated minute-scale-duration loadings method was developed for the monitoring and the modelling of the viscoelastic properties of cement based materials in three laboratories: the Université Libre de Bruxelles (ULB), the Institut Français des Sciences et Technologies des Transports, de l'Aménagement et des Réseaux (IFSTTAR) and the Technische Universität Wien (TUWIEN). At ULB and Ifsttar, the main application of this test was to monitor and to model accurately the viscoelastic behavior at very early age (Boulay et al. 2012, 2013, 2014; Delsaute et al. 2012, 2013, 2014, 2016a, b, c, 2017; Delsaute and Staquet 2014; Staquet et al. 2014). In order to apply repeated loadings since the final setting time, a TSTM device was used at ULB. At TUWIEN, the new testing methodology was used for the identification of creep properties at the microstructural scale and also to study the water transfer between aggregate and cement paste during hydration (Ausweger et al. 2019; Göbel et al. 2018a, b; Irfan-ul-Hassan et al. 2016, 2017; Karte et al. 2015; Königsberger et al. 2016).

6.5.2 Implementation of Non-destructive Methods on a TSTM Device

To characterize experimentally the internal restriction of the cement paste caused by the aggregate, Pirskawetz et al. (2012) have developed a TSTM device which combines non-destructive methods. Ultrasound velocity method and acoustic emission have been integrated to the TSTM mold. These methods allow monitoring continuously the damage process in hardening material. Ultrasound velocity method characterizes the development of the microstructure during hardening while acoustic emission has been used to observe and detect the formation and the development of micro-cracks. The TSTM device was designed for cement paste and mortar (Fig. 6.21). The central part of the TSTM has a length of 34 cm and a cross section of 4 cm × 4 cm. Ultrasonic sensors are placed at the end of both extremities while the 3 acoustic emission sensors are put on the upper part of the measuring part.

It should be noticed that the first TSTM device designed for cement paste scale was developed in Delft University of Technology for the study of the development of the autogenous strain and the development of shrinkage-induced stresses (Lura 2003; Zhou et al. 2006). This device has a cross section of 7.5 mm × 7.5 mm and a measurement length of 37.5 mm.

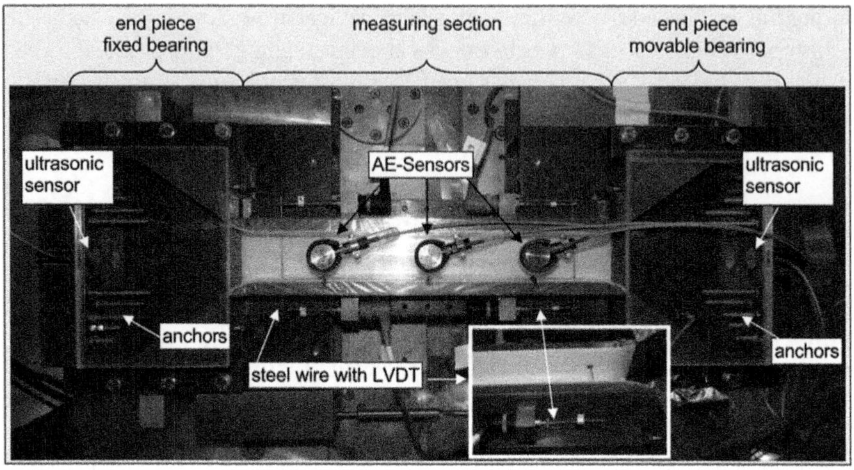

Fig. 6.21 TSTM device combining acoustic emission and ultrasound velocity method (Pirskawetz et al. 2012)

6.5.3 Testing Concrete in Its Plastic State

TSTM device were generally designed for the study of the behavior of cement based materials during the hardening phase. However, concrete elements, with a large exposed surface, are sensitive to cracking when concrete is still in a plastic state. Different curing procedures were developed to reduce the cracking risk of plastic concrete. One way to prevent cracking consists periodically in applying a fog spray on the exposed surface. Such method is responsible of wetting and drying cycles and thus causes variation of capillary pressure inside concrete. This corresponds to cyclic loadings of concrete during its plastic state. In order to simulate this situation, Khan (2018) and Khan et al. (2017) have developed a TSTM device which is enable to measure the tensile strength, and stress relaxation properties of plastic concrete as well as its resistance to cyclic loading till the end of setting. The device is shown in Fig. 6.22. The whole test setup is supported by a beam. The measurement length is 100 mm and the total length of the sample is 618 mm. One LVDT sensor is placed on the top of the specimen in the central part for the measurement of the concrete deformation. Concrete was tested in a climate control room.

6.5.4 Influence of Cyclic Loading/Displacement on the Hardening Process of Grout Material

The foundation of offshore wind turbines in a seabed is a major technical challenge for engineers. To insure the stability of the wind turbine, the ring between the steel

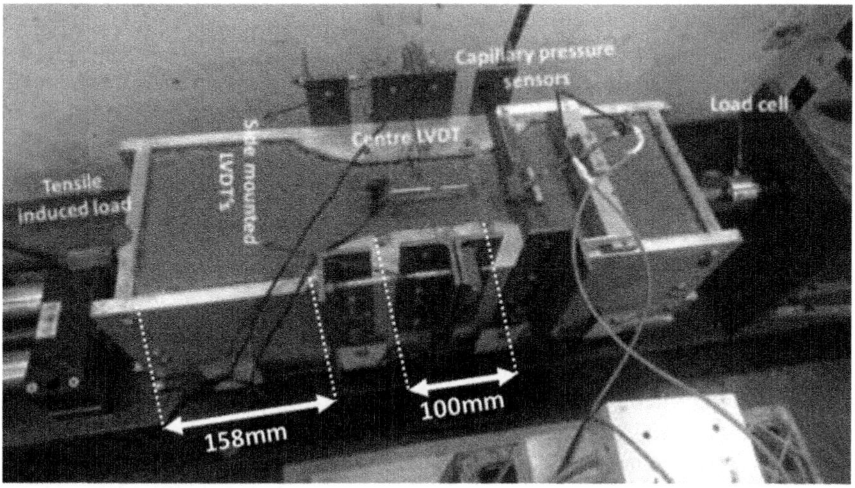

Fig. 6.22 TSTM device developed to study the influence of tensile cyclic loading and relaxation on plastic concrete (Khan 2018)

monopile and the seabed is filled with a grout. Since casting, the grouted connections have to carry the loadings resulting from waves acting on the monopile. As illustrated in Fig. 6.23, this leads to cycling loading of the grout and further to potential local damage in form of micro-cracks if the associated stress/strength ratio is high enough. The damage might accumulate over time and might eventually lead to a significant reduction of the local mechanical strength and local durability properties of the grout. In addition, the application of cyclic loadings can lead to irreversible plastic strain which can induce the formation of a gap at the interface between the monopile and the grouted connections (Lohaus et al. 2015). In order to point out the influence of early age movements and deformational restraints on the performance of a grouted connection, Delsaute (2019) and Delsaute et al. (2018) have developed a new testing methodology in which a TSTM device has been used. The test protocol was developed in order to reproduce the displacement/loading imposed by the waves on the grout

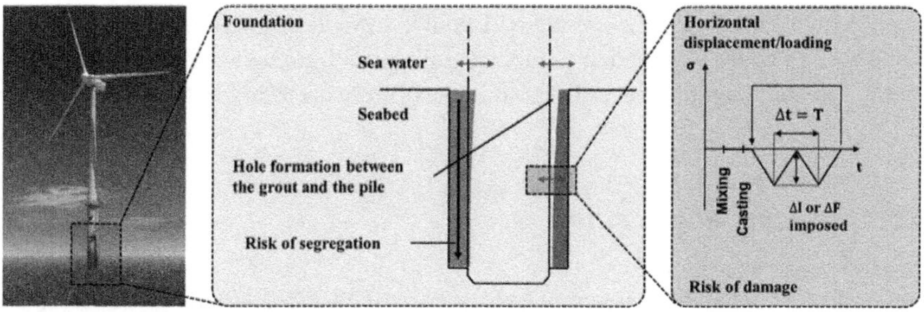

Fig. 6.23 Loading case induced by the cyclic movement of the wave on the grout (Delsaute 2019)

annulus during the construction process. Firstly, cyclic displacements are applied to the sample since an age of 1 h till the moment that a force threshold is reached (around the initial setting). Then cyclic loadings are applied on the sample. At this moment, displacement sensors were activated in order to monitor the displacement of the anchored rod in the straight part of the sample. From this moment, the recording of the displacement was used to compute the E-modulus for each loading and unloading. At an age of 24 h the cyclic loadings are stopped. At the end of the test, the prismatic part of the TSTM and dummy specimens were removed from the mold and were cut in several cubes of 10 cm side for the determination of the compressive strength. Compressive strength from both specimens are compared in order to define a potential strength reduction induced by the cyclic loading. Results of the E-modulus obtained by means of cyclic loading are then compare to the E-modulus obtained by means of repeated loadings in order to evaluate the potential damage of the grout. Finally, the free strains of the loaded specimen and the dummy specimen are also compared in order to quantify the irreversible strain induced by the application of cyclic loadings. From this new experimental protocol, new indicators were developed in order to evaluate the performance of grout materials under cyclic displacement and loading since casting.

6.5.5 Structural Scale

At the Technical University of Graz, a TSTM device has been elaborated at the structural scale. Such device was created for the development of a mechanically consistent design model for crack width concrete in restrained concrete members (Turner et al. 2015). For this purpose, the sample has a dog bone shape with a total length of 3.7 m and a cross section of 25 cm $\times$ 25 cm in the central part. The stiffness of the whole frame has been designed to simulate restrained shrinkage of reinforced concrete structure. In addition, hydraulic cylinders have been placed at one end of the steel columns and are used to impose displacement to the concrete member in order to simulate load variation induced by the seasonal temperature change. At the other end of the columns, a load cell was placed to monitor the force induced by the restrained of the free deformation of the concrete member and by the displacement imposed with the hydraulic cylinders. Finally, variable insulation system has been developed in order to simulate the thermal history of concrete element induced by hydration heat. Complete details about the test setup are given in the Chap. 7.

6.5.6 Degree of Restraint

Generally, in concrete structures, only a part of the free strains of concrete are restrained. 2D or 3D linear elastic calculation can be performed in order to evaluate the degree of restriction of a concrete element by the structure [see e.g. (Schlicke and Matiašková 2019)]. Thus, the degree of restraint in early age concrete structure depends on the location of the concrete element and the general configuration of the structure (Klausen 2016). At NTNU, Klausen et al. (2015) have updated a TSTM device in order to control the degree of restraint R (between 0 and 100%) of a concrete sample. Also, the test protocol was adapted. When a measured length change in the TSTM exceeds a displacement threshold χ, a displacement is automatically applied to cancel partially the deformation and corresponds to $-R \cdot \chi$. The concrete sample can shrink again and the load is kept constant till a new length change of χ is reached. Then another partial compensation cycle ($-R \cdot \chi$) begins. At Tsinghua University, Zhu et al. (2017, 2018) have developed a multiple TSTM system named "Synchronous closed loop federated control TSTM system". Four TSTM systems are placed in a same room which is controlled in temperature and relative humidity. The degree of restraint of each TSTM system can vary between 0 and 100% and the mold of each TSTM can be detached from the test rig. Such configuration allows casting the concrete directly on a construction site.

6.6 Conclusions

In this chapter, a succinct state-of-the-art of passive and active restrained shrinkage test method for the evaluation of the risk of cracking of cementitious materials under restrained condition has been presented. In the 1990s, a new experimental concrete testing concept has been designed: the Temperature Stress Testing Machine (TSTM). The device is able to monitor the mechanical behaviour of concrete since setting time under restrained conditions. Actually, the restrained shrinkage test set up designed with a TSTM system takes into account all the parameters affecting the stresses development inside the concrete sample: the degree of restriction, the relaxation of stresses, the evolution of the strength and the stiffness … On basis of this kind of test, several parameters can be quantified such as the elastic deformation, free and restrained shrinkages and also, creep and relaxation.

Between 1990 and 2010, less than fifteen laboratories worldwide have developed or acquired this kind of experimental equipment. All devices bear some similarities but major differences remain in the test set up designs (e.g. sample dimension, measurement length, instrumentation, thermal regulation), in the testing processes (test start, deformation threshold) and also in the scale of the material used (cement paste, mortar or concrete). Based on the development of several TSTM devices, a revisited TSTM system has been designed at ULB for monitoring the mechanical properties of concrete in tension and in compression since setting time. Several updates have been operated during this last decade in order to improve the device for the study of the restrained shrinkage and to extend its use to other applications. The design of this system includes a thermal regulation in contact with all sides of the concrete sample and, especially, for the very early age, displacement sensors without contact to limit the measurement artefacts on the concrete. A dummy mold was added to the revisited TSTM system for monitoring the free shrinkage and the thermal deformation.

So, the direct measurement of the creep deformation and the relaxation stresses is possible with the simultaneous use of the first and the dummy mold. The test method for measuring the early age properties of concrete must also avoid an early cracking of the concrete sample. To do so, the test method starts after the end of setting time of concrete when the stiffness of the studied material is sufficient.

This test device can offer much to the characterization of cement-based materials. Several types of tests can be performed according to the intended application. The TSTM device is able to simulate experimentally the situation of a concrete member (e.g. temperature, degree of restraint) which makes the TSTM a key tool for the study of the thermo-mechanical response of cement based materials since very early age.

References

Altoubat, S. A., & Lange, D. A. (2001). Creep, shrinkage, and cracking of restrained concrete at early age. *ACI Materials Journal, 98*(4), 323–331.

Altoubat, S. A., & Lange, D. A. (2002). Grip-Specimen interaction in uniaxial restrained tests. In *Concrete: Materials science to applications* (pp. 189–204), edited by ACI.

Aly, T., & Sanjayan, J. G. (2008). Shrinkage cracking properties of slag concretes with one-day curing. *Magazine of Concrete Research, 60*(1), 41–48.

Ausweger, M., Binder, E., Lahayne, O., Reihsner, R., Maier, G., Peyerl, M., et al. (2019). Early-age evolution of strength, stiffness, and non-aging creep of concretes: Experimental characterization and correlation analysis. *Materials, 12*(2), 207.

Baesens, P., & Delsaute, B. (2010). Comportement de Bétons Au Jeune Âge En Conditions de Déformations Libres et Restreintes. Université Libre de Bruxelles.

Banthia, N., Azzabi, M., & Pigeon, M. (1993). Restrained shrinkage cracking in fibre-reinforced cementitious composites. *Materials and Structures, 26*(7), 405–413.

Banthia, N., Azzabi, M., & Pigeon, M. (1995). Restrained shrinkage tests on fiber reinforced cementitious composites. *ACI Special Publication, 155,* 137–152.

Banthia, N., Yan, C., & Mindess, S. (1996). Restrained shrinkage cracking in fiber reinforced concrete: A novel test technique. *Cement and Concrete Research, 26*(1), 9–14.

Benboudjema, F., Carette, J., Delsaute, B., Honorio de Faria, T., Knoppik, A., Lacarrière, L., et al. (2019). Mechanical properties. In E. M. R. Fairbairn and M. Azenha (Eds.), *Thermal cracking of massive concrete structures—State of the art report of the rilem technical committee 254-CMS* (pp. 69–114).

Bendimerad, A. Z., Delsaute, B., Roziere, E., Staquet, S., & Loukili, A. (2020). Advanced techniques for the study of shrinkage-induced cracking of concrete with recycled aggregates at early age. *Construction and Building Materials.*

Bentur, A. (2003a). Evaluation of early age cracking characteristics in cementitious systems. *Materials and Structures, 36*(257), 183–190.

Bentur, A. (2003b). Terminology and definitions. In A. Bentur (Ed.), *Early age cracking in cementitious systems—Report of RILEM technical committee 181-EAS—Early age shrinkage induced stresses and cracking in cementitious systems* (pp. 13–15). RILEM Publications SARL.

Bjontegaard, O. (1999). Thermal dilation and autogenous deformation as driving forces to self-induced stresses in high-performance concrete.

Bloom, R., & Bentur, A. (1995). Free and restrained shrinkage of normal and high-strength concretes. *ACI Materials Journal, 92*(2), 211–217.

Boulay, C., & Colson, A. (1981). Un Extensomètre à Béton Éliminant l'influence Des Déformations Transversales Sur La Mesure Des Déformations Longitudinales. *Materials and Structures, 14*(79), 35–38.

Boulay, C., Crespini, M., Delsaute, B., & Staquet, S. (2012). Monitoring of the creep and the relaxation behaviour of concrete since setting time, Part 1 : Compression. In *Strategies for sustainable concrete structures* (p. 10). Aix-en-Provence.

Boulay, C., Staquet, S., Azenha, M., Deraemaeker, A., Crespini, M., Carette, J., et al. (2013). Monitoring elastic properties of concrete since very early age by means of cyclic loadings, ultrasonic measurements, natural resonant frequency of composite beam (EMM-ARM) and with smart aggregates. In *Proceedings of the 8th International Conference on Fracture Mechanics of Concrete and Concrete Structures, FraMCoS 2013*.

Boulay, C., Staquet, S., Delsaute, B., C, J., C, Michela, Yazoghli-Marzouk, O., et al. (2014). How to monitor the modulus of elasticity of concrete, automatically since the earliest age? *Materials and Structures, 47*(1–2), 141–155.

Bourchy, A. (2018). Relation Chaleur d'hydratation Du Ciment : Montée En Température et Contraintes Générées Au Jeune Âge Du Béton. Université Paris-Est.

Briffaut, M., Benboudjema, F., Torrenti, J. M., & Nahas, G. (2011a). A thermal active restrained shrinkage ring test to study the early age concrete behaviour of massive structures. *Cement and Concrete Research, 41*(1), 56–63.

Briffaut, M., Benboudjema, F., Torrenti, J. M., & Nahas, G. (2011b). Numerical analysis of the thermal active restrained shrinkage ring test to study the early age behavior of massive concrete structures. *Engineering Structures, 33*(4), 1390–1401.

Carette, J., Delsaute, B., & Staquet, S. (2018a). Estimating the stress development in early age concrete with mineral additions from coupled measurements. In *Interdisciplinary Approaches for Cement-based Materials and Structural Concrete: Synergizing Expertise and Bridging Scales of Space and Time* (pp. 171–176).

Carette, J., Joseph, S., Cizer, Ö., & Staquet, S. (2018b). Decoupling the autogenous swelling from the self-desiccation deformation in early age concrete with mineral additions: Micro-macro observations and unified modelling. *Cement and Concrete Composites, 85,* 122–132.

CEOS.fr. (n.d.). CEOS.Fr Comportement et Evaluation Des Ouvrages Spéciaux Vis-à-Vis de La Fissuration et Du Retrait." Retrieved January 23, 2019. https://www.ceosfr.irex.asso.fr/en/.

Charron, J. P. (2003). Contribution à l'étude Du Comportement Au Jeune Âge Des Matériaux Cimentaires En Conditions Des Déformations Libre et Restreinte. Ph.D. thesis, Université Laval.

Darquennes, A. (2009). Comportement Au Jeune Âge de Bétons Formulés à Base de Ciment Au Laitier de Haut Fourneau En Condition de Déformations Libre et Restreinte. Ph.D. thesis, Université Libre de Bruxelles.

Darquennes, A., Staquet, S., & Espion, B. (2006). Shrinkage of slag cement concrete in free and restrained conditions. In O. M. Jensen, P. Lura, & K. Kovler (Eds.), *International RILEM Conference on Volume Changes of Hardening Concrete: Testing and Mitigation* (p. 10).

Delsaute, B. (2016). New approach for monitoring and modelling of the creep and shrinkage behaviour of cement pastes, mortars and concretes since setting time. Université Libre de Bruxelles (BATir) and Université Paris Est (IFSTTAR).

Delsaute, B. (2019). Influence of cyclic movement on the hardening process of grout: case of offshore wind turbine installation. In G. Pijaudier-cabot, P. Grassl, & C. La Borderie (Eds.), *10th International Conference on Fracture Mechanics of Concrete and Concrete Structures, FraMCoS-X* (p. 11). Anglet.

Delsaute, B., Boulay, C., Granja, J., Carette, J., Azenha, M., Dumoulin, C., et al. (2016a). Testing concrete E-modulus at very early ages through several techniques: An inter-laboratory comparison. *Strain, 52*(2), 91–109.

Delsaute, B., Boulay, C., & Staquet, S. (2016b). Creep testing of concrete since setting time by means of permanent and repeated minute-long loadings. *Cement and Concrete Composites, 73,* 75–88.

Delsaute, B., Carette, J., & Staquet, S. (2013). Monitoring of the creep and the relaxation at very early age: Complementary results on the CEOS concrete. In *VIII International Conference on Fracture Mechanics of Concrete and Concrete Structures (IA-FramCOS-8)* (pp. 453–458).

Delsaute, B., Furnémont, R., Königsberger, M., & Staquet, S. (2018). Influence de Mouvement Cyclique Sur Le Durcissement de Coulis: Cas Des Éoliennes Offshores. In *19ème édition des Journées Scientifiques (RF)* ^{2}*B (Regroupement Francophone pour la Recherche et la Formation sur le Béton* (p. 10). Bayonne, France.

Delsaute, B., Hamami, A., Rozière, E., Staquet, S., & Loukili, A. (n.d.). Autogenous and drying shrinkage induced stresses in early age CEM I concrete – influence of the nature and the porosity of gravel. In *Cement and Concrete Composites*.

Delsaute, B., & Staquet, S. (2014). Early age creep and relaxation modelling of concrete under tension and compression. In *CONMOD 2014: Proceedings of the RILEM International Symposium on Concrete Modelling* (pp. 12–14).

Delsaute, B., & Staquet, S. (2017). Decoupling thermal and autogenous strain of concretes with different water/cement ratios during the hardening process. *Advances in Civil Engineering Materials, 6*(2).

Delsaute, B., & Staquet, S. (2019). Development of strain-induced stresses in early age concrete composed of recycled gravel or sand. *Journal of Advanced Concrete Technology, 17*(6), 319–334.

Delsaute, B., Staquet, S., & Boulay, C. (2012). Monitoring of the creep and the relaxation behaviour of concrete since setting time, Part 2 : Traction. In *Strategies for Sustainable Concrete Structures* (p. 10). Aix-en-Provence.

Delsaute, B., Staquet, S., & Boulay, C. (2014). Early age creep and relaxation behavior of concrete under tension and compression. In *Concrete Innovation Conference CIC 2014* (p. 10). Oslo.

Delsaute, B., Torrenti, J.-M., & Staquet, S. (2016c). Monitoring and modeling of the early age properties of the vercors concrete. In *TINCE 2016* (p. 12). Paris.

Delsaute, B., Torrenti, J. M., & Staquet, S. (2017). Modeling basic creep of concrete since setting time. *Cement and Concrete Composites, 83,* 239–250.

Dong, W., Yuan, W., Zhou, X., & Wang, F. (2018). The fracture mechanism of circular/elliptical concrete rings under restrained shrinkage and drying from top and bottom surfaces. *Engineering Fracture Mechanics, 189,* 148–163.

Göbel, L., Königsberger, M., Osburg, A., & Pichler, B. (2018a). Viscoelastic behavior of polymer-modified cement pastes: insight from downscaling short-term macroscopic creep tests by means of multiscale modeling. *Applied Sciences, 8*(4), 487.

Göbel, L., Osburg, A., & Pichler, B. (2018b). The mechanical performance of polymer-modified cement pastes at early ages: Ultra-short non-aging compression tests and multiscale homogenization. *Construction and Building Materials, 173,* 495–507.

Grazia, T. (1999). Comportement Des Bétons Au Jeune Âge. Ph.D. thesis, Université Laval.

He, Z., Zhou, X., & Li, Z. (2004). New experimental method for studying early-age cracking of cement-based materials. *ACI Materials Journal, 101*(1), 50–56.

Hossain, A. B., & Weiss, J. (2004). Assessing residual stress development and stress relaxation in restrained concrete ring specimens. *Cement and Concrete Composites, 26*(5), 531–540.

Hossain, A. B., & Weiss, J. (2006). The role of specimen geometry and boundary conditions on stress development and cracking in the restrained ring test. *Cement and Concrete Research, 36*(1), 189–199.

Igarashi, S. I., Bentur, A., & Kovler, K. (2000). Autogenous shrinkage and induced restraining stresses in high-strength concretes. *Cement and Concrete Research, 30*(11), 1701–1707.

Irfan-ul-Hassan, M., Königsberger, M., Reihsner, R., Hellmich, C., & Pichler, B. (2017). How water-aggregate interactions affect concrete creep: Multiscale analysis. *Journal of Nanomechanics and Micromechanics, 7*(4).

Irfan-ul-Hassan, M., Pichler, B., Reihsner, R., & Hellmich, Ch. (2016). Elastic and creep properties of young cement paste, as determined from hourly repeated minute-long Quasi-Static tests. *Cement and Concrete Research, 82,* 36–49.

Jaafri, R., Samouh, H., Roziere, E., Alam, S. Y., Wisniewski, V., & Loukili, A. (2019). Experimental and numerical analysis of curling behavior of natural hydraulic lime—Cement based mortars. *Cement and Concrete Research, 117,* 1–15.

Ji, G. M., Kanstad, T., & Bjøntegaard, Ø. (2018). Calibration of material models against TSTM test for crack risk assessment of early-age concrete containing fly ash. *Advances in Materials Science and Engineering, 9,* 1–11.

Kamen, A., Denarié, E., Sadouki, H., & Brühwiler, E. (2008). Thermo-mechanical response of UHPFRC at early age—Experimental study and numerical simulation. *Cement and Concrete Research, 38*(6), 822–831.

Kanavaris, F., Azenha, M., Soutsos, M., & Kovler, K. (2019). Assessment of behaviour and cracking susceptibility of cementitious systems under restrained conditions through ring tests: A critical review. *Cement and Concrete Composites, 95,* 137–153.

Karte, P., Hlobil, M., Reihsner, R., Dörner, W., Lahayne, O., Eberhardsteiner, J., et al. (2015). Unloading-based stiffness characterisation of cement pastes during the second, third and fourth day after production. *Strain, 51*(2), 156–169.

Khan, M.Y. (2018). *The tensile material properties of plastic concrete and the influence on plastic cracking*. Stellenbosch University.

Khan, M. Y., Kolawole, J. T., Boshoff, W. P., & Combrinck, R. (2017). Influence of relaxation and cyclic loading on the tensile material properties of plastic concrete. In S. Staquet & D. G. Aggelis (Eds.) *2nd International RILEM/COST Conference on Early Age Cracking and Serviceability in Cement-based Materials and Structures (EAC2)* (pp. 379–384). RILEM Publications S.A.R.L.

Kishi, T., & Lin, Z. (2008). A tentative experimental evaluation on early-age creep. In *Creep, shrinkage and durability mechanics of concrete and concrete structures* (pp. 285–291).

Klausen, A. B. E. (2016). *Early age crack assessment of concrete structures: Experimental investigation of decisive parameters*. NTNU.

Klausen, A. E., Kanstad, T., & Bjøntegaard, Ø. (2015). Updated temperature-stress testing machine (TSTM): Introductory tests, calculations, verification, and investigation of variable fly ash content. In *CONCREEP 2015: Mechanics and Physics of Creep, Shrinkage, and Durability of Concrete and Concrete Structures—Proceedings of the 10th International Conference on Mechanics and Physics of Creep, Shrinkage, and Durability of Concrete and Concrete Structure*.

Königsberger, M., Irfan-ul-Hassan, M., Pichler, B., & Hellmich, C. (2016). Downscaling based identification of nonaging power-law creep of cement hydrates. *Journal of Engineering Mechanics, 142*(12), 04016106.

Kovler, K. (1994). Testing system for determining the mechanical behaviour of early age concrete under restrained and free uniaxial shrinkage. *Materials and Structures, 27*(6), 324–330.

Kovler, K., & Bentur, A. (2009). *Cracking sensitivity of normal- and high-strength concrete*.

Kovler, K., Sikuler, J., & Bentur, A. (1993). Restrained shrinkage tests of fibre-reinforced concrete ring specimens: Effect of core thermal expansion. *Materials and Structures, 26*(4), 231–237.

Lohaus, L., Cotardo, D., Werner, M., Schaumann, P., & Kelma, S. (2015). Experimental and numerical investigations of grouted joints in monopiles subjected to early-age cycling. *Journal of Ocean and Wind Energy, 2*(4).

Lokhorst, S. J. (1998). *Deformational behaviour of concrete influenced by hydration related changes of the microstructure*.

Lura, P. (2003). *Autogenous deformation and internal curing of concrete*. Ph.D. thesis, Delft University of Technology.

Melo Neto, A. A., Cincotto, M. A., & Repette, W. L. (2007). *Desenvolvimento de Metodologia e Equipamento Para a Medida Da Retração Restringida*. Brazil.

Paillere, A. M., Buil, M., & Serrano, J. J. (1989). Effect of fiber addition on the autogenous shrinkage of silica fume concrete. *ACI Materials Journal, 86*(2), 139–144.

Pirskawetz, S., Weise, F., & Fontana, P. (2012). Analysis of early-age cracking of cementitious materials by combination of various non destructive testing methods. In *2nd International Conference on Microstructural-related Durability of Cementitious Composites* (pp. 350–359).

Rilem 42-CEA. (1981). Properties of set concrete at early ages state-of-the-art-report. *Matériaux et Constructions, 14*(6), 399–450.

RILEM TC 119-TCE. (1997). Avoidance of thermal cracking in concrete at early ages—Recommendations. *Materials and Structures, 30*, 451–461.

Schlicke, D., Matiašková, L. (2019). Advanced computational methods versus analytical and empirical solutions for determining restraint stresses in bottom-restrained walls. *Journal of Advanced Concrete Technology, 17*, 335–349.

Schlitter, J. L., Barrett, T., & Weiss, W. J. (2010a). Restrained shrinkage behavior due to combined autogenous and thermal effects in mortars containing super absorbent polymer (Sap). In *International RILEM Conference on Use of Superabsorbent Polymers and Other New Additives in Concrete* (August).

Schlitter, J. L., Senter, A. H., Bentz, D. P., Nantung, T., & Weiss, W. J. (2010b). A dual concentric ring test for evaluating residual stress development due to restrained volume change. *Journal of ASTM International, 7*(9), 103118.

Schöppel, K., Plannerer, M., & Springenschmid, R. (1994). Determination of restraint stresses and of material properties during the hydration of concrete with temperature-stress testing machine. In *RILEM Proceedings 25, Thermal Cracking in Concrete at Early Ages*.

Springenschmid, R., Breitenbücher, R., & Mangold, M. (1994). Development of the cracking frame and the temperature-stress testing machine. In *RILEM Proceedings 25, Thermal Cracking in Concrete at Early Ages* (pp. 137–144).

Staquet, S., Azenha, M., Boulay, C., Delsaute, B., Carette, J., Granja, J., et al. (2014). Maturity testing through continuous measurement of E-modulus: An inter-laboratory and inter-technique study. In *Proceedings of ECO-CRETE International Symposium on Sustainability* (p. 8).

Staquet, S., Delsaute, B., Fairbairn, E. M. R., Torrent, R., Knoppik, A., Ukrainczyk, N., et al. (2019). *Mixture Proportioning for Crack Avoidance* (vol. 27).

Turcry, P. (2004). Retrait et Fissuration Des Bétons Autoplaçans : Influence de La Formulation. Université de Nante.

Turner, K., Schlicke, D., & Tue, N. V. (2015). Restraint and crack width development during service life regarding hardening caused stresses. In *Proceeding of the fib Symposium* (pp. 1–8). Copenhagen, Denmark.

Vaysburd, A. M., Emmons, P. H., Bissonnette, B., & Pigeon, M. (2001). Some aspects of evaluating cracking sensitivity of repair materials. In K. Kovler & A. Bentur (Eds.), *RILEM Proceedings PRO 23 Early Age Cracking in Cementitious Systems—EAC'01* (pp. 169–185). RILEM Publications S.A.R.L.

Weiss, J., Lura, P., Rajabipour, F., & Sant, G. (2008). Performance of shrinkage-reducing admixtures at different humidities and at early ages. *ACI Materials Journal, 105*(5), 478–486.

Weiss, W. J., & Shah, S. P. (2002). Restrained shrinkage cracking: the role of shrinkage reducing admixtures and specimen geometry. *Materials and Structures, 35*(246), 85–91.

Weiss, J., Yang, W., & Shah, S. (2000). Influence of specimen size/geometry on shrinkage cracking of rings. *Journal of Engineering Mechanics, 126*(3), 233–242.

Yokoyama, K., Hiraishi, S., Kasai, Y., & Kishitani, K. (1994). Shrinkage and cracking of high-strength concrete and flowing concrete at early ages. In *ACI Special Publication* (vol. 148, pp. 243–258).

Zhou, X., Dong, W., & Oladiran, O. (2014). Experimental and numerical assessment of restrained shrinkage cracking of concrete using elliptical ring specimens. *Journal of Materials in Civil Engineering, 26*(11), 04014087.

Zhou, J., Ye, G., Schlangen, E., & Van Breugel, K. (2006). Autogenous deformation of portland cement paste blended with blast furnace slag measured by mini-TSTM. In *International RILEM Conference on Volume Changes of Hardening Concrete: Testing and Mitigation* (pp. 367–374, vol. C).

Zhu, H., Li, Q., Hu, Y., & Ma, R. (2018). Double feedback control method for determining early-age restrained creep of concrete using a temperature stress testing machine. *Materials, 11*(7), 1079.

Zhu, H., Li, Q., & Yu, H. (2017). Self-developed testing system for determining the temperature behavior of concrete. *Materials, 10*(4), 419.

Chapter 7
Adjustable Restraining Frames for Systematic Investigation of Cracking Risk and Crack Formation in Reinforced Concrete Under Restrained Conditions

Dirk Schlicke, Katrin Hofer and Nguyen Viet Tue

Abstract The Institute of Structural Concrete at Graz University of Technology developed adjustable restraining frames (ARFs) in order to enable systematic investigations of the behaviour of reinforced concrete under restrained conditions. The focus was hereby set on the following aspects:

- holistic experimental simulation of the stress history due to imposed deformations including cracking and crack opening by time, as well as
- experimental determination of cracking patterns to be expected in externally restrained and ordinarily reinforced thick members (reinforcement is located near the surface whereby the core is not affected by reinforcement; $A_c \gg A_{c,eff}$).

The results of these experiments give profound insights on the stress history, the cracking risk, the crack pattern and further crack opening to be expected due to the complex interplay of thermal, hygric and mechanical behaviour of concrete. Special attention was given to the restraining condition as well as the present reinforcement. In particular, the most important results of the whole campaign were:

- confirmation and additional findings regarding the effect of viscoelastic behaviour on the hardening-induced stress history,
- clarification of the behaviour of reinforced concrete when it comes to superimposition of hardening-induced stresses with further stresses due to imposed deformations during service life and
- experimental verification of the existence and effectiveness of secondary cracks in thick members on the limitation of the crack width of the primary crack.

Besides the outlined findings and their application in guidelines for crack risk assessment (ÖBV in Analytical design of watertight structures with optimized concrete (in German). Austrian society for construction technology, Vienna, Austria, 2018) as well as crack width control and jointless design (BAW in Merkblatt

---

D. Schlicke (✉) · K. Hofer · N. V. Tue
Institute of Structural Concrete, Graz University of Technology, Graz, Austria
e-mail: dirk.schlicke@tugraz.at

K. Hofer
Federal Waterways Engineering and Research Institute (BAW), Karlsruhe, Germany

M. Serdar et al. (eds.), *Advanced Techniques for Testing of Cement-Based Materials*, Springer Tracts in Civil Engineering, https://doi.org/10.1007/978-3-030-39738-8_7

früher Zwang – Rissbreitenbegrenzung für frühen Zwang in massiven Wasserbauwerken. Referat Massivbau der BAW, Karlsruhe, 2010; BAW in Merkblatt Zwang bei fugenlosen Wasserbauwerken. Referat Massivbau der BAW, Karlsruhe, 2019), these experiments provide also clear and transparent data for the verification of thermo-mechanical calculation models for simulation of hardening-induced stress histories, (e.g. Schlicke in Mindestbewehrung für zwangbeanspruchten Beton. Graz University of Technology, Graz, 2014; Heinrich in effiziente Berechnung viskoelastischer Spannungen in gezwängten Bauteilen. Graz University of Technology, Graz, 2018; Jędrzejewska in Constr Build Mater 174, 2018) as well as nonlinear calculation models for simulation of secondary cracking in thick concrete members.

Keywords Restraining frames · Restraint stressing · Cracking risk · Crack width development · Superposition of hardening-induced stresses with restraint stressing during service life

7.1 Introduction

Concrete shows distinct volume changes due to cement hydration, drying, creep and thermal dilation. Under restrained conditions, in which concrete members cannot deform freely, restraint stresses occur due to these imposed deformations. Temporal occurrence of herewith caused cracking is subject to a complex interplay of thermal, hygric and mechanical behaviour of the concrete as well as the restraining condition. Crack width and further crack opening during service life depend also on the present reinforcement and the bond.

In addition, thick members with ordinary reinforcement near the surface show a deviant mode of cracking due to restraint as the usually assumed successive cracking of a reinforced concrete tie in current theories, (e.g. EC2 2015 or *fib* 2010) Despite the possibility of surface cracking, macroscopic cracking in thick members starts also with a primary crack. With regard to the very low reinforcement ratio in thick concrete members, the influence of the reinforcement on the creation of the next primary crack is very small and it can be assumed on the safe side, that the next primary crack occurs according to the restraining condition, as explained in (Schlicke 2014; Schlicke and Tue 2015). These primary cracks have a certain distance according to geometrical conditions of the restraining condition and separate the member into almost independent parts. Further cracking between the primary cracks is still possible, but this secondary cracking is now caused by the activation of reinforcement in the primary crack. As illustrated in Fig. 7.1, the secondary cracks are restricted to the vicinity of the primary crack and do not propagate through the whole cross section. A profound understanding of this so-called secondary cracking and its effectiveness on the limitation of the crack width of the primary crack is of crucial importance for an economical determination of minimum reinforcement for crack width control as well as for the jointless design of concrete structures (Turner 2017).

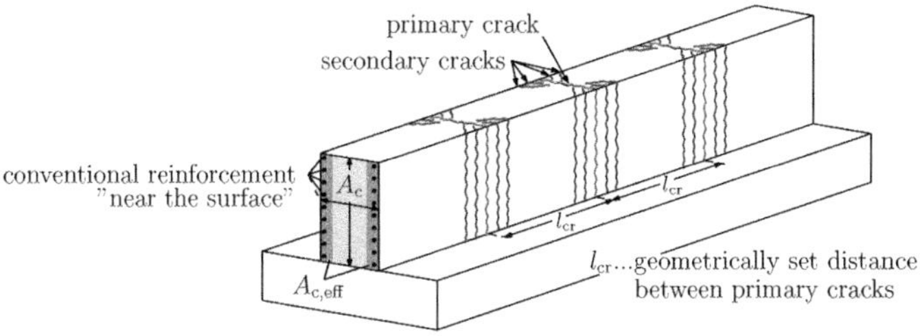

Fig. 7.1 Schematic illustration of geometrically set primary cracks in a bottom-restrained wall and secondary cracks due to reinforcement in the vicinity of the primary cracks

For a systematic and holistic investigation of this behaviour with all its particular effects the Institute of Structural Concrete at Graz University of Technology developed adjustable restraining frames for reinforced concrete (ARFs). On the one hand, the experimental setup provides a continuous monitoring of stressing due to hydration in the concrete since setting. And after hydration, a realistic experimental superimposition of the prior imposed hardening-induced stressing with further restraint stressing during service life is enabled by an integrated extension mechanism in the frame. On the other hand, the ARFs can be modified in a way that they allow experimental studies of cracking patterns to be expected in conventionally reinforced thick members with reinforcement near the surface.

This chapter contains a comprehensive description of the design and functionality of the experimental setup. Besides, conditions and properties of the experimental programme as well as the most important insights are presented. Several parts of the herewith outlined specifications and results can also be found in other contributions of the authors; however, they had never been published in such a comprehensive and detailed form as in the present chapter.

7.2 Motivation and Concept

The basic motivation of this research was the provision of experiments by which the behaviour of reinforced concrete with imposed deformations can be studied as close to real member behaviour as possible. The study is twofold. In the first step, a holistic investigation of restrained stress histories and herewith caused crack formation was carried out. The holistic approach refers to a time-discrete simulation beginning from hardening-induced stresses over the superimposition with additionally imposed deformations representative for service life. And in order to enable targeted parametric analysis of this complex behaviour, two identical frames were built and used in parallel for comparative studies. Besides, and since these investigations were conducted on slender concrete ties, the particular cracking behaviour of thick members was investigated in the second step.

For the hardening-induced stress history the aim was to provide a partial restraining condition which is given passively in order to implicitly account for the changing degree of restraint due to the evolution of stiffness of the hardening concrete. Besides, the imposed deformations of the hardening phase should be given exclusively by internally occurring strains of the hardening concrete itself (thermal dilation due to hydration heat, and additional deformations due to shrinkage and viscoelasticity) in order to ensure a physically correct coupling of thermal and mechanical behaviour with respect to maturity. Occurring deformations and stresses had to be monitored since setting which required a casting directly into the frames.

After the hardening phase, it was desired to impose additional deformations, representative for service life, without affecting the existing stresses due to hardening. For this purpose, the frames had to have the possibility to move the head girders apart from each other as well as to lock the system afterwards in the new deformed stage.

The test specimens in these series were reinforced concrete ties with a square-shaped cross section and symmetrically located reinforcement in the corners. And with the reinforcement in the specimen the experiment could be conducted beyond cracking with detailed monitoring of the development of the remaining restraint forces after cracking and the crack width opening in the course of time.

Next to the comparative tests on ties, one of the ARFs was modified to conduct further deformation-controlled tension tests on plate-shaped test specimens representing a horizontal slice of a thick wall with reinforcement near the surface. As outlined before, these tests should give important insights on the specific cracking patterns to be expected in thick members. Therefore, it was very important that these tests were deformation controlled to consider the influence of the decreasing member stiffness on the actual restraint force.

Figure 7.2 gives a summarized illustration of both experimental settings with the ARFs with regard to the respective static system, a picture of the setup and a schematic illustration of the force evolution according to imposed deformations.

The self weight of the specimen was carried in the ARFs by an auxiliary frame construction providing four vertical supporting points over the length of the specimen. By this, any effects of additional bending due to self weight of the specimens was excluded.

7.3 Technical Specification of the Frames

7.3.1 General Setup and Functionality

The ARFs were developed with regard to the functionality of different prior developed experimental setups aiming at determination of stress histories in restrained concrete, in particular *Springenschmid Cracking Frame* and *Temperature-stress test machines* (TSTMs) as presented e.g. in Springenschmidt et al. (1994). Of course, current advancements in the application and further development of TSTMs as well

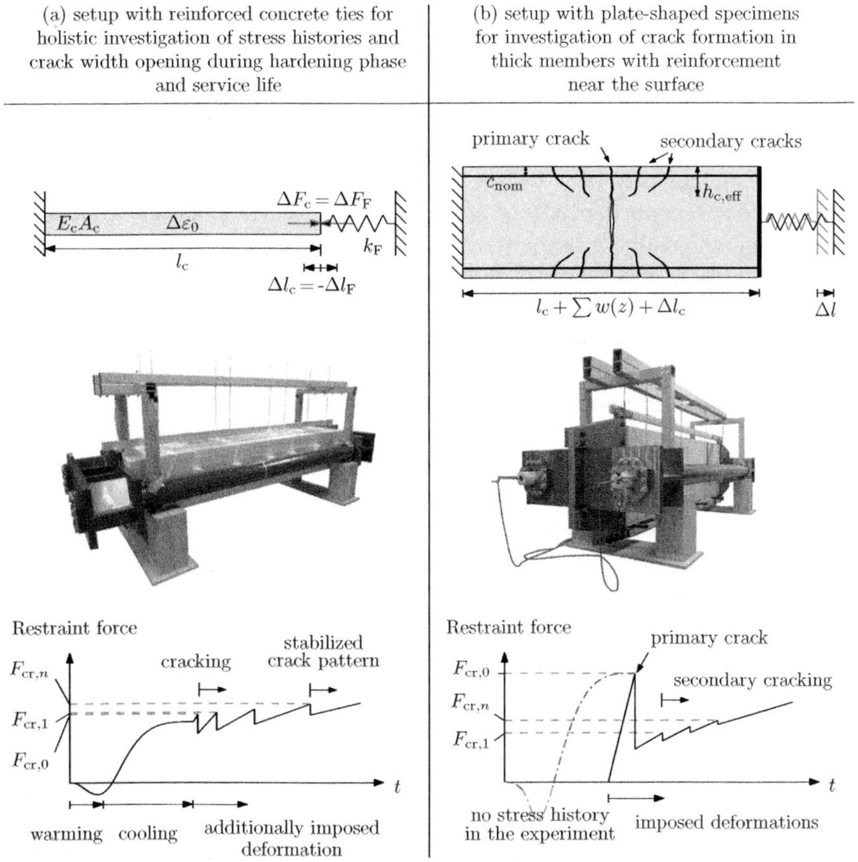

Fig. 7.2 TU Graz adjustable restraining frames (photographs D. Schlicke)

as the development of specific devices was considered during the project, in particular this refers to the works of (Springenschmidt et al. 1994; Thielen and Hintzen 1994; Bjøntegaard and Sellevold 2004; Klausen et al. 2015; Westman and Emborg 1994; Staquet et al. 2012; Bentur and Kovler 2003). The ring test method, as comprehensively summarized in Kanavaris et al. 2019, was not further regarded for this research. It is simply less suitable for detailed investigations of temperature-dominated stress histories with change of sign in stresses (initial expansion with compression in the warming and subsequent contraction due to cooling until pronounced tension at temperature equalization), as well as for monitoring of restraint-induced crack opening of reinforced concrete. Besides, restraining ring tests have usually very thin specimens with rather small opportunities to produce temperature-dominated stress histories solely with the heat of hydration of the concrete, they produce always non-uniform stress distributions in the specimen and it is problematic to precisely determine the stiffness of the restraining condition for detailed analysis.

As a basic requirement of measuring the hydration-induced stress history in concrete, the stress recording in the ARFs starts also with the beginning of setting. And similar to the cracking frame but in contrast to TSTMs, the restraining condition in the ARFs is realized fully passively. Such passive restraining condition implicitly provides a time-dependent partial restraining condition as it is to be expected in real cases of restrained members. In fact, the restraining condition at setting is always 100% and decreases in dependence of the evolving elastic modulus in the hardening concrete and the stiffness of the restraining condition. In case of the experimental setup of the ARFs the degree of restraint develops from 100% down to 65%. Of course, this behaviour can be simulated with TSTMs as well, however, in a purely passive restraining condition, viscoelastic strains interact only within this partial restraining condition, so that the effect of relaxation is implicitly given within the setup of the ARFs. In contrary to this, the simulation of partial restraining conditions with TSTM's requires a mechanical imitation of the restraining condition and concerns exist that this can cause an unnatural drift in the resulting stress history under partial restraint.

In contrast to *Springenschmid Cracking Frame* and conventional TSTM's, the test specimens of the ARFs are reinforced in order to enable the conduction of the experiment beyond cracking of the specimen. The reinforcement ratio in the measurement length of the ARF test specimens for stress history amounts 1% and 2% in the anchorage regions at the edges respectively. By this, a representative reinforcement ratio for conventional reinforced concrete is provided in the cracked state of the specimen. Restraining frame experiments with reinforced concrete bars can also be found elsewhere, e.g. in Faria et al. 2016, but these experiments work with rather slender specimens and solely in combination with an active control. The focus of these experiments is predominantly set on the material behaviour during drying, however, the aimed fully passive restraining of the hardening phase with the ARFs cannot be achieved with them.

Subsequently after the hardening phase, it was desired to impose additional deformations representative for service life without affecting the prior stresses due to hardening. The aim of this research was a detailed investigation of the stress history of predominantly restrained concrete members with pronounced hardening-induced stresses. Detailed investigations on the superposition of stresses during the service life of reinforced concrete members can also be found elsewhere, e.g. in (Gomes et al. 2018; Schnell et al. 2017; Berger and Feix 2018). However, these experiments focus strongly on the reduction of restraint stresses due to crack formation and crack opening caused by loading which is clearly to be distinguished from the interplay of restraint stresses only. For the simulation of the interplay of hardening-induced restraint stresses with imposed deformations during service life, one head girder of the ARFs was therefore designed with the ability to move it on the longitudinal support beam with externally attachable hydraulic cylinders. In the hardening phase with passive restraining condition the head girder was rigidly screwed to the support beam providing full stiffness. Only after hydration and before moving the head girder, the connection was unscrewed and the head girder could be decoupled from the support beam without any loss of stresses in the test specimen. This is possible

Fig. 7.3 View of the individual steel components of one ARF (photographs LKI)

because the head girder is sitting in vertical direction on the support beam. And in horizontal direction the contact area between head girder and supporting beam is compressed at this stage due to the tension in the test specimen. The great challenge for the experiments with the ARFs was to realize this movable connection without weakening the passive restraining condition. The passive restraining condition needs a high stiffness and thus, the contact surfaces between head girder and supporting beam required a very low roughness (ISO Rz 6.3) and also very small tolerances of the fit of the movable parts were required (ISO F8).

Figure 7.3 shows the two head girders and the two supporting beams of one ARF before assembling. Attention should be paid to the solid supporting beams with diameter of 230 mm and the highly precise prepared surfaces of the one edge of the beam as well as the holes in one head girder to provide a movable connection without any play.

The head girders were produced with 60 mm sheets and fillet welding with pre-conditioned surfaces between the contact surfaces in between the single sheets. After welding and coating the required holes for the connection with the supporting beams were drilled. The detailed connections between supporting beams and head girders are shown in Fig. 7.4 in the final state. The moveable connection at one side was simply enabled by inserting the surface-conditioned end of the supporting beam into the precise hole of the moveable head girder. After that, supporting beam and head

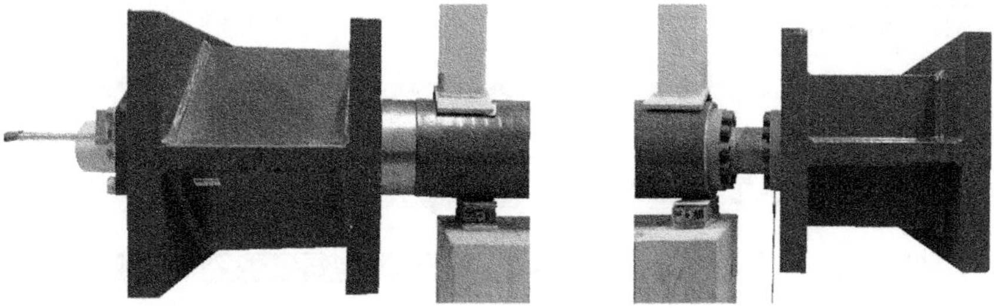

(a) moveable head girder with cylinder (b) load cell and frictionless support

Fig. 7.4 Constructive details of the head girder connections (photographs D. Schlicke)

girder were screwed together with a head plate, which is visible in the photographs of Fig. 7.2. Besides, these head plates have also a hole in its centre for later attachment of the cylinders, see Fig. 7.4a). On the other side of the frame, the connection between head girder and supporting beam is fully stiff. And in between this rigid connection very stiff load cells (GTM RF 500 kN) were included for monitoring the force directly in the frame, see Fig. 7.4b). The opportunity of monitoring the force indirectly on basis of strain measurements attached to the supporting beam was deliberately not done due to concerns regarding the influence of the ambient temperature on the measurement results as well as the stability of such sensors in the long-term.

The ARFs were put on wooden slats and wedge mounts for precise positioning. Hereby, the support was located below the supporting beams and at the position, where the loads of the auxiliary frame for carrying the self weight of the specimens was introduced. In addition, Teflon plates were placed between wedge mounts and supporting beam. Altogether, all these measures provided a frictionless support enabling a perfect force coupling between the specimen and the frame. Figure 7.5 exemplifies the outlined functionality of the ARFs in the passive and in the active phase schematically.

The experiments showed that the frames provide a suitable restraining condition in the horizontal plane. However, the experiments showed also that the head girders are sensible against rotation to some extent. Thus, special care was needed when cracking started only at one side of the specimen (top or bottom side) and the significant change of bending stiffness in the specimen caused a rotation of its ends. In such cases the

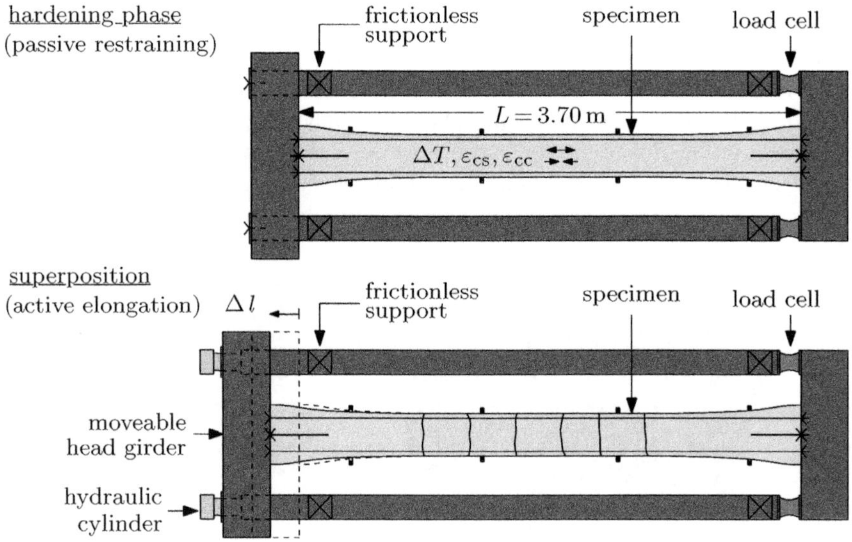

Fig. 7.5 Schematic illustration of the functionality of the ARFs with concrete ties

rotation had to be corrected by tightening or loosening the additional supports of the specimen by adjusting the auxiliary frame.

The experiments were monitored with various measurements consisting of:

- force measurement in the frames with load cells (Type GTM RF 500 kN),
- overall deformation measurement of the specimens using LVDTs with a measuring length spanning between both head girders (3.70 m),
- concrete deformations with locally applied LVDTs and a measurement length of 10 cm,
- crack width development with high strain transducers (DD1) applied to the surface with a measuring length of 2 cm as well as
- temperature in the concrete using thermo-elements Type K (one sensor in the core and one in the edge of the cross section).

The measurements were taken continuously with a measurement frequency of 1/300 Hz (every 5 min.). Only in times in which deformations were externally imposed with the cylinders the measurement frequency was increased to 2 Hz.

7.3.2 Passive Frame for Simulation of Hardening-Induced Stress History

The passive frame for simulation of hardening-induced stress history is the most basic setting of the frame. Head girders and supporting beams are rigidly connected and a concrete tie is casted directly into the frame. The connection between the frame and the concrete tie was simply provided by a screw connection of the continuous reinforcement as well as additional starter bars at the dog bone-shaped ends of the concrete tie, see Fig. 7.5. Before casting, formwork as well as reinforcement were precisely placed in the centre axis of the frame and the reinforcement was prestressed with ~20 MPa to minimize the sag of the reinforcement to an acceptable value. After casting, the formwork was closed on top with an identical wooden plate as it was used at bottom and side surfaces. Besides, the specimen was insulated with an XPS layer as well as a foil cover. The whole procedure is illustrated with Fig. 7.6.

Although, the cross section of the concrete tie is rather slender, the formwork and XPS insulation around it could cause a hydration-induced temperature history in the tie which is comparable with mass concrete. E.g. the general setting with a concrete cross section of 25×25 cm^2 and 2 cm wooden formwork as well as an insulation of 30 cm XPS showed a temperature history which is representative for a mass concrete member with an effective thickness of ~1.0 to 1.5 m. After cooling down, the specimen was stripped as shown in Fig. 7.7a). Hereby, foil, XPS layer and formwork were fully removed except very small stripes of formwork below the suspension in order to maintain an undistorted support of the self weight. In some cases, the specimen was stripped much earlier, at the time of the maximum temperature, in order to study the effect of a rapid temperature shock.

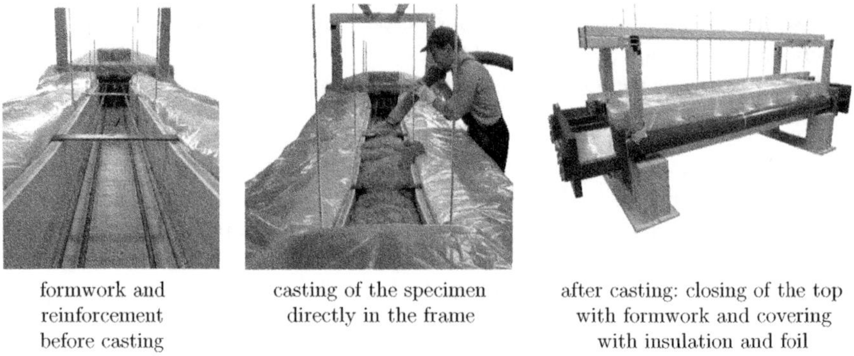

Fig. 7.6 Preparation, casting and insulation for holistic monitoring of hardening-induced stresses since setting (photographs D. Schlicke)

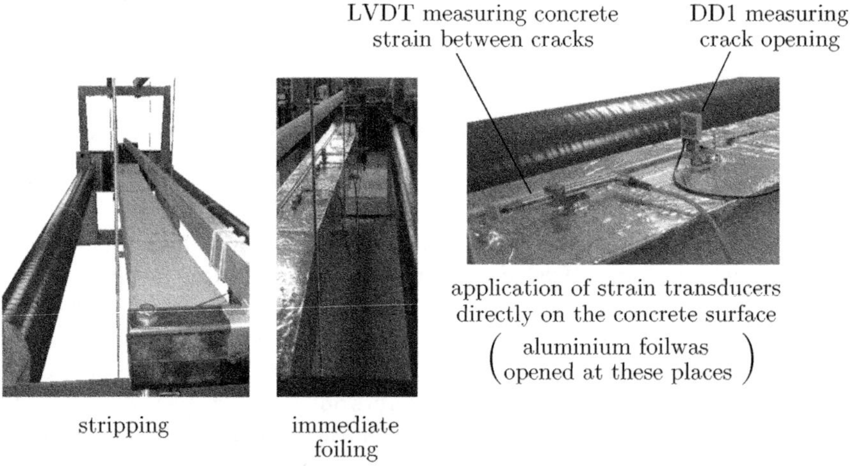

Fig. 7.7 Stripping, immediate foiling with aluminium foil and application of strain sensors at the surface of the specimen (photographs D. Schlicke and K. Hofer)

Independent of the stripping time, all specimens were equipped after stripping with additional deformation sensors along the specimen length in order to monitor the development of the crack widths and the concrete strain between cracks, see Fig. 7.7. In detail, the development of the crack widths was measured with 2 cm long 'DD1 high accuracy strain transducers' which were applied over the cracks whereas the development of the concrete strain between cracks was measured with 15 cm long LVDTs. Besides, some specimens were sealed after stripping with aluminium foil in order to prevent the specimens from drying, which is also shown in Fig. 7.7.

Altogether, the stress-independent deformation of the hardening concrete tie consists of the occurring thermal dilation and autogenous shrinkage. Besides, stress-dependent deformation develops according to the stress history and its timely viscoelastic behaviour. Drying shrinkage, however, becomes solely relevant at later stages after hardening and only in cases in which the specimens had not been sealed with aluminium foil after stripping.

The hardening-induced stressing of the test specimen can be reliably concluded from the measurements in the load cells. On basis of the force equilibrium in the system it can be formulated at any point in time:

$$\sum F_{\mathrm{ARF}} = -\sigma^{\mathrm{I}}_{\mathrm{c,max}} \cdot A_{\mathrm{c}} \cdot (1 + \alpha_{\mathrm{E}}\rho_{\mathrm{s}}) = -\sigma^{\mathrm{II}}_{\mathrm{s}} \cdot A_{\mathrm{s}} = -[\sigma_{\mathrm{c}}(x) \cdot A_{\mathrm{c}} + \sigma_{\mathrm{s}}(x) \cdot A_{\mathrm{s}}] \tag{7.1}$$

where:

$\sum F_{\mathrm{ARF}}$—sum of forces in the two load cells in the ARF,
$\sigma^{\mathrm{I}}_{\mathrm{c,max}}$—maximum concrete stresses in the uncracked part of the specimen (this part of the equation is only valid as long as the stabilized crack pattern is not reached),
A_{c}—cross sectional area of the concrete in the specimen,
α_{E}—ratio of modulus of elasticity ($E_{\mathrm{s}}/E_{\mathrm{c}}$),
ρ_{s}—reinforcement ratio in the cross section ($A_{\mathrm{s}}/A_{\mathrm{c}}$),
$\sigma^{\mathrm{II}}_{\mathrm{s}}$—steel stresses directly in a crack,
A_{s}—cross sectional area of present reinforcement,
$\sigma_{\mathrm{c}}(x)$—concrete stresses at a certain location x in the specimen and
$\sigma_{\mathrm{s}}(x)$—steel stresses at a certain location x in the specimen.

At the same time, the compatibility of deformations can be formulated on basis of the imposed deformations and the deformation response of both the frame and the specimen. The solution is not trivial and requires a time-discrete analysis with special regard to the evolution of the elastic modulus and the highly significant viscoelastic behaviour of the hardening concrete. For the purpose of a better understanding, however, the solution is given firstly in a general way without the context of timely dependence. It holds:

$$\varepsilon_{0,\mathrm{p}} \cdot L = \Delta L_{\mathrm{real}} - \sum w - \int_0^L \frac{\sigma_{\mathrm{c}}(x)}{E_{\mathrm{c}}} \mathrm{d}x \tag{7.2}$$

where:

$\varepsilon_{0,\mathrm{p}}$—imposed deformation due to hardening concrete in the passive phase; i.a. consisting of thermal dilation due to hydration, shrinkage strain and viscoelastic strain,
L—length of the specimen,
ΔL_{real}—real deformation occurring in the frame (free deformation due to partial restraining condition),

$\sum w$—sum of occurring crack widths along the specimen and
E_c—elastic modulus of the concrete.

On basis of the force equilibrium in Eq. (7.1) and the compatibility of deformations shown in Eq. (7.2) the stressing in the ARFs can be directly determined by means of another important state parameter of restrained concrete members, namely the degree of restraint. In particular, the degree of restraint is a factor representing the amount of imposed deformations which will be restrained. In case of the ARFs it can be determined with:

$$a = \frac{1}{1 + \frac{EA_m}{L} \cdot \frac{1}{k_F}} \tag{7.3}$$

where:

a—degree of restraint,
EA_m—mean axial stiffness of the specimen with respect to the present modulus of elasticity as well as cracking and
k_F—equivalent spring stiffness of the frame.

For the ARFs, in which the stressing of the specimen is directly connected to the sum of forces in the load cells according to Eq. (7.1), the resulting forces to an imposed deformation have a size of:

$$\sum F_{ARF} = \left(\varepsilon_{0,p} + \frac{\sum w}{L}\right) \cdot EA_m \cdot a \tag{7.4}$$

As mentioned before, the occurring stress histories in restrained concrete members and also in the specimens of the ARFs are strongly time-dependent for two main reasons. One reason is the evolution of the modulus of elasticity by time so that the size of the stress response to a certain restrained deformation increment changes throughout the hardening phase. For instance, a deformation increment at an early concrete age with low modulus of elasticity causes only low stresses, whereas the imposition of the same deformation increment at a later concrete age with higher modulus of elasticity causes a higher stress response. And the other reason is the pronounced viscoelastic behaviour of concrete depending on the occurring stress history. In terms of time, these additional viscoelastic deformations are depending on the age of concrete at stressing as well as on the duration of the stressing. The detailed determination of viscoelastic strains in hardening concrete is not trivial but also no essential requirement for the general understanding of the functionality of the ARFs. Thus, this contribution considers the viscoelastic strains implicitly within the imposed deformation without further discussion on their actual determination. Further detailed information on the actual consideration of viscoelastic behaviour in the analysis of the experiments are given in Schlicke (2014), Heinrich (2018).

Finally, and with regard to the time-dependence of stress histories in restrained concrete members, the real stressing in the ARFs can be determined with:

$$\sum F_{\mathrm{ARF}}(t_{\mathrm{n}}) = \int_{t_0}^{t_{\mathrm{n}}} \left(\Delta\varepsilon_{0,\mathrm{p}}(t) + \frac{\sum_{i=0}^{\mathrm{n}} \Delta \mathrm{w}_i(t)}{L} \right) \cdot EA_{\mathrm{m}}(t) \cdot a(t)\mathrm{d}t \tag{7.5}$$

where:

t_0—time at setting of concrete,
t_{n}—given point in time and
n—number of existing cracks.

Equation (7.5) is composed in a way that increments of the imposed deformation and the change of crack width are superimposed with the currently present axial stiffness of the specimen as well as the currently present degree of restraint. The axial stiffness of the member affects the degree of restraint directly, as indicated in Eq. (7.3). Of course, axial stiffness and degree of restraint could be merged in Eq. (7.5) but they are considered separately for the purpose of a clear formulation. Either way, both parameters depend on the present modulus of elasticity of the concrete as well as on the stage of cracking and they change contrarily. In general, the increasing modulus of elasticity increases the axial stiffness but decreases the degree of restraint, whereas cracking decreases the axial stiffness and increases the degree of restraint.

Before cracking, concrete stresses can be assumed to be uniformly distributed along the specimen length so that axial stiffness and degree of restraint change throughout the time only due to the evolution of the elastic modulus. After cracking, however, concrete stresses are not uniformly distributed along the specimen length anymore and steel stresses increases significantly in the cracks. The axial stiffness depends now predominantly on the steel strain in the cracks and the tension stiffening effect of the concrete between the cracks. A computational verification of the experimental observation in the cracked state requires at least the knowledge about the distribution of steel strains along the member length. Unequivocal measurements of the stress distribution along the member length could not be provided in the past test series. However, in (Turner 2017) it was shown on basis of detailed analysis of the experiments with respect to the observed crack widths that the deformation compatibility can be roughly verified with the average bond strength and the tension stiffening factor according to EC2 (2015).

Altogether, the stress history in the ARFs can be characterized as following. After casting and before setting of concrete, the axial stiffness is negligible small and the degree of restraint is 1.0. In this phase, all occurring strains are fully restrained without occurrence of any significant force in the ARF. Directly with the setting of concrete the occurring deformations are immediately restrained, whereby the degree of restraint decreases proportionally with the evolution of the elastic modulus. Later, if critical tensile stresses are reached and cracking occurs, the degree of restraint increases according to the decreasing axial stiffness of the specimen. In case of the investigated hardening ties, the degree of restraint decreases from 100% at setting to 65% when the evolution of elastic modulus shows no significant changes anymore and the specimen is uncracked. In case of cracking, however, the axial stiffness of the test specimen decreases, so that the degree of restraint increases again. For

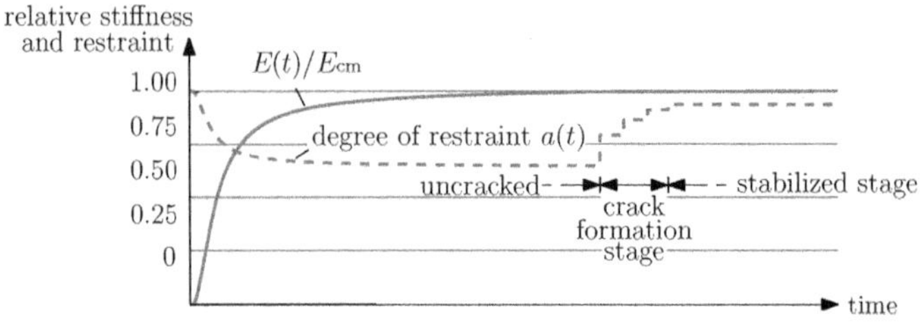

Fig. 7.8 Evolution of the degree of restraint in the experiment

the reinforcement ratio of 1% an increase of the degree of restraint up to 95% in the stabilized crack pattern was observed in the experiments. Figure 7.8 gives a schematic illustration of this behaviour.

In general, externally restrained members show a direct correlation between the increase of the degree of restraint and the decrease of forces when cracking occurs. The same can be observed in the experiments. As long as the ARF is kept in the passive stage and cracking occurs rather at a single point in time without significant changes of imposed deformations at this time, the change of the degree of restraint can be derived directly from the reduction of forces in the system due to cracking. It holds:

$$a\left(t^{\mathrm{II}}\right) = \frac{\sum F_{\mathrm{ARF}}\left(t^{\mathrm{I}}\right)}{\sum F_{\mathrm{ARF}}\left(t^{\mathrm{II}}\right)} \cdot a\left(t^{\mathrm{I}}\right) \leq 1 \tag{7.6}$$

where:

t^{I}—time directly before cracking and
t^{II}—time directly after cracking.

7.3.3 *Activation of the Frame for Superimposition of Hardening-Induced Stresses with Additional Stresses During Service Life*

After temperature equalization in the passive phase, the actual stress state of the specimen was superimposed with additional deformations typical for service life. For this superimposing the passivity of the frame can be halted temporarily without losing any stresses in the specimen. Only the screw connection between the movable head girder and the supporting beam was opened, but this does not change the present restraining condition of the specimen since this connection is compressed at that stage. Then, the hydraulic cylinders were attached at the head girder and a relative

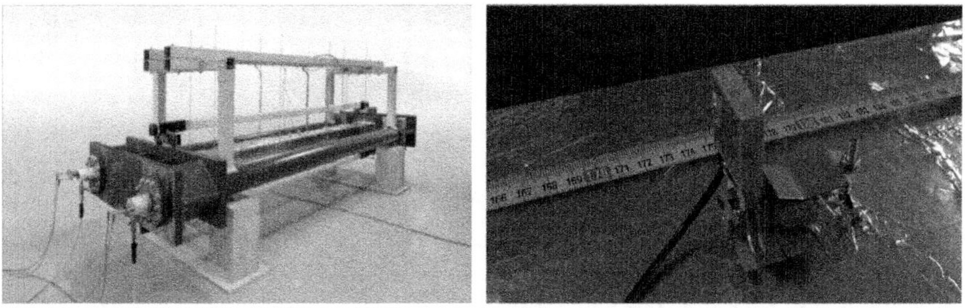

Fig. 7.9 ARF setup for long-term observation, left: application of hydraulic cylinders for imposition of additional deformations, right: crack width monitoring with DD1 high accuracy strain transducers (Photographs BAW and K. Turner)

displacement is induced between head girder and supporting beam, see Figs. 7.4 and 7.9. With this modification, the deformation impact is increased according to the targeted change in the system, namely:

1. increase of stressing in a still uncracked specimen
2. creation of a first crack or
3. creation of additional cracks until the stabilized cracking pattern is reached.

Afterwards, the ARF is locked again and the frame works passively again.

For the further analysis, the externally imposed deformations with the hydraulic cylinders can be considered the same way as the internally imposed deformations in the passive phase, except that they have to be taken into account as fully restrained deformation at the time of their imposition. However, all further (internally) occurring deformations after the system is locked again are working against the remaining degree of restraint. Altogether, the following can be formulated:

$$F_{\mathrm{ARF}}(t_{\mathrm{n}}) = \int\limits_{t_0}^{t_{\mathrm{n}}} \left(\Delta\varepsilon_{0,\mathrm{p}}(t) + \Delta\varepsilon_{\mathrm{cs},1}(t) + \Delta\varepsilon_{\mathrm{cc},1}(t) - \frac{\Delta L_{\mathrm{cyl}}(t)}{L \cdot a(t)} + \frac{\sum_{i=0}^{\mathrm{n}} \Delta \mathrm{w}_i(t)}{L} \right) \cdot EA_{\mathrm{m}}(t) \cdot a(t)\mathrm{d}t \tag{7.7}$$

where:

$\Delta\varepsilon_{\mathrm{cs},1}(t)$—shrinkage strain in the concrete occurring after hardening phase,
$\Delta\varepsilon_{\mathrm{cc},1}(t)$—viscoelastic strain in the concrete occurring due to stress change according to externally imposed deformations and
$\Delta L_{\mathrm{cyl}}(t)$—actively imposed elongation of the specimen with hydraulic cylinders (representative for imposed deformations during service life).

As indicated in Eq. (7.7), the new status of the system (here expressed with the force in the ARF) can be determined with a clear distinction between the two phases of imposed deformation due to hardening and imposed deformation due to external conditions. In contrast to the generally in practice accepted assumption, that ongoing creep has a significant influence on the remaining stressing from the hardening phase (Grube 1991; VDZ 2014), the experiments with the ARFs indicated that the time of such superimposition is of minor importance as long as the concrete can be considered as hardened concrete. The reason is a very small relaxation of the restraint force after temperature equalization. Further information and first explanation attempts can be found in Schlicke et al. (2015), Schlicke and Dorfmann (2017).

In the long-term, the stress history and the crack width development depend both on the local and global difference between steel and concrete strain along the specimen. The strain difference in turn is influenced by the bond properties which impair under sustained and/or repeated loading. Next to the actively imposed elongation by hydraulic cylinders in the experiment, the drying shrinkage contributes to the imposed deformations in all cases without sealing with aluminium foil after stripping.

7.3.4 Modified Frame for Investigation of Cracking in Thick Members

The modified frame for investigation of cracking in thick members is basically working like the frame with the tie in the active mode when additional deformations are imposed externally with the hydraulic cylinders. Only differences are:

1. plate-shaped specimen instead of a tie with quadratic cross section,
2. additional adaptors at the head girders for mounting of the specimen in the frame,
3. stressing is caused without simulation of hardening-induced stresses and
4. no long-term monitoring of the development of the crack width after loading.

The aim of these experiments was the experimental simulation of the cracking pattern to be expected in thick concrete members, as illustrated in Fig. 7.1. Therefore, a plate-shaped specimen was designed which enables detailed investigations of one crack system (one primary crack + secondary cracks) in a thick wall. The thickness of the specimen represents a horizontal cut of one layer of longitudinal reinforcement in the wall and the length is representative for the distance between the geometrically set primary cracks, so that all parts of the wall affecting the crack openings in one crack system are represented. Figure 7.10 illustrates the theoretical location of the specimen in the wall.

Similar experiments have been conducted in the past by Fellmann and Menn (1984) and Helmus (1990), however, these experiments were force-controlled and provide only limited information on the development of the restraining force during the cracking process. Thus, the experiments with the modified frames were deformation-controlled. An overview of the experimental setup and its functionality is given in Fig. 7.11.

A basic requirement of the experiment was ensuring an even cross sections at both free edges of the plate. Therefore, massive steel adaptor constructions were used to mount the specimen into the frame. They consisted of a massive outer part around

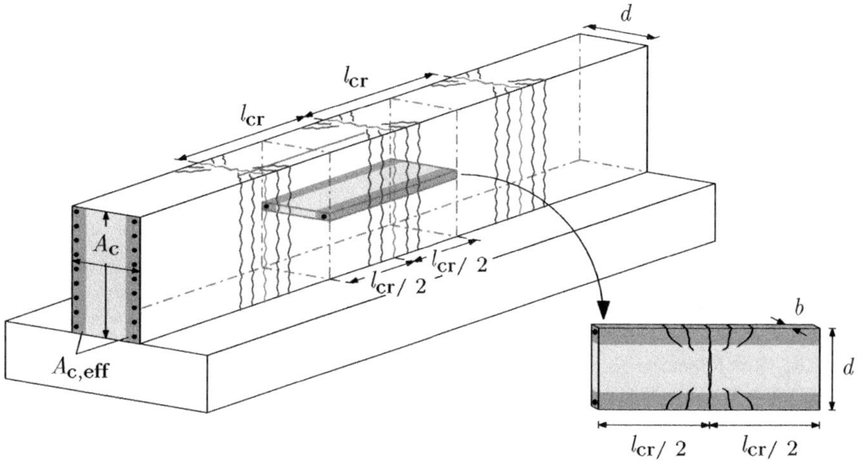

Fig. 7.10 Location of the plate-shaped specimen in the wall

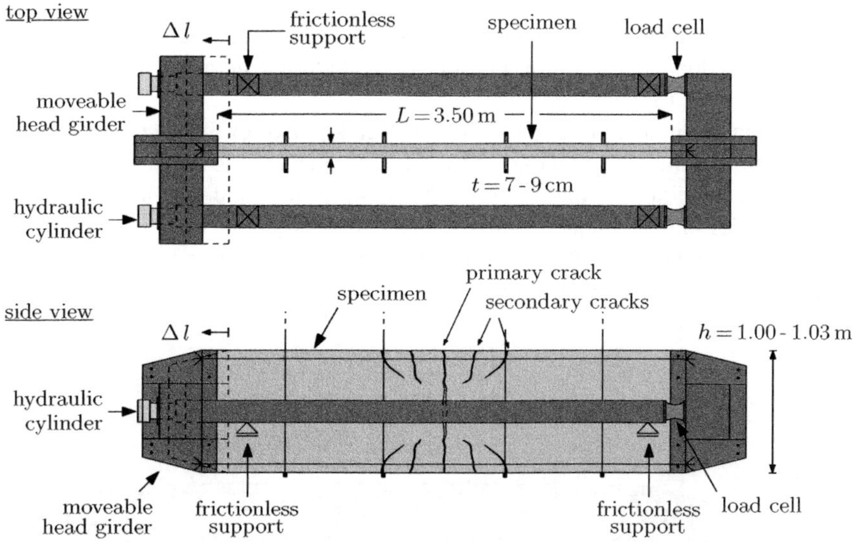

Fig. 7.11 Schematic illustration of the functionality of the modified ARF for experimental investigation of cracking patterns to be expected in thick members with reinforcement near the surface

the head girder and an inner slat which was anchored to the plate with the continuous horizontal reinforcement at top and bottom of the vertically aligned plate as well as additional horizontal starter bars in between, see Fig. 7.12.

Overall, the presented construction provided adequate maintaining of even cross sections at both free edges of the specimen during the experiment, without affecting

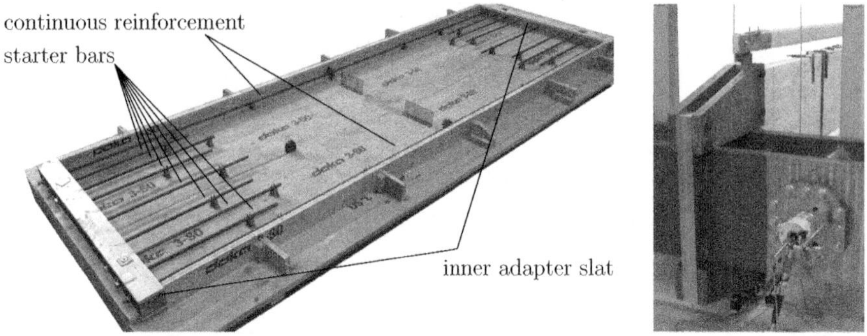

Fig. 7.12 Reinforcement arrangement of the plate-shaped specimen and construction for mounting the specimen into the frame

the area in which the crack formation takes place. However, the secondary cracking of the specimen did usually not occur perfectly symmetrically over the length and the height. It was realized, that the beginning of secondary cracking at only one side (top or bottom) comes along with a significant change of bending stiffness in the specimen which caused a rotation of the head girders. In such cases the rotation had to be corrected by tightening or loosening the additional supports of the auxiliary frame, see Fig. 7.13.

Furthermore, it should be noted that the free surfaces of the specimen in plate direction (horizontal cut surfaces of the specimen in the wall) do not exactly represent the support condition of the conceived layer when it would still be in the wall. Practically, it is almost impossible to overcome this disagreement in the experiment, however, the influence on the cracking pattern itself is minor and the influence on the restraining force can be determined with comparative computational studies on basis of the experimental results.

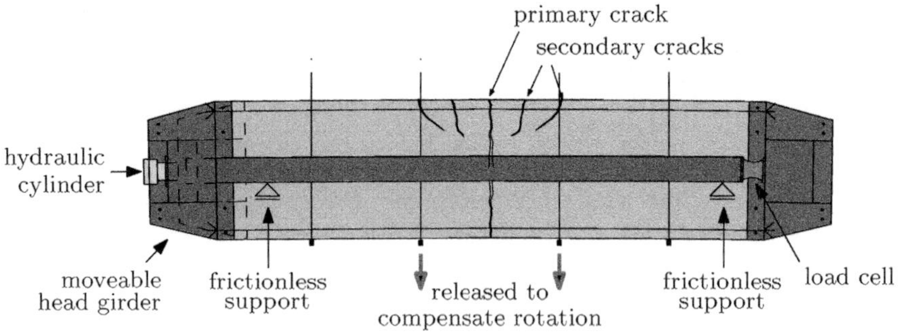

Fig. 7.13 Sensitivity of the head girders against rotation and countermeasures

7.4 Selected Results

7.4.1 General Remarks on the Testing Programme Conducted So Far

Since October 2012 until now, a comprehensive testing programme was conducted with the ARFs. Altogether, this includes five comparative test series on the restraint stress history to be expected in mass concrete and 3 tests on the cracking pattern to be expected in thick members. All tests were performed with a common concrete composition for mass concrete with a strength class C35/45. The concrete mix is given in Table 7.1. Besides, comprehensive material testing was performed in parallel to this testing programme in order to enable a profound analysis of the experiments. In detail, this refers to measurements on the hydration heat release, the evolution of mechanical properties (compressive strength, tensile strength and elastic modulus), autogenous shrinkage and drying shrinkage on dummy specimens of the same cross sectional dimensions as the specimens as well as targeted tensile creep tests, given in Schlicke (2014), Turner (2017), Schlicke et al. (2015), Schlicke and Dorfmann (2017).

The specimens were all reinforced with reinforcement steel $f_{yk} = 500$ N/mm^2 and a diameter of 14 mm. The concrete ties contained in the measurement length 4 Ø 14 mm with one rebar in each corner and a cover of 40 mm to each side, whereas the plate-shaped specimens contained in the measurement length 2 Ø 14 mm, one at each side with a cover of 38 mm.

With a total duration of up to twelve months per series, the five series on the restraint stress history gave insights on the evolution of hardening-induced stresses and the risk of early age cracking as well as the continuation of this stress history in the service life with respect to cracking and crack opening. The tests on the cracking pattern, however, were performed in one go without further investigations on the crack width development in the long-term. Selected results and insights were presented in the following.

Table 7.1 Concrete properties

Parameter	Value (kg/m^3)
CEM III/A 32.5 N Holcim	300
Water	145
Sand and Aggregates: 0-2/2-8/8-16	620/720/630
BV ViscoCrete-1051 PCE	3

7.4.2 Hardening-Induced Stress History and Risk of Early Age Cracking

Figure 7.14 shows the temperature and stress evolution of the five series during the hardening phase until temperature equalization. Each series consist of two specimens which were cast at the same time and run in parallel. In general, the series can be distinguished in comparative studies on the influence of the duration of the cooling down phase presented in Fig. 7.14 (left) and comparative studies on the crack formation and crack width development presented in Fig. 7.14 (right).

Qualitatively seen, the experimentally simulated temperature and stress histories comply with the expected results. The temperature increase due to hydration heat has a size of ~20 °C which is representative for an effective thickness of ~1.0 to 1.5 m. Also, the undisturbed cooling phase is realistic for such thickness. Only in order to safe time during the experiments, the specimens with continued insulation were stripped slightly before temperature equalization.

Quantitatively seen, however, the experiments regarding the influence of the duration of the cooling down phase indicated a controversial behaviour compared to what was generally expected. In detail the following observations could be made from series 1 and 2 shown in Fig. 7.14 (left):

1. The duration of the cooling down phase did not significantly reduce the size of the occurring tensile stresses. The results of both series show almost the same tensile stresses at the end, even though the duration of the cooling down phase of

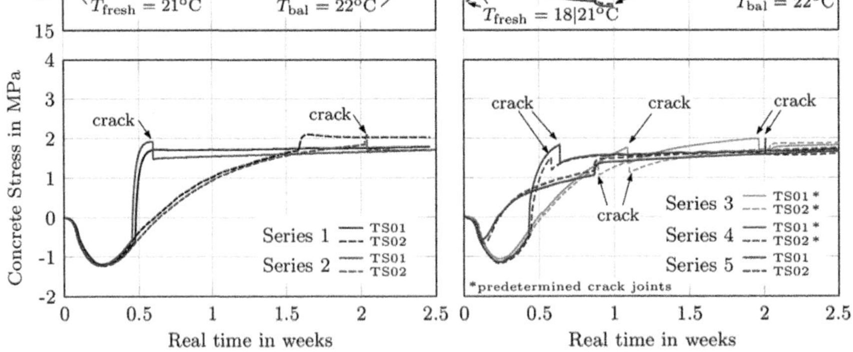

Fig. 7.14 Hardening induced temperature (top) and stress (bottom) histories

the specimens within one series differed by more than one week. According to the theory, the slower cooling should lead to lower stresses since the time-dependent stress reduction due to viscoelasticity (relaxation) has more time to act. The hypothesis that the expected difference was compensated by the evolution of the elastic modulus (early stripping with smaller elastic modulus but less creep may lead to the same stresses as late stripping with bigger elastic modulus but more creep) cannot be confirmed by the accompanying testing of the elastic modulus.

2. The early stripping and associated early occurrence of high tensile stresses did not significantly affect the risk of cracking. According to the theory, the specimens with early stripping in series 1 and 2 should be subjected to a higher risk of cracking. Their tensile strength is less evolved at the early occurrence of high tensile stresses than in case of the two specimens with late occurrence of high tensile stresses. However, both specimens of series 1 remained uncracked whereas both specimens of series 2 cracked. Therefore, it appears for cases with a pronounced warming and associated maturity effect that the small variation of mechanical properties between the series may even have a bigger effect than the time of stripping.
3. The created tensile stresses after temperature equalization show no significant reduction with ongoing time. According to theory, the time-dependent viscoelastic behaviour should cause a relief of stresses when no further strains are imposed, see e.g. Grube (1991) and VDZ (2014). Although, this relaxation can clearly be found in the phase with compression stresses between 42 and 60 h, when the temperature is only slightly but still increasing meanwhile the compressive stresses were already visibly reduced by ongoing viscoelasticity under compression. On the contrary, this behaviour cannot be obtained in tension. The hypothesis that such reduction in tension is compensated by parallel drying cannot be confirmed by the accompanying shrinkage dummy.

The results of series 3–5 shown in Fig. 7.14 (right) do partly confirm the above outlined observations. However, these three series were designed in order to conduct comparative studies on the crack formation and crack width development. Therefore, the temperature histories in these series are always identical between the two specimens of one series. The differences to series 1 and 2 are metal sheets for predetermination of cracking at desired positions in series 3 and 4. And series 5 stands out because no thermal insulation was used at all.

Altogether, series 3 to 5 do not provide additional insights on the hardening-induced stress history and risk of early age cracking. These series were mainly needed for investigations of the superimposition of hardening-induced stressing with further imposed deformations representative for service life. But it is noteworthy that reducing the temperature maximum in series 5 did not lead to a significant decrease of restraint stresses after temperature equalization compared to series 1 to 4.

7.4.3 Superimposition of Hardening-Induced Stressing with Further Imposed Deformations Representative for Service Life

Theoretical considerations of pronounced viscoelastic material properties in young concrete would support the conclusion of a significant relaxation of hardening-induced stresses with ongoing time. On the contrary, the created tensile stresses in the experiments do not show any significant reduction after temperature equalization; they even remain almost to a full extent. And this applies in the same way to uncracked specimens as to cracked ones.

For a better understanding of this behaviour, a comparative computational analysis was performed in Turner (Turner 2017) based on the calculation model in Schlicke (2014) and Heinrich (2018). In this analysis, the series 1 and the series 5 were computationally simulated, whereby different case studies with and without viscoelasticity as well as with and without drying shrinkage (according to the conditions in the experiment) were performed. The result is shown in Fig. 7.15.

Generally seen, the agreement between measurements and computational results under consideration of viscoelasticity and drying shrinkage is acceptable. A small disagreement exists for series 1 after temperature equalization, which could be explained with the very high tensile stresses in the magnitude of $f_{ctk,0.05}$ at that time. This high utilization may cause non-linear viscoelastic effects which were not taken into account here.

Nonetheless, this comparative study confirms the very limited effect of relaxation of hardening-induced stresses after temperature equalization. Even if drying is not

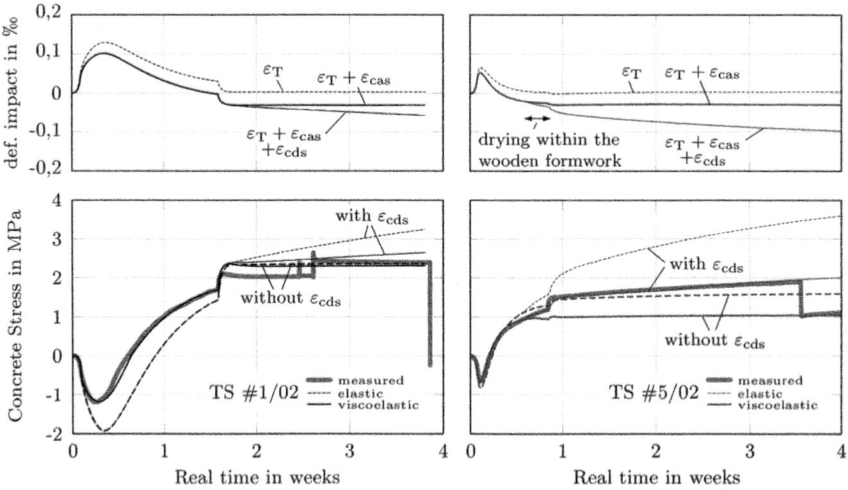

Fig. 7.15 Influence of drying shrinkage and viscoelasticity on the stress history after temperature equalization, cf. Turner (2017)

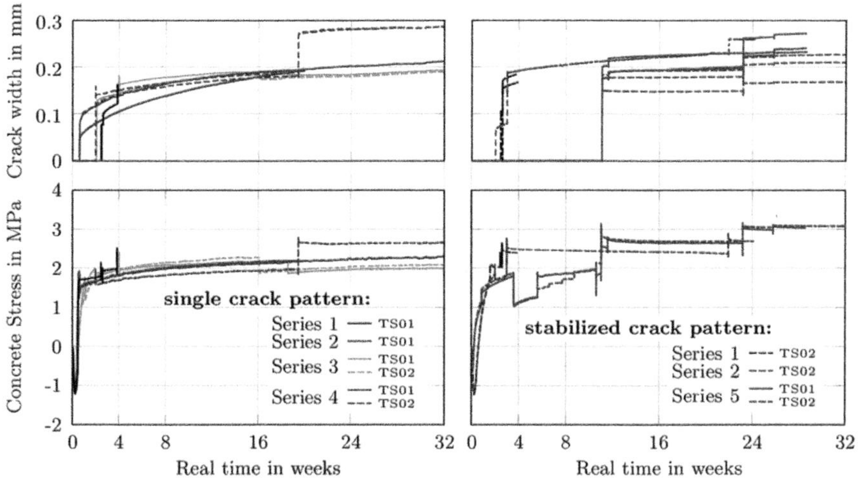

Fig. 7.16 Development of the restraint force and crack opening due do sustained restraining

taken into consideration, the initially imposed stresses do not decrease in a visible manner. It should be noted, that the achieved agreement depends profoundly on the applied material model for viscoelasticity according to Schlicke (2014), which has a very small long-term component for creep in tension.

Altogether, the common assumption of a significant reduction of hardening-induced stresses after temperature equalization cannot be confirmed, neither with experimental results, nor with the computational analysis. This conclusion is also viable for a longer time frame, as shown in Fig. 7.16. A safe design of structures with strict requirements on crack avoidance or crack width control require therefore a combined consideration of imposed deformations originating from the hardening phase as well as during service life.

7.4.4 Crack Opening During Service Life

The combined action of hardening-induced stresses with further restraint stresses during service life makes it very likely that cracking occurs under service, especially in case of exposed structures. After cracking, the stressing is reduced since parts of the imposed and restrained deformation were released in form of crack width. Nonetheless, time-dependent effects in the transfer length due to bond creep and local drying shrinkage as well as additionally imposed deformations during service life, e.g. due to seasonal temperature variations and/or overall drying of the structure, cause an increase of the crack width. Such behaviour was also simulated with the ARFs by opening the frames without affecting the hardening-induced stress state and imposing additional deformations with the hydraulic cylinders. Afterwards the system was locked again.

The additionally imposed deformations lead to first or further cracking in the specimens since the stress level after hardening has already reached approx. 80% of the concrete's tensile strength. In some specimens the impact was increased until an equivalent cooling of 45 °C and the crack pattern was stabilized. The obtained results for the development of the restraint force (given as concrete stresses in the uncracked parts) and the associated crack width development are given in Fig. 7.16, whereby it was distinguished between specimens which remained still in the crack formation stage (left) and specimens with stabilized crack patterns (right).

In the viewed time, the stress histories and crack width developments are subject to drying shrinkage and changing bond properties, which counteracts regarding the restraint stresses.

In specimens with single cracks the contraction due to drying shrinkage is dominant and leads to continuously increasing tensile stresses and crack widths. In the specimens of series 3 this results in further cracking after 16 and 18 weeks. In a stabilized crack pattern, the local effects in the bond length (bond creep) are more influential and reduce the axial member stiffness and the restraint stresses accordingly. Despite the declining steel stress the crack widths increase.

In the specimens of Series 5 the upper stress level of 3.2 MPa was re-induced at different points of time. It showed that the impairing of the bond properties slows down by time, however, the bond stress is always higher than 70% of the short-term property which confirms the current design code assumption to be on the safe side.

7.4.5 Evolution of the Restraint Force During Cracking in Thick Members with Reinforcement Near the Surface

Figure 7.17 shows the evolution of the restraint force during the crack formation process in thick members with reinforcement near the surface. In detail, the experimental results determined with three experiments as well as the theoretical result without influences of the experimental set-up are shown. At first, the restraint force is built up with imposed deformations in one go with a speed of ~10 μm/min. until the primary crack occurs. Subsequently, the primary crack opens meanwhile the restraint force decreases until a new equilibrium is found. Afterwards, the imposition of deformations is continued and the restraint force is increased again, whereby the stiffness of the specimen is significantly smaller after formation of the primary crack. Nonetheless, the stresses in the effective concrete zone increase comparably fast so that secondary cracks occur now on a lower force level than in the uncracked state ($F_{cr,0}$). In summary, all three experiments absorbed an imposed deformation of ~−65‰ by formation of one primary crack and two pairs of secondary cracks,

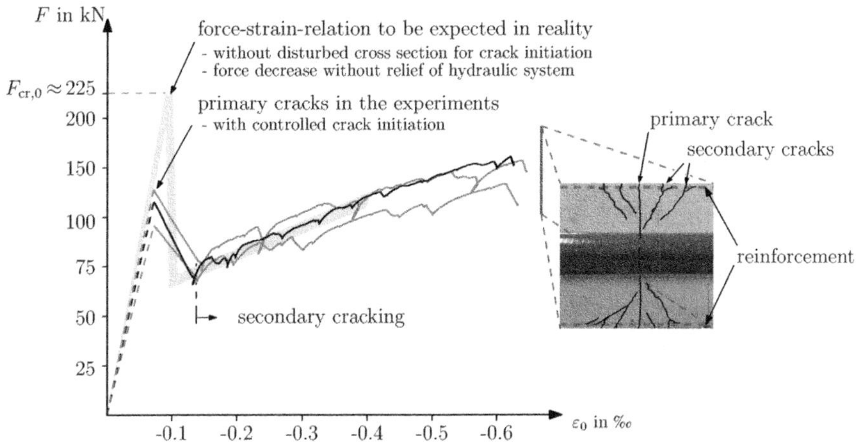

Fig. 7.17 Experimentally simulated evolution of the restraint force during cracking and final crack pattern in a thick cross section with reinforcement near the surface

whereby the cracking force reached a magnitude of ~66% of the cracking force of the undisturbed cross section.

The real stress history due to hydration was not particularly simulated during the experiment since this would had been enormously more complicated without gaining any beneficial effect on the test result. As indicated in Fig. 7.2, these types of thick members have a comparably late time of macro-cracking in which the mechanical properties of the concrete are far developed. When the cracking force is reached at such stage, the prior stress history is not important for the further crack formation process.

Besides, the primary cracks in the experiments occurred already at ~55% of the theoretical cracking force of the undisturbed cross section. The reason was the targeted initiation of the primary cracks at half length of the specimen by inclusion of metal sheets in the cross section at this location. In reality, the location of the primary crack is geometrically set, as outlined in Sect. 7.1, but in the experiment, it was important to create the primary crack in the symmetry of the specimen. After cracking, the restraint forces in the experiments decrease slower as compared with the theoretical decrease. The reason is the relief of the hydraulic system in the experiment when the restraint force decreases. This imposes additional deformations on the specimen. In reality, this influence does not exist and the decrease of the restraint force is much more abrupt. However, the stiffness of the experiment after primary cracking is similar to the theoretical one and the forces increase similarly as soon as the imposed deformations due to the hydraulic relief are compensated.

One key insight of the experiments with the modified frames is the evolution of the restraint force during cracking in thick members with reinforcement near the surface. Further insights are the cracking pattern to be expected as well as the temporal development of crack widths of the primary crack as well as the secondary cracks. These results will be presented elsewhere in the near future.

7.5 Conclusion and Outlook

This chapter presented an experimental setup for the holistic simulation of the stress history in restrained RC members with special regard of the hydration-induced stress development since setting and a continuous superposition with additionally imposed deformations representative for service life of exposed structures. By this, important parameters such as concrete technology, thermal treatment and crack formation itself can be studied systematically. The major insights of the past experiments are:

- Thermal control after maximum temperature did not influence the hardening induced restraint positively. Therefore, the effect of thermal treatment in order to prevent early through-cracks might be overestimated in building practice. However, this does not apply to the risk of surface cracking caused by high temperature gradients when removing the formwork at a time of high temperatures in the core of the member.
- Common simplifications in design practice neglect the superposition of early deformation impacts and later imposed deformations during service life. These simplifications are based on a significant decrease of hardening-induced restraint stresses after temperature equalization which could not be confirmed in the experiments. This insight is particularly important for the design of jointless concrete structures, which require for that reason a combined consideration of all deformation impacts during the life time.
- Especially watertight constructions are often optimized for preventing early crack formation with measures of concrete technology and specific construction methods. Nevertheless, those constructions can be subjected to high tensile stresses after hardening. If there is the possibility of additional deformation impacts in service life, the higher crack risk at later age should be considered within design.

Besides, the experimental setup was modified in a second step in order to simulate the restraint-induced crack formation in conventionally reinforced thick concrete members with reinforcement near the surface. The major insights of these experiments are:

- Thick concrete members with conventional reinforcement near the surface show a specific mode of cracking consisting of a primary crack, which is caused by the restraining condition, and secondary cracking, which is created by the activation of the reinforcement in the vicinity of the primary crack.
- The force level of secondary cracking is significantly lower than the restraint force causing the primary crack, which was detectable by conducting these experiments with externally imposed deformations instead of a force control.

- The deformation compatibility along the member is achieved by the sum of crack widths of the primary crack and the secondary crack. By this, the crack widths at the surface are effectively limited by distributing the imposed deformations over several cracks. On the contrary, the width of the primary crack increases significantly in the core of the member to which secondary cracking does not propagate.
- A clear understanding of the mechanical background of this cracking enables an effective reinforcement design for control of cracking in thick concrete members on basis of the deformation compatibility.

The experimental programme with these frames will be used in the near future for targeted in-depth studies of the effect of viscoelastic properties of concrete during the hydration-induced stress history and in the transition phase of construction process and service life. Besides, the modified setup will be used for a better description of the crack development in the primary crack when secondary cracks are formed. In particular, it is planned to monitor the zone of cracking with digital image correlation as well as equipping the reinforcement with distributed fibre optical measurements for determining the steel strain distribution along the cracking zone.

Acknowledgements The cooperation with the Federal Waterways Engineering and Research Institute (BAW) in Karlsruhe, Germany was scientifically and financially significant for the project's success and we thank Dipl.-Ing. Rainer Ehmann and Dr.-Ing. Jörg Bödefeld for the prosperous collaboration. The authors would also like to thank the team of the Laboratory for Structural Engineering at Graz University of Technology (LKI) for their great support.

References

BAW-Merkblatt. (2010). Merkblatt früher Zwang – Rissbreitenbegrenzung für frühen Zwang in massiven Wasserbauwerken. Referat Massivbau der BAW, Karlsruhe.

BAW-Merkblatt. (2019). Merkblatt Zwang bei fugenlosen Wasserbauwerken. Referat Massivbau der BAW, Karlsruhe (Gelbdruck).

Bentur, A., & Kovler, K. (2003). Evaluation of early age cracking characteristics in cementitious systems. *Materials and Structures, 36,* 183–190.

Berger, J., & Feix, J. (2018). Constraint forces from imposed deformations. Structural Concrete (online version of record before inclusion in an issue) 10 p. https://doi.org/10.1002/suco.201800123.

Bjøntegaard, Ø., & Sellevold, E. J. (2004). The temperature-stress testing machine (TSTM): Capabilities and limitations. In J. Weiss, K. Kovler, J. Marchand, & S. Mindess (Eds.), *International RILEM Symposium Advances in Concrete Through Science and Engineering, RILEM proceeding* 48. e-ISBN: 2912143926.

EN 1992-1-1:2015. (2015). *EUROCODE 2: Design of concrete structures—Part 1-1: General rules and rules for buildings.*

Faria, R., Leitão, L., Teixeira, L., Azenha, M., & Cusson, D. (2016). A structural experimental technique to characterize the viscoelastic behaviour of concrete under restrained deformations. *Strain, 53,* 1305–1475. https://doi.org/10.1111/str.12216.

Fellmann, W., & Menn, C. (1984). Zugversuche an Stahlbetonscheiben. Technical Report No. 7604-1. Institute of Structural Engineering, ETH Zürich.

fib Model Code 2010 (2013), fib Model Code for Concrete Structures 2010.
Gomes, J., Azenha, M., Granja, J., Faria, R., Sousa, C., Zahabizadeh, B., et al. (2018). Proposal of a test set up for simultaneous application of axial restraint and vertical loads to slab-like specimens: Sizing principles and application. In *International RILEM/COST conference Interdisciplinary Approaches for Cement-based Materials and Structural Concrete. SynerCrete'18*, Funchal (pp. 303–308). https://doi.org/10.5281/zenodo.1405563.
Grube, H. (1991). *Ursachen des Schwindens von Beton und Auswirkungen auf Betonbauteile, Schriftenreihe der Zementindustrie, Heft 52/ 1991*. Düsseldorf: Verlag Bau + Technik.
Heinrich P. J. (2018), Effiziente Berechnung viskoelastischer Spannungen in gezwängten Bauteilen. Ph.D. thesis, Graz University of Technology. (in German).
Helmus, M. (1990). *Mindestbewehrung zwangbeanspruchter dicker Stahlbetonbauteile, DAfStb Heft 412*. Berlin: Deutscher Ausschuss für Stahlbeton.
Jędrzejewska, A., et. al. (2018). COST TU1404 benchmark on macroscopic modelling of concrete and concrete structures at early age: Proof-of-concept stage. *Construction and Building Materials, 174*. ISSN 0950-0618.
Kanavaris, F., Azenha, M., Soutsos, M., & Kovler, K. (2019). Assessment of behaviour and cracking susceptibility of cementitious systems under restrained conditions through ring tests: A critical review. *Cement and Concrete Composites, 95*, 137–153. https://doi.org/10.1016/j.cemconcomp.2018.10.016.
Klausen, A. E., Kanstad, T., Bjøntegaard, Ø. (2015). Updated temperature-stress testing machine (TSTM): Introductory tests, calculations, verification, and investigation of variable fly ash content. In *Mechanics and Physics of Creep, Shrinkage and Durability of Concrete and Concrete Structures. CONCREEP-10*, Vienna. https://doi.org/10.1061/9780784479346.086.
ÖBV Bulletin. (2018). *Analytical design of watertight structures with optimized concrete (in German)*. Vienna, Austria: Austrian society for constructuion technology.
Schlicke, D. (2014). Mindestbewehrung für zwangbeanspruchten Beton. Ph.D. thesis (2nd Ed.) Graz University of Technology.
Schlicke, D., & Dorfmann, E. (2017). Influence of gradual imposition of tensile stresses on associated viscoelastic behaviour. In *RILEM/COST Conference EAC-02*, Brussels, Belgium.
Schlicke, D., & Tue, N. V. (2015). Minimum reinforcement for crack width control in restrained concrete members considering the deformation compatibility. *Structural Concrete, 16*, 221–232. https://doi.org/10.1002/suco.201400058.
Schlicke, D., Turner, K., & Tue, N. V. (2015). Decrease of tensile creep response under realistic restraint conditions in structures. In: *Mechanics and Physics of Creep, Shrinkage and Durability of Concrete and Concrete Structures. CONCREEP-10*, Vienna. https://doi.org/10.1061/9780784479346.158.
Schnell, J., Albert, A., & Dridiger, A. (2017). Ressourcenschonende Reduktion von Bewehrungsstahl in Hochbaudecken, TU Kaiserslautern, ForschungsberichtBBSR/ZukunftBAU (SWD-10.08.18.7-14.17).
Springenschmidt, R., Breitenbucher, R., & Mangold M. (1994). Development of the cracking frame and temperature stress testing machine. In R. Springenschmidt (Ed.), *International RILEM Symposium Thermal Cracking in Concrete at Early-Age, RILEM proceedings* 25, Munich (pp. 137–144). London, UK: E&FN Spon.
Staquet, S., Delsaute, B., Darquennes, A., & Espion, B. (2012). Design of a revisited TSTM system for testing concrete since setting time under free and restraint conditions. In *Crack Control of Mass Concrete and Related Issues concerning Early-Age of Concrete Structures* (pp. 99–110). RILEM Publications SARL.
Thielen, G., & Hintzen, W. (1994). Investigation of concrete behaviour under restraint with a Temperature-Stress testing machine. In R. Springenschmid (Ed.), *International RILEM Symposium Thermal cracking in concrete in early ages, RILEM Proceedings 25*, Munich (pp. 135–152). London, UK: E&FN Spon.
Turner, K. (2017). Ganzheitliche Betrachtung zur Ermittlung der Mindestbewehrung für fugenlose Wasserbauwerke. Ph.D. thesis, Graz University of Technology (in German).

VDZ Zement-Merkblatt "Risse in Beton" Nr. B18, Verein Deutscher Zementwerke e.V., Düsseldorf, 2014.

Westman, G., & Emborg, M. (1994). *Development of a relaxation frame*. Nordic Concrete Research, Publication No.15.2/94, Oslo (pp. 89–96).

Printed by Printforce, the Netherlands